Generalized, Linear, and Mixed Models

Generalized, Linear, and Mixed Models

Second Edition

Charles E. McCulloch

University of California
Department of Epidemiology and Biostatistics
San Francisco, CA

Shayle R. Searle

Cornell University
Department of Biological Statistics and Computational Biology
Ithaca, NY

John M. Neuhaus

University of California
Department of Epidemiology and Biostatistics
San Francisco, CA

WILEY

A JOHN WILEY & SONS, INC., PUBLICATION

Library of Congress Cataloging-in-Publication Data:

McCulloch, Charles E.
   Generalized, linear, and mixed models / Charles E. McCulloch, Shayle R. Searle, John M.
Neuhaus. — 2nd ed.
       p. cm.
   Includes bibliographical references and index.
   ISBN 978-0-470-07371-1 (cloth)
1. Linear models (Statistics) I. Searle, S. R. (Shayle R.), 1928– II. Neuhaus, John M. III. Title.
   QA279.M3847 2008
   519.5'35—dc22
                                                                    2008002724

10 9

List of Chapters

Contents

Preface

For the Second Edition of *Generalized, Linear and Mixed Models* we incorporated advances of the intervening years since the First Edition. Chapter 8, the former Chapter 7, includes more extensive development of longitudinal data models, especially for non-normally distributed outcomes. There are three new chapters: Chapter 9 extracts and expands on the First Edition's treatment of marginal models, Chapter 10 describes methods of incorporating multiple outcomes (especially of different distributional types), and Chapter 12 explores the consequences of departures from assumptions and suggests remedies when assumptions are violated. Models that incorporate all the features of the book, e.g., generalized, linear, mixed models are assumption laden and it is important to delineate which assumptions are crucial, ways to diagnose them, and possible remedies. In addition, the rest of the book has been lightly revised.

Although this edition is expanded, the text can still be used as originally intended. The additional chapters serve to expand the basic offerings and show a wider variety of ways in which mixed models can be used and some of the practical issues in implementation. Chapters 1 through 3 are the introductory chapters. Chapter 1 lays out the basic terminology and gives a of number examples. Chapters 2 and 3 cover the central ideas of the entire book in simple scenarios (one-way classification and single predictor regression). Chapters 4 through 7 form the "meat" of the book and describe the main classes of models. The rest of the Chapters cover more specialized topics and can be read or taught in a relatively independent manner, as interest allows, following Chapters 4 through 7. Chapters 8 through 11 cover variations on the main models (longitudinal data models, models incorporating multivariate responses, and nonlinear models) and Chapters 12 through 14 cover issues (departures from assumptions, prediction and computing) that largely cross-cut the topics of previous chapters.

As before, the emphasis is on the applications of these models and the assumptions necessary for valid statistical inference. The focus is not on the details of data analysis nor on the use of statistical software, though we do briefly mention some examples.

San Francisco, CA

Charles E. McCulloch

Ithaca, NY

Shayle R. Searle

May 2008

John M. Neuhaus

Preface to the First Edition

The last thirty or so years have been a time of enormous development of analytic results for the linear model (LM). This has generated extensive publication of books and papers on the subject. Much of this activity has focused on the normal distribution and homoscedasticity. Even for unbalanced data, many useful, analytically tractable results have become available. Those results center largely around analysis of variance (ANOVA) procedures, and there is abundant computing software which will, with wide reliability, compute those results from submitted data.

Also within the realm of normal distributions, but permitting heterogeneity of variance, there has been considerable work on linear mixed models (LMMs) wherein the variance structure is based on random effects and their variance components. Algebraic results in this context are much more limited and complicated than with LMs. However, with the advent of readily available computing power and the development of broadly applicable computing procedures (e.g., the EM algorithm) we are now at a point where models such as the LMM are available to the practitioner. Furthermore, models that are nonlinear and incorporate non-normal distributions are now feasible. It is to understanding these models and appreciating the available computing procedures that this book is directed.

We begin by reviewing the basics of LMs and LMMs, to serve as a starting point for proceeding to generalized linear models (GLMs), generalized linear mixed models (GLMMs) and some nonlinear models. All of these are encompassed within the title "Generalized, Linear, and Mixed Models."

The progress from easy to difficult models (e.g. from LMs to GLMMs) necessitates a certain repetition of basic analysis methods, but this is appropriate because the book deals with a variety of models and the application to them of standard statistical methods. For example, maximum likelihood (ML) is used in almost every chapter, on models that get progressively more difficult as the book progresses. There is, indeed, purposeful concentration on ML and, very noticeably, an (almost complete) absence of analysis of variance (ANOVA) tables.

Although analysis of variance methods are quite natural for fixed effects linear models with normal distributions, even in the case of linear mixed models with normal distributions they have much less appeal. For example, with unbalanced data from mixed models, it is not clear what the "appropriate" ANOVA table should be. Furthermore, from a theoretical viewpoint, any such table represents an over-summarization of data: except in special cases, it does not contain sufficient statistics and therefore engenders a loss of information and efficiency. And these deficiencies are aggravated if one tries to generalize analysis of variance to models based on non-normal distributions such as, for example, the Poisson or binomial. To deal with these we therefore concentrate on ML procedures.

Although ML estimation under non-normality is limited in yielding analytic results, we feel that its generality and efficiency (at least with large samples) make it a natural method to use in today's world. Today's computing environment compensates for the analytic intractability of ML and helps makes ML more palatable.

As prelude to the application of ML to non-normal models we often show details of using it on models where it yields easily interpreted analytic results. The details are lengthy, but studying them engenders a confidence in the ML method that hopefully carries over to non-normal models. For these, the details are often not lengthy, because there are so few of them (as a consequence of the model's inherent intractability) and they yield few analytic results. The brevity of describing them should not be taken as a lack of emphasis or importance, but merely as a lack of neat, tidy results. It is a fact of modern statistical practice that computing procedures are used to gain numerical information about the underlying nature of algebraically intractable results. Our aim in this book is to illuminate this situation.

The book is intended for graduate students and practicing statisticians. We begin with a chapter in which we introduce the basic ideas of fixed and random factors and mixed models and briefly discuss general methods for the analysis of such models. Chapters 2 and 3 introduce all the main ideas of the remainder of the book in two simple contexts (one-way classifications and linear regression) with a minimum of emphasis on generality of results and notation. These three chapters could form the core of a quarter course or, with supplementation, the basis of a semester-long course for Master's students. Alternatively, they could be used to introduce generalized mixed models towards the end of a linear models class.

Chapters 4, 5, 6 and 8 cover the main classes of models (linear, generalized linear, linear mixed, and generalized linear mixed) in more generality and breadth. Chapter 7 discusses some of the special features of longitudinal data and shows how they can be accommodated within LMMs. Chapter 9 presents the idea of prediction of realized values of random effects. This is an important distinction introduced by considering models containing random effects. Chapter 10 covers computing issues, one of the main barriers to adoption of mixed models in practice. Lest the reader think that everything can be accommodated under the rubric of the generalized linear mixed model, Chapter 11 briefly mentions nonlinear mixed models. And the book ends with two short appendices, M and S, containing some pertinent results in matrices and statistics.

For students with some training in linear models, the first 10 chapters, with light emphasis on Chapters 1 through 4 and 6, could form a "second" course extending their linear model knowledge to generalized linear models. Of course, the book could also be used for a semester long course on generalized mixed models, although in-depth coverage of all of the topics would clearly be difficult.

Our emphasis throughout is on modeling and model development. Thus we provide important information about the consequences of model assumptions, techniques of model fitting and methods of inference which will be required for data analysis, as opposed to data analysis itself. However, to illustrate the concepts we do also include analysis or illustration of the techniques for a variety of real data sets.

The chapters are quite variable in length, but all of them have sections, subsections and sub-subsections, each with its own title, as shown in the Table of Contents. At times we have sacrificed the flow of the narrative to make the book more accessible as a reference. For example, Section 2.1d is basically a catalogue of results with titles that make retrieval more straightforward, particularly because those titles are all listed in the table of contents.

Ithaca, NY   Charles E. McCulloch
September 2000   Shayle R. Searle

Chapter 1

INTRODUCTION

1.1 MODELS

a. Linear models (LM) and linear mixed models (LMM)

In almost all uses of statistics, major interest centers on averages and on variation about those averages. For more than sixty years this interest has frequently manifested itself in the widespread use of analysis of variance (ANOVA), as originally developed by R. A. Fisher. This involves expressing an observation as a sum of a mean plus differences between means, which, under certain circumstances, leads to methods for making inferences about means or about the nature of variability. The usually-quoted set of circumstances which permits this is that the expected value of each datum be taken as a linear combination of unknown parameters, considered as constants; and that the data be deemed to have come from a normal distribution. Thus the linear requirement is such that the expected value (i.e., mean), μ_{ij}, of a random observation y_{ij} can be, for example, of the form $\mu_{ij} = \mu + \alpha_i + \beta_j$ where μ, α_i and β_j are unknown constants — unknown, but which we want to estimate. And the normality requirement would be that y_{ij} is normally distributed with mean μ_{ij}. These requirements are the essence of what we call a *linear model*, or LM for short. By that we mean that the model is linear in the parameters, so "linearity" also includes being of the form $\mu_{ij} = b_0 + b_1 x_{1ij} + b_2 x_{2ij}^2$, for example, where the xs are known and there can be (and often are) more than two of them.

A variant of LMs is where parameters in an LM are treated not as constants but as (realizations of) random variables. To denote this different meaning we represent parameters treated as random by Roman rather than Greek letters. Thus if the αs in the example were to be considered random, they would be denoted by as, so giving $\mu_{ij} = \mu + a_i + \beta_j$. With the βs remaining as constants, μ_{ij} is then a mixture of random

1

and constant terms. Correspondingly, the model (which is still linear) is called a *linear mixed model*, or LMM. Until recently, most uses of such models have involved treating random a_is as having zero mean, being homoscedastic (i.e., having equal variance) with variance σ_a^2 and being uncorrelated. Additionally, normality of the a_is is usually also invoked.

There are many books dealing at length with LMs and LMMs. We name but a few: Graybill (1976), Seber (1977), Arnold (1981), Hocking (1985), Searle (1997), and Searle et al. (1992).

b. Generalized models (GLMs and GLMMs)

The last twenty-five years or so have seen LMs and LMMs extended to *generalized linear models* (GLMs) and to *generalized linear mixed models* (GLMMs). The essence of this generalization is two-fold: one, that data are not necessarily assumed to be normally distributed; and two, that the mean is not necessarily taken as a linear combination of parameters but that some function of the mean is. For example, count data may follow a Poisson distribution, with mean λ, say; and $\log \lambda$ will be taken as a linear combination of parameters. If all the parameters are considered as fixed constants the model is a GLM; if some are treated as random it is a GLMM.

The methodology for applying a GLM or a GLMM to data can be quite different from that for an LM or LMM. Nevertheless, some of the latter is indeed a basis for contributing to analysis procedures for GLMs and GLMMs, and to that extent this book does describe some of the procedures for LMs (Chapter 4) and LMMs (Chapter 6); Chapters 5 and 7 then deal, respectively, with GLMs and GLMMs. Chapters 2 and 3 provide details for the basic modeling of the one-way classification and of regression, prior to the general cases treated later.

1.2 FACTORS, LEVELS, CELLS, EFFECTS AND DATA

We are often interested in attributing the variability that is evident in data to the various categories, or classifications, of the data. For example, in a study of basal cell epithelioma sites (akin to Abu-Libdeh et al., 1990), patients might be classified by gender, age-group and extent of exposure to sunshine. The various groups of data could be summarized in a table such as Table 1.1.

The three classifications, gender, age, and exposure to sunshine,

Table 1.1: A Format for Summarizing Data

Gender	Low Exposure to Sunshine			High Exposure to Sunshine		
	Age Group			Age Group		
	A	B	C	A	B	C
Male						
Female						

Table 1.2: Summarizing Exam Grades

Gender	English			Geology		
	Section			Section		
	A	B	C	A	B	C
Male						
Female						

which identify the source of each datum are called *factors*. The individual classes of a classification are the *levels* of a factor (e.g., male and female are the two levels of the factor "gender"). The subset of data occurring at the "intersection" of one level of every factor being considered is said to be in a *cell* of the data. Thus with the three factors, gender (2 levels), age (3 levels) and sunshine (2 levels), there are $2 \times 3 \times 2 = 12$ cells.

Suppose that we have student exam grades from each of three sections in English and Geology courses. The data could be summarized as in Table 1.2, similar to Table 1.1. Although the layout of Table 1.2 has the same appearance as Table 1.1, sections in Table 1.2 are very different from the age groups of Table 1.1. In Table 1.2 section A of English has no connection to (and will have different students from) section A of Geology; in the same way neither are sections B (or C) the same in the two subjects. Thus the section factor is *nested* within the subject factor. In contrast, in Table 1.1 the three age groups are the same for both low and high exposures to sunshine. The age and sunshine factors are said to be *crossed*.

In classifying data in terms of factors and their levels, the feature of interest is the extent to which different levels of a factor affect the variable of interest. We refer to this as the *effect* of a level of a factor

on that variable. The effects of a factor are always one or other of
two kinds, as introduced in Section 1.1 in terms of parameters. First
is the case of parameters being considered as fixed constants or, as we
henceforth call them, *fixed effects*. These are the effects attributable to
a finite set of levels of a factor that occur in the data and which are
there because we are interested in them. In Table 1.1 the effects for all
three factors are fixed effects.

The second case corresponds to what we earlier described as parameters being considered random, now to be called *random effects*. These
are attributable to a (usually) infinite set of levels of a factor, of which
only a random sample are deemed to occur in the data. For example, four loaves of bread are taken from each of six batches of bread
baked at three different temperatures. Since there is definite interest
in the particular baking temperatures used, the statistical concern is to
estimate those temperature effects; they are fixed effects. No assumption is made that the temperature effects are random. Indeed, even if
the temperatures themselves were chosen at random, it would not be
sensible to assume that the temperature *effects* were random. This is
because temperature is defined on a continuum and, for example, the
effect of a temperature of $450.11°$ is almost always likely to be a very
similar to the effect of a $450.12°$ temperature. This nullifies the idea of
temperature effects being random.

In contrast, batches are not defined on a continuum. They are real
objects, just as are people, or cows, or clinics and so, depending on the
circumstance, it can be perfectly reasonable to think of their effects as
being random. Moreover, we can do this even if the objects themselves
have not been chosen as a random sample — which, indeed, they seldom
are. So we assume that batch effects are random, and then interest in
them lies in estimating the variance of those effects. Thus data from
this experiment would be considered as having two sources of random
variation: batch variance and, as usual, error variance. These two
variances are known as *variance components*: for linear models their
sum is the variance of the variable being observed.

Models in which the only effects are fixed effects are called *fixed
effects models*, or sometimes just *fixed models*. And those having (apart
from a single, general mean common to all observations) only random
effects are called *random effects models* or, more simply, *random models*.
Further examples and properties of fixed effects and of random effects
are given in Sections 1.3 and 1.4.

Figure 1.1: Examples of Interaction and No Interaction.

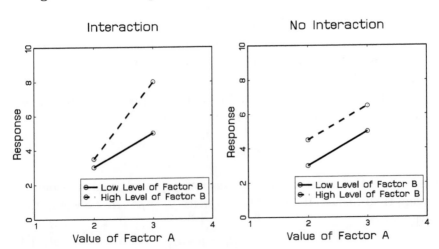

When there are several factors, the effect of the combination of two or more of them is called an *interaction effect*. In contrast, the effect of a single factor is called a *main effect*. The concept of interaction is as follows: If the change in the mean of the response variable between two levels of factor A is the same for different levels of factor B, we say that there is no interaction; but if that change is different for different levels of B, we say that there is an interaction. Figure 1.1 illustrates the two cases.

Details of the application of analysis of variance techniques to LMs and LMMs depend in many cases on whether every cell (as defined above) of the data has the same number of observations. If that is so the data are said to be *balanced data*; if not, the data are *unbalanced data*, in which case they are either *all-cells-filled data* (where every cell contains data) or *some-cells-empty data* (where some cells have no data). Implications of these descriptions are discussed at length in Searle et al. (1992, Sec. 1.2).

1.3 FIXED EFFECTS MODELS

Fixed effects and random effects have been specified and described in general terms. We now illustrate the nature of these effects using real-life examples and emphasizing the particular properties of random effects.

a. Example 1: Placebo and a drug

Diggle et al. (1994) describe a clinical trial to treat epileptics with
the drug Progabide. We ignore the baseline period of the experiment,
and consider a response which is the number of seizures after patients
were randomly allocated to either the placebo or the drug. If y_{ij} is
the number of seizures experienced by patient j receiving treatment i
($i = 1$ for placebo and $i = 2$ for Progabide), a possible model for y_{ij}
could be based upon starting from the expected value

$$\mathrm{E}[y_{ij}] = \mu_i,$$

where μ_i is the mean number of seizures expected from someone receiving treatment i. If we wanted to write $\mu_i = \mu + \alpha_i$ we would then
have

$$\mathrm{E}[y_{ij}] = \mu_i = \mu + \alpha_i, \tag{1.1}$$

where μ is a general mean and α_i is the effect on the number of seizures
due to treatment i.

In this model of the expected value of y_{ij}, each μ_i (or μ and each α_i)
is considered as a fixed unknown constant, the magnitudes of which we
wish, in some general sense, to estimate; that is, we want to estimate
μ_1, μ_2, and $\mu_1 - \mu_2$. And having estimated that difference we would
want to test if it is less than zero (i.e., to test if the drug is reducing the
number of seizures). In doing this the μ_is (or the α_is) correspond to
the two different treatments being used. They are the only two being
used, and in using them there is no thought for any other treatments.
This is the concept of fixed effects. We consider just the treatments
being used and no others, and so the effects are called *fixed effects*.

The manner in which data are obtained always affects inferences that
can be drawn from them. We therefore describe a sampling process
pertinent to this fixed effects model. The observed data are envisaged
as being one possible set of data involving these same two treatments
that could be derived from repetitions of a clinical trial, repetitions for
which a different sample of people receiving each treatment would be
used. This would lead on each occasion to a set of data that would
be two random samples, one from a population of possible data having
mean μ_1, and another from a population having μ_2.

The all-important feature of fixed effects is that they are deemed to
be constants representing the effects on the response variable y of the
various levels of the factor concerned, in this case the two treatments,

placebo and drug. These treatments are the levels of the factor of particular interest, chosen because of interest in those treatments in the trial. But they could just as well be different fertilizers applied to a corn crop, different forage crops grown in the same region, different methods used in a manufacturing process, different drugs given for the same illness, and so on. The possibilities are legion, as are the varieties of models and their complexities. We offer two more brief examples.

b. **Example 2: Comprehension of humor**

A recent study (Churchill, 1995) of the comprehension of humor ("Did you get it?") involved showing three types of cartoons (visual only, linguistic only, and visual-linguistic combined) to two groups of adolescents (normal and learning disabled). Motivated by this study, suppose the adolescents record scores of 1 through 9, with 9 representing extremely funny and 1 representing not funny at all. Then with $\bar{y}_{ij\cdot}$ being the mean score from showing cartoon type i to people in group j, a suitable start for a model for $\bar{y}_{ij\cdot}$ could be

$$\mathrm{E}[\bar{y}_{ij\cdot}] = \mu_{ij} = \mu + \alpha_i + \beta_j, \tag{1.2}$$

where μ is a general mean, α_i is the effect on comprehension due to cartoon type i ($= 1$, 2 or 3) and β_j is the effect due to respondents being in adolescent group j ($= 1$ or 2). Because each of the same three cartoon types is shown to each of the two adolescent groups, this is an example of two crossed factors, cartoon type and adolescent group. Furthermore, since the three cartoon types and the two groups of people have been chosen as the only types and groups being considered, the α_is and β_js are fixed effects corresponding to the three types and two groups. They are the specific features of interest in this study, and under no circumstances can they be deemed to have been chosen randomly from a larger array of types and groups. Thus the α_is and β_js are fixed effects. This is just a simple extension of Example 1 which has one factor with two levels. In Example 2 there are two factors: one, type of cartoon, with three levels, and another, group of people, with two levels. After estimating effects for these levels we might want to make inferences about the extent to which the visual-linguistic cartoons were comprehended differently from the average of the visual and linguistic ones; and we would also want to test if the learning-disabled adolescents differed in their cartoon comprehension from the non-disabled adolescents. (In actual practice this study involved 8 different cartoons within each type, and several people in each group; these two factors, cartoon-within-type of cartoon, and people-within-group, had also to be taken into account.)

c. Example 3: Four dose levels of a drug

Suppose we have a clinical trial in which a drug (e.g., Progabide of Example 1) is administered at four different dose levels. For y_{ij} being the datum for the jth person receiving dose i we could start with

$$\mathrm{E}[y_{ij}] = \mu_i = \mu + \alpha_i . \tag{1.3}$$

This is just like (1.1), only where $i = 1$, 2, 3 or 4, corresponding to the four dose levels. The μ_is (and α_is) are fixed effects because the four dose levels used in the clinical trial are the only dose levels being studied. They are the doses on which our attention is fixed. This is exactly like Example 1 which has only two fixed effects whereas here there are four, one for each dose level. And after collecting the data, interest will center on differences between dose levels in their effectiveness in reducing seizures.

No matter how many factors there are, if they are all fixed effects factors and the fixed effects are combined linearly as in (1.1), (1.2) and (1.3), the model is called a fixed model; or, more generally just a linear model (LM). And there can, of course, also be fixed effects in nonlinear models.

1.4 RANDOM EFFECTS MODELS

a. Example 4: Clinics

Suppose that the clinical trial of Example 3 was conducted at 20 different clinics in New York City. Consider just the patients receiving the dose level numbered 1. The model equation for y_{ij}, which represents the jth patient at the ith clinic, could then be

$$\mathrm{E}[y_{ij}] = \mu + a_i, \tag{1.4}$$

with $i = 1$, 2, $\ldots$, 20 for the 20 clinics. But now pause for a moment. It is not unreasonable to think of those clinics (as do Chakravorti and Grizzle, 1975) as a random sample of clinics from some distribution of clinics, perhaps all the clinics in New York City.

Note that (1.4) is essentially the same algebraically as (1.3), save for having a_i in place of α_i. However, the underlying assumptions are different. In (1.3) each α_i is a fixed effect, the effect of dose level i on the number of seizures; and dose level i is a pre-decided treatment of interest. But in (1.4) each a_i is the effect on number of seizures of the observed patient having been in clinic i; and clinic i is just one clinic, the one from among the randomly chosen clinics that happened to be numbered i in the clinical trial. The clinics have been chosen randomly with the object of treating them as a representation of the population of all clinics in New York State, and inferences can and will be made about that population. This is a characteristic of random effects: they can be used as the basis for making inferences about populations from which they have come. Thus a_i is a *random effect*. As such, it is, indeed, a random variable, and the data will be useful for making an inference about the variance of those random variables; i.e., about the magnitude of the variation among clinics; and for predicting which clinic is likely to have the best reduction of seizures.

b.   Notation

As indicated briefly in Section 1.1 we adopt a convention of μ for a general mean and, for purposes of distinction, Greek letters for fixed effects and Roman for random effects. Thus for fixed effects equations (1.1) and (1.3) have α_i, and (1.2) has α_i and β_j; but (1.4) has a_i for random effects.

– i.   *Properties of random effects in LMMs*

With the a_is being treated as random variables, we must attribute probabilistic properties to them. There are two that are customarily employed; first, that all a_is are independently and identically distributed (i.i.d.); second, that they have zero mean, and then, that they all have the same variance, σ_a^2. We summarize this as

$$a_i \sim \text{ i.i.d. } (0, \sigma_a^2) \quad \forall \quad i. \tag{1.5}$$

This means that

$$\begin{aligned}
\mathrm{E}[a_i] &= 0 \ \forall \ i, & (1.6)\\
\mathrm{var}(a_i) &= \mathrm{E}[(a_i - \mathrm{E}[a_i])^2] = \mathrm{E}[a_i^2] = \sigma_a^2 & (1.7)
\end{aligned}$$

and

$$\operatorname{cov}(a_i, a_k) \;=\; 0 \text{ for } i \neq k. \tag{1.8}$$

There are, of course, properties other than these that could be used such as the covariances of (1.8) being non-zero.

– ii.   *The notation of mathematical statistics*

A second outcome of treating the a_is as random variables is that we must consider $E[y_{ij}] = \mu + a_i$ of (1.4) with more forethought, because it is really a mean calculated conditional on the value of a_i. To describe this situation carefully, we revert for a moment to the standard mathematical statistics notation which uses capital letters for random variables and lowercase letters for realized values. Since a random variable appears on the right-hand side of (1.4) the more precise way of writing (1.4) is

$$E[Y_{ij}|A_i = a_i] = \mu + a_i, \tag{1.9}$$

from which, when the realized value of A_i is not known, we write

$$E[Y_{ij}|A_i] = \mu + A_i. \tag{1.10}$$

In fact, of course, (1.10) is the basic result from which (1.9) is the special case. And from (1.10) we get the standard result

$$E[Y_{ij}] = E_A\,[E[Y_{ij}|A_i]] = E_A[\mu + A_i] = \mu + E_A(A_i) = \mu \tag{1.11}$$

because we are taking $E_A[A_i]$, which is the expectation of A_i over the distribution of A, as being zero.

Note that assuming $E_A[A_i] = 0$ involves no loss of generality to the results (1.9), (1.10) or (1.11). This is because if instead of $E_A[A_i] = 0$ we took $E_A[A_i] = \tau$, say, then (1.11) would become

$$E[Y_{ij}] = \mu + E_A[A_i] = \mu + \tau = \mu', \text{ say.}$$

And then (1.10) would be

$$E[Y_{ij}|A_i] = \mu + A_i = \mu + \tau + A_i - \tau = \mu' + A_i'$$

for $A_i' = A_i - \tau$ with $E_A[A_i'] = \tau - \tau = 0$; and so (1.10) is effectively unaffected. Similarly, (1.9) would become

$$E[Y_{ij}|A_i = a_i] = \mu + a_i = \mu + \tau + a_i - \tau = \mu' + a_i',$$

which is (1.9) with μ' and a_i' in place of μ and a_i, respectively, and the form of (1.10) and (1.11) is retained.

Finally, if there is interest in the particular level of the random effect (e.g., in knowing how the ith clinic differs from the average) then we will be interested in predicting the realized value a_i. On the other hand, if interest lies in the population from which A_i is drawn, we will be interested in $\text{var}(A_i) = \sigma_a^2$.

From now on, for notational convenience, we judiciously ignore the distinction between a random variable and its realized value and let a_i do double duty for both; likewise for y_{ij}.

– iii. *Variance of* y

Having defined $\text{var}(a_i) = \sigma_a^2$, we now consider $\text{var}(y_{ij})$ by first considering the variation that remains in the data after accounting for the random factors. If the data were normally distributed we would typically define a residual error $y_{ij} - \text{E}[y_{ij}|a_i]$ and to it attribute a normal distribution. Equivalently we could simply assert that

$$y_{ij}|a_i \sim \text{ i.i.d. } N(\text{E}[y_{ij}|a_i], \sigma^2). \tag{1.12}$$

Either approach works perfectly well when assuming normality. But for non-normal cases (1.12) is more sensible.

In Example 1 each y_{ij} is a count of the number of seizures. It is therefore quite natural to think that y_{ij} should follow a Poisson model, and assert that

$$y_{ij}|a_i \sim \text{ i.i.d. Poisson}(\text{E}[y_{ij}|a_i]). \tag{1.13}$$

In doing this the "residual" variation is encompassed in the conditional distribution which in (1.13) is taken to be Poisson. If we tried to attribute a distribution to the residual $y_{ij} - \text{E}[y_{ij}|a_i]$ it would be much less natural since, for example, $y_{ij} - \text{E}[y_{ij}|a_i]$ may not even take on integer values. Thus we would have an awkward-to-deal-with distribution.

– iv. *Variance and conditional expected values*

To obtain $\text{var}(y)$ we will often use the formula which relates variance to conditional expected values in order to partition variability:

$$\text{var}(y) = \text{var}(\text{E}[y|u]) + \text{E}[\text{var}(y|u)].$$

For the case of the homoscedastic linear model (1.4) through (1.7), this gives the usual *components of variance* breakdown:

$$\begin{aligned}
\sigma_y^2 \;=\; \mathrm{var}(y_{ij}) \;&=\; \mathrm{var}(\mathrm{E}[y_{ij}|a_i]) + \mathrm{E}[\mathrm{var}(y_{ij}|a_i)] && (1.14)\\
&=\; \mathrm{var}(\mu + a_i) + \mathrm{E}[\sigma^2]\\
&=\; \sigma_a^2 + \sigma^2. && (1.15)
\end{aligned}$$

A similar formula holds for covariances:

$$\mathrm{cov}(y, w) = \mathrm{cov}_u(\mathrm{E}[y|u], \mathrm{E}[w|u]) + \mathrm{E}_u[\mathrm{cov}(y, w|u)], \qquad (1.16)$$

a derivation of which is to be found in Searle et al. (1992, p. 462). Applying this to the homoscedastic linear model gives

$$\begin{aligned}
\mathrm{cov}(y_{ij}, y_{ij'}) \;&=\; \mathrm{cov}(\mu + a_i, \mu + a_i) + \mathrm{E}[0]\\
&=\; \sigma_a^2. && (1.17)
\end{aligned}$$

Thus σ_a^2 is the intra-class covariance, that is, the covariance between every pair of observations in the same class; and $\sigma_a^2/(\sigma_a^2 + \sigma^2)$ is the intra-class correlation coefficient.

c.   Example 5: Ball bearings and calipers

Consider the problem of manufacturing ball bearings to a specified diameter that must be achieved with a high degree of accuracy. Suppose that each of 100 ball bearings is measured with each of 20 micrometer calipers, all of the same brand. Then a suitable model equation for y_{ij}, the diameter of the ith ball bearing measured with the jth caliper, could be

$$\mathrm{E}[y_{ij}] = \mu + a_i + b_j. \qquad (1.18)$$

This is another example of two crossed factors as in Example 2, with the same model equation as in (1.2) except that the symbols a_i and b_j are used rather than α_i and β_j. Thus it is the equation of a different model because a_i and b_j are random effects corresponding, respectively, to the 100 ball bearings being considered as a random sample from the production line, and to the 20 calipers that are being considered as a random sample of calipers from some population of available calipers. Hence in (1.18) each a_i and b_j is treated in the same manner as a_i is treated in Example 4, with the additional property of stochastic independence of the a_is and b_js; thus

$$\mathrm{cov}(a_i, b_j) = 0. \qquad (1.19)$$

In this case, inferences of interest will be those concerning the magnitudes of the variance among ball bearings and the variance among calipers.

1.5 LINEAR MIXED MODELS (LMMs)

a. Example 6: Medications and clinics

Another example of two crossed factors is to suppose that all four dose levels of Example 3 were used in all 20 clinics of Example 4, such that in each clinic each patient was randomly assigned to one of the dose levels. If y_{ijk} is the datum for patient k on dose level j in clinic i, then a suitable model equation for $\mathrm{E}[y_{ijk}]$ would be

$$\mathrm{E}[y_{ijk}] = \mu + a_i + \beta_j + c_{ij}, \tag{1.20}$$

where a_i, β_j and c_{ij} are effects due to clinic i, dose j and clinic-by-dose interaction, respectively. Since, as before, the doses are the only doses considered, β_j is a fixed effect. But the clinics that have been used were chosen randomly, and so a_i is a random effect. Then, because c_{ij} is an interaction between a fixed effect and a random effect, it is a random effect, too. Thus the model equation (1.20) has a mixture of both fixed effects, the β_js, and random effects, the a_is and c_{ij}s. It is thus called a *mixed model*. It incorporates problems relating to the estimation of both fixed effects and variance components. Inferences of interest will be those concerning the effectiveness of the different doses and the variability (variance) among the clinics.

In application to real-life situations, mixed models have broader use than random models, because so often it is appropriate (by the manner in which data have been collected) to have both fixed effects and random effects in the same model. Indeed, every model that contains a μ is a mixed model, because it also contains unexplained variation, and so automatically has a mixture of fixed effects and random elements. In practice, however, the name *mixed model* is usually reserved for any model having both fixed effects (other than μ) and random effects, as well as the customary unexplained variation.

b. Example 7: Drying methods and fabrics

Devore and Peck (1993) report on a study for assessing the smoothness of washed fabric after drying. Each of nine different fabrics were subjected to five methods of drying (line drying; line drying after brief machine tumbling; line drying after tumbling with softener; line drying with air movement; and machine drying). Clearly, method of drying is a fixed effects factor. But how about fabric? If those nine fabrics were specifically chosen as being the only fabrics under consideration, then fabric is a fixed factor. But if the nine fabrics just happened to be the fabrics occurring in a family wash, then it might be reasonable to think of those fabrics as just being a random sample of fabrics from some population of fabrics — and fabric would be a random effect.

Notice that it is what we think is the nature of a factor and of its levels occurring in the data that determines whether a factor is to be called fixed or random. This is discussed further in Section 1.6. As in many mixed models, inference is directed to differences between fixed effects and to the magnitude of the variance among random effects.

c. Example 8: Potomac River Fever

A study of Potomac River Fever in horses was conducted by sampling horses from social groups of horses within 522 farms in New York State (Atwill et al., 1996). The social groups were defined by whether the animals tended to be kept together (i.e., in the same barn or pasture). Breed, gender, and types of animal care (e.g., stall cleaning and frequency of spraying flies) are some of the fixed effects factors that might need to be reckoned with. But farm, and equine social group nested within farm, are clearly random factors.

d. Regression models

Customary regression analysis is based on model equations (for three predictor variables, e.g.) of the form

$$E[y_i] = \beta_0 + \beta_1 x_{1i} + \beta_2 x_{2i} + \beta_3 x_{3i},$$

where the βs are considered to be fixed constants which one estimates from data on y and the xs. But sometimes it is appropriate to think of some βs as being random. When this is so the model is often called a *random coefficients model*.

e. Longitudinal data

A common use of mixed models is in the analysis of longitudinal data, which are defined as data collected on each subject (broadly inter-

preted) on two or more occasions. Methods of analysis have typically
been developed for the situation where the number of occasions is small
compared to the number of subjects. Experiments with longitudinal
data are widely used for at least three reasons: (1) to increase sensitivity by making within-subject comparisons, (2) to study changes
through time, and (3) to use subjects efficiently once they are enrolled
in a study.

The decision as to whether a factor should be fixed or random in a
longitudinal study is often made on the basis of which effects vary with
subjects. That is, subjects are regarded as a random sample of a larger
population of subjects and hence any effects that are not constant for
all subjects are regarded as random.

For example, suppose we are testing a blood pressure drug at each
of two doses and a control dose (dose $= 0$) for each subject in our
study. Individuals clearly have different average blood pressures, so
our model must have a separate intercept for each subject. Similarly,
the response of each subject to increasing dosage of the drug might
vary from subject to subject, so we would model the slope for dose
separately for each subject. To complete our model, we might also
assume that blood pressure changes gradually with a subject's age,
measured at the beginning of the study. If we let y_{ij} denote the blood
pressure measurement taken on the ith subject (of age x_i) on occasion
j at dose d_{ij}, we could then model $\mathrm{E}[y_{ij}]$ as

$$\mathrm{E}[y_{ij}] = a_i + b_i d_{ij} + \gamma x_i. \qquad (1.21)$$

Since the a_i and b_i are specific to the ith subject, they would be declared
random factors. Since γ is the same for all subjects, it is declared fixed.

If we are interested in the overall population response to the drug
we can separate overall terms from the terms specific to each subject.
To do so we rewrite (1.21) as

$$\mathrm{E}[y_{ij}] = (\alpha + a_i') + (\beta + b_i')d_{ij} + \gamma x_i, \qquad (1.22)$$

where $a_i' = a_i - \alpha$ and $b_i' = b_i - \beta$, with α and β being averages over the
population of subjects and are thus fixed effects. On the other hand,
a_i' and b_i' are subject-specific deviations from these overall averages and
so are treated as random effects with means zero and variances σ_a^2 and
σ_b^2. Chapter 8 is devoted to the analysis of longitudinal data.

f.   Example 9: Osteoarthritis Initiative

The OAI (Osteoarthritis Initiative — www.oai.ucsf.edu) was a large study of the determinants of osteoarthritis of the knee. Osteoarthritis causes problems ranging from stiffness and mild pain to severe joint pain and even disability. Persons aged 45 and above at high risk for developing knee osteoarthritis were followed yearly for four years. One of the measured outcomes was a numeric score measuring the degree of knee pain. Let y_{ij} be the pain score for the ith person on jth occasion and let x_{ij} be their age on that occasion. We might hypothesize the model

$$y_{ij} = \beta_0 + b_{0i} + (\beta_1 + b_{1i})x_{ij} + \epsilon_{ij}, \tag{1.23}$$

in which $(\beta_1 + b_{1i})$ is the trend in pain with age for person i.

g.   Model equations

Notice in the preceding examples that equations (1.1), (1.2), (1.3), (1.4), (1.18) and (1.20) are all described as *model equations*. Many writers refer to such equations as models; but this is not correct, because description of a model demands not only a model equation but also explanation of the nature of the terms in such an equation. For example, (1.1) and (1.4) are essentially the same model *equation*, $\mathrm{E}[y_{ij}] = \mu + \alpha_i$ and $\mathrm{E}[y_{ij}] = \mu + a_i$, but the models are not the same. The α_i is a fixed effect but the a_i is a random effect; and this difference, despite the sameness of the model equations, means that the models are different and the analysis of data in the two cases is accordingly different. Moreover, models being different but with the same right-hand sides of their model equation applies not just to whether effects are fixed or random, but can also apply to models that are even more different. For example, $\mu + \alpha_i$, which occurs so often in analysis of variance style models can also occur on the right-hand side of a model equation in a binomial model.

1.6   FIXED OR RANDOM?

Equation (1.2) for modeling cartoon types and groups of people (normal and disabled) is indistinguishable from (1.18) for modeling ball bearings and calipers. But the complete models in these cases are different because of the interpretation attributed to the effects: in the one case, fixed, and in the other, random. In these and the other examples most of the effects are clearly fixed or random; thus drugs and methods of drying are fixed effects, whereas clinics and farms are random effects. But such clear answers to the question "fixed or random?" are not necessarily the norm. Consider the following example.

a. Example 10: Clinical trials

A multicenter clinical trial is designed to judge the effectiveness of a new surgical procedure. If this procedure will eventually become a widespread procedure practiced at a number of clinics, then we would like to select a representative collection of clinics in which to test the procedure and we would then regard the clinics as a random effect.

However, suppose we change the situation slightly. Now assume that the surgical procedure is highly specialized and will be performed mainly at a very few referral hospitals. Also assume that all of those referral hospitals are enrolled in the trial. In such a case we cannot regard the selected clinics as a sample from a larger group of clinics and we will be satisfied with making inferences only to the clinics in the study. We would therefore treat clinic as a fixed effect.

Is is clear that we could envision situations that are intermediate between the "treat clinics as random" scenario and the "treat clinics as fixed" scenario and making the decision between fixed and random would be very difficult. Thus it is that the situation to which a model applies is the deciding factor in determining whether effects are to be considered as fixed or random.

b. Making a decision

Sometimes, then, the decision as to whether certain effects are fixed or random is not immediately obvious. Take the case of year effects, for example, in studying wheat yields: are the effects of years on yield to be considered fixed or random? The years themselves are unlikely to be random, for they will probably be a group of consecutive years over which data have been gathered or experiments run. But the effects on yield may reasonably be considered random, subject, perhaps, to correlation between yields in successive years.

In endeavoring to decide whether a set of effects is fixed or random, the context of the data, the manner in which they were gathered and the environment from which they came are the determining factors. In considering these points the important question is: are the levels of the factor going to be considered a random sample from a population of values which have a distribution? If "yes" then the effects are to be considered as random effects; if "no" then, in contrast to randomness, we think of the effects as fixed constants and so the effects are considered as fixed effects. Thus when inferences will be made about a distribution of effects from which those in the data are considered to be a random sample, the effects are considered as random; and when inferences are going to be confined to the effects in the model, the effects are considered fixed.

Another way of putting it is to ask the questions: "Do the levels of a factor come from a probability distribution?" and "Is there enough information about a factor to decide that the levels of it in the data are like a random sample?" Negative answers to these questions mean that one treats the factor as a fixed effects factor and estimates the effects of the levels; and treating the factor as fixed indicates a more limited scope of inference. On the other hand, affirmative answers mean treating the factor as a random effects factor and estimating the variance component due to that factor. In that case, when there is also interest in the realized values of those random effects that occur in the data, then one can use a prediction procedure for those values.

It is to be emphasized that the assumption of randomness does not carry with it the assumption of normality. Often this assumption *is* made for random effects, but it is a separate assumption made subsequent to that of assuming effects are random. Although many estimation procedures for variance components do not require normality, if distributional properties of the resulting estimators are to be investigated then normality of the random effects is often assumed.

For any factor, the decision tree shown in Figure 1.2 has to be followed in order to decide whether the factor is to be considered as fixed or random. Consider using Figure 1.2 for Example 7 of Section 1.5b where the two factors of interest are five methods of drying and nine different fabrics. To the question atop Figure 1.2, for methods of drying the answer is clearly "no." The five methods cannot be thought of as coming from a probability distribution. But for the other factor, fabrics, it might seem quite reasonable to answer that question with "Yes, the nine fabrics used in the experiment can be thought of as coming from a probability distribution insofar as their propensity for drying is concerned." Thus methods of drying would be treated as fixed effects and fabrics as random. On the other hand, suppose the nine fabrics were nine mixtures of Orlon and cotton being manufactured by one company for a shirt maker. Then those nine fabrics would be the only fabrics of interest to their manufacturer — and in no way would they be thought of as coming from a distribution of fabrics. So they would be treated as fixed.

Figure 1.2: Decision tree for deciding fixed versus random

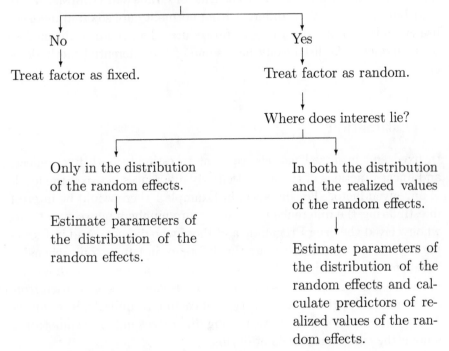

Is it reasonable to assume that levels of the factor come from a probability distribution?

No

Treat factor as fixed.

Yes

Treat factor as random.

Where does interest lie?

Only in the distribution of the random effects.

Estimate parameters of the distribution of the random effects.

In both the distribution and the realized values of the random effects.

Estimate parameters of the distribution of the random effects and calculate predictors of realized values of the random effects.

1.7 INFERENCE

The essence of statistical analysis has three parts: (1) collection of data, (2) summarizing data, and (3) making inferences. Data get considered as samples from populations, and from data one makes inferences about populations. These inferences might well be termed conclusions supported by probability statements. In contrast to conclusions derived by deductions as being rock-solid and immutable, one might say that conclusions drawn from inference are conclusions diluted by probability statements. In any case, that is where the use of statistics usually leads us. We therefore briefly summarize goals we aim for when using inference, and some of the methods involved in doing that. Goals in the use of statistics are principally of three kinds: estimating (including confidence intervals), testing and predicting.

Inference in traditional linear models is based largely on least squares estimation for fixed effects, on analysis of variance sums of squares for estimating variances of random effects, and on normality assumptions for making tests of hypotheses, and for calculating confidence intervals, best predictors and prediction intervals. These procedures will always have their uses for LMs and LMMs. But for GLMs and GLMMs, where distribution assumptions different from normality are so often invoked, use is made of a broader range of procedures than those traditional to linear models. We list briefly here some of the inferential methodologies.

a. Estimation

In fixed and in mixed models we want to estimate both fixed effects and linear functions of them, particularly differences between the levels of any given factor. For instance, in Example 1 there would be interest in estimating the difference in mean seizure numbers between patients who received the drug Progabide and those who did not. And in Example 2 we would want to estimate the difference in humor comprehension between normal and learning-disabled people; and also the difference in the average of the visual-only plus verbal-only types of cartoon from the visual-and-verbal-combined type of cartoon. Similarly, in Example 7 we would be interested in estimating differences in fabric smoothness among the various methods of drying.

When random effects are part of a model we often want to estimate variances of the effects within a factor — and of covariances too, to the extent that they are part of the specification of the random effects. Thus in Example 4 estimating the variance of the clinic effects would be important because it would be an estimate of the variability within the entire population of clinics, not just within the 20 clinics used in the study. And in mixed models, as well as estimating fixed effects we also want to estimate variances of random effects just as in random models. Thus in Example 6 we would estimate differences in seizure numbers as between the various Progabide dose levels; and also estimate the variance among clinics.

– i.   *Maximum likelihood (ML)*

The primary method of estimation we consider throughout this book is *maximum likelihood* (ML). If $\mathbf{y}$ is the data vector and $\boldsymbol{\theta}$ the vector of parameters in the distribution function of $\mathbf{y}$, we can represent that function as $f(\mathbf{y}|\boldsymbol{\theta})$, meaning that for some given value of $\boldsymbol{\theta}$ it is the density function of $\mathbf{y}$. But for $\boldsymbol{\theta}$ being simply the representation of any one of the possible values of $\boldsymbol{\theta}$ we could also rewrite the density function as $L(\boldsymbol{\theta}|\mathbf{y}) = f(\mathbf{y}|\boldsymbol{\theta})$, which is called the *likelihood*. This is a function of $\boldsymbol{\theta}$ and ML is the process of finding that value of $\boldsymbol{\theta}$ which maximizes $L(\boldsymbol{\theta}|\mathbf{y})$. For mathematically tractable density functions this process can be quite straightforward, yielding a single, algebraic expression for the maximizing $\boldsymbol{\theta}$ as a function of $\mathbf{y}$. But for difficult functions it can demand iterative numerical methods and may not always yield a single value for the maximizing $\boldsymbol{\theta}$. Naturally, this presents problems when applying ML to real data.

– ii.   *Restricted maximum likelihood (REML)*

A method related to ML is *restricted* (or *residual*) *maximum likelihood* (REML), which involves the idea of applying ML to linear functions of $\mathbf{y}$, say $\mathbf{K}'\mathbf{y}$, for which $\mathbf{K}'$ is specifically designed so that $\mathbf{K}'\mathbf{y}$ contains none of the fixed effects which are part of the model for $\mathbf{y}$. So in ML, replace $\mathbf{y}$ with $\mathbf{K}'\mathbf{y}$ and one has REML. Historically REML was derived only for the case of linear mixed models (Patterson and Thompson, 1971) but has been generalized to nonlinear models (e.g., Schall, 1991).

Two valuable consequences of using REML are first, that variance components are estimated without being affected by the fixed effects. This means that the variance estimates are invariant to the values of the fixed effects. Second, in estimating variance components with REML, degrees of freedom for the fixed effects are taken into account implicitly, whereas with ML they are not. The simplest example of this is in estimating σ^2 from normally distributed data y_i, which we denote as $y_i \sim \mathcal{N}(\mu, \sigma^2)$ for $i = 1, 2, \ldots, n$. With $\bar{y} = \sum_{i=1}^{n} y_i / n$ and $S_{yy} = \sum_{i=1}^{n}(y_i - \bar{y})^2$, the REML estimator of σ^2 is $S_{yy}/(n-1)$ whereas the ML estimator is S_{yy}/n. Further examples appear in Sections 2.1 and 3.1, and a full discussion of ML and REML is given in Chapter 6. Derivation of the preceding results involving S_{yy} is covered in Exercise E 1.6.

Beyond LMMs, the ideas of REML can be generalized directly to non-normal models where, in limited cases, a linear function of **y** can be constructed to contain none of the fixed effects. However, for non-normal and nonlinear models, other, alternative "definitions" of REML have been put forth. Two examples are the solutions of equations that equate quadratic forms of predicted random effects with their expected values, and the maximization of a likelihood after "integrating out" the fixed effects.

Whichever definition is adopted, note that REML does nothing about estimating fixed effects. This is because all REML methods are designed to be free of the fixed effects portion of a model.

– iii.   *Solutions and estimators*

Estimators such as ML or REML are found by maximizing a function of the parameters (the likelihood or restricted likelihood) within the bounds of the parameter space. In general this problem, the maximization of a nonlinear function within a constrained region, is quite difficult. For many models considered in this book, we will be able to do little more than point the reader in the direction of numerical methods for finding the numerical *estimates* for a particular data set. For other models we can be more explicit.

In cases where explicit, closed-form solutions exist for the maximizing values, an oft-successful method for finding those solutions is to differentiate the likelihood or restricted likelihood and set the derivatives equal to zero. From the resulting equations, solutions for the parameter symbols might well be thought of as ML or REML estimators, in which case they would be denoted by a "hat" or tilde over the parameter symbol. However, this is not a fail-safe method because, in some cases, solutions to these estimating equations may not be in the parameter space. For example, in some cases of ML or REML a solution for an estimated variance is such that it is possible for it to be negative; that is, it is possible for data to be such that the solution is a negative value (e.g., Section 2.2). Since, under ML and REML, negative estimates of positive parameters are not acceptable, we will often denote solutions with a dot above the parameter (e.g., $\dot{\sigma}_a^2$). We then proceed to adjust them (in accord with established procedures) to yield the ML estimators which will be denoted in the usual way with a "hat" above the parameter symbol. Thus for ν denoting a variance we can have

$$\nu \;\; = \;\; \text{parameter}$$

$$\dot{\nu} \;=\; \text{solution} \qquad (1.24)$$

$$\hat{\nu} \;=\; \text{estimator.}$$

– iv. *Bayes theorem*

Even for fixed effects one school of thought in statistics is to assume that they are random variables with a distribution $\Pi(\boldsymbol{\theta})$. This is called the *prior distribution* of $\boldsymbol{\theta}$. Then, because

$$f(\mathbf{y},\,\boldsymbol{\theta}) \;=\; f(\mathbf{y}|\boldsymbol{\theta})\Pi(\boldsymbol{\theta}) = \Pi(\boldsymbol{\theta}|\mathbf{y})f(\mathbf{y}), \qquad (1.25)$$

we have

$$\Pi(\boldsymbol{\theta}|\mathbf{y}) \;=\; \frac{f(\mathbf{y}|\boldsymbol{\theta})\Pi(\boldsymbol{\theta})}{f(\mathbf{y})} = \frac{f(\mathbf{y}|\boldsymbol{\theta})\Pi(\boldsymbol{\theta})}{\int f(\mathbf{y}|\boldsymbol{\theta})\Pi(\boldsymbol{\theta})d\boldsymbol{\theta}}. \qquad (1.26)$$

This is called the *posterior density* of $\boldsymbol{\theta}$; and the mean of $\boldsymbol{\theta}|\mathbf{y}$ derived from this density is an often-used estimator of $\boldsymbol{\theta}$ — a Bayes estimator.

Suppose that $\Pi(\boldsymbol{\theta})$ and $\Pi(\boldsymbol{\theta}|\mathbf{y})$ themselves involve parameters φ so that

$$f(\mathbf{y}) = \frac{f(\mathbf{y}|\boldsymbol{\theta})\Pi(\boldsymbol{\theta}|\varphi)}{\Pi(\boldsymbol{\theta}|\mathbf{y},\varphi)} \qquad (1.27)$$

from (1.26). There can be cases where φ can be estimated as a function of y from (1.27) using, for example, marginal maximum likelihood. Then, on using those estimates in (1.26), the $\mathrm{E}[\boldsymbol{\theta}|\mathbf{y}]$ derived from that adaptation of $f(\boldsymbol{\theta}|\mathbf{y})$ is known as an empirical Bayes estimate of $\boldsymbol{\theta}$. This is essentially the approach we follow for random effects in later chapters, assigning them a distribution such as a normal distribution, and the best predicted values described in Chapter 13 are essentially empirical Bayes estimates.

– v. *Quasi-likelihood estimation*

In many problems of statistical estimation we know some detail of the distribution governing the data, but may be unwilling to specify it exactly. This precludes the use of maximum likelihood, which requires exact specification of the distribution in order to construct the likelihood. The idea of quasi-likelihood, developed by Wedderburn (1974), addresses this concern. This is a method of estimation that requires only a model for the mean of the data and the relationship between the mean and the variance, yet in many cases retains the full, or nearly full, efficiency of maximum likelihood. Since the input is minimal, the method is robust to misspecification of finer details of the model. Section 5.6 contains further details.

– vi.   *Generalized estimating equations*

To capture some of the beneficial aspects of quasi-likelihood estimation in the context of correlated data models, Zeger and Liang (1986) and Liang and Zeger (1986) developed *generalized estimating equations* methods (GEEs). These methods are robust in the presence of misspecification of the variance-covariance structure of data. Estimates using GEEs are often easier to compute than maximum likelihood estimates but may have different interpretations and magnitudes than MLEs. GEEs are discussed in Section 9.3.

b.   Testing

Insofar as testing is concerned one's usual interests are to test hypotheses about the parameters (and/or functions of them) which have been estimated as described above. With fixed effects, we test hypotheses of the form that differences between levels of a factor are zero, or occasionally that they equal some pre-decided constant. And for random effects a useful hypothesis is that a variance component is zero — or, occasionally, that it equals some pre-decided value.

Ancillary to all these cases we often also want to use parameter estimates to establish confidence intervals for parameters, or for combinations of them.

– i.   *Likelihood ratio test (LRT)*

Traditional analysis of variance methodology (under normality assumptions) leads to hypothesis tests involving F-statistics which are ratios of mean squares. These statistics can also be shown to be an outcome of the *likelihood ratio test* (LRT), first propounded by Neyman and Pearson (1928). This in its general form can be applied much more broadly than to traditional ANOVA, and is therefore useful for GLMs and GLMMs. That general form can be described as follows. Let $\hat{\boldsymbol{\theta}}$ be the maximizing $\boldsymbol{\theta}$ over the complete range of values of each element of $\boldsymbol{\theta}$. Similarly, let $\hat{\boldsymbol{\theta}}_0$ be the maximizing $\boldsymbol{\theta}$, limited (restricted or defined) by some hypothesis H pertaining to some elements of $\boldsymbol{\theta}$; and let $L(\hat{\boldsymbol{\theta}}_0)$ be the value of the likelihood using $\hat{\boldsymbol{\theta}}_0$ for $\boldsymbol{\theta}$. Then the likelihood ratio is $L(\hat{\boldsymbol{\theta}}_0)/L(\hat{\boldsymbol{\theta}})$; and it leads to a test statistic for the hypothesis H.

– ii.   *Wald's procedure*

Another very general procedure for developing a hypothesis test, known as *Wald's test* (Wald, 1941), is that if $\hat{\boldsymbol{\theta}}$ is an estimate of $\boldsymbol{\theta}$ and $\mathbf{I}(\boldsymbol{\theta})$ is the information matrix for $\hat{\boldsymbol{\theta}}$, then $(\hat{\boldsymbol{\theta}} - \boldsymbol{\theta}_*)'[\mathbf{I}(\boldsymbol{\theta}_*)]^{-1}(\hat{\boldsymbol{\theta}} - \boldsymbol{\theta}_*)$ is a test statistic for the hypothesis $H : \boldsymbol{\theta} = \boldsymbol{\theta}_*$; and it has, under some conditions, approximately a χ_p^2 distribution (p being the order of $\boldsymbol{\theta}$). For LMs this is exactly χ_p^2, or gets modified to exactly $\mathcal{F}$ when estimates are used in $\mathbf{I}(\boldsymbol{\theta})$. And for $p = 1$, the signed square root of this quadratic form in $\hat{\boldsymbol{\theta}}$ has approximately a normal distribution with zero mean and unit variance.

c.   **Prediction**

Finally, there is prediction. When dealing with a random effects factor the random effects occurring in the data are realizations of a random variable. But they are unobservable. Nevertheless, in many situations we would like to use the data to put some sort of numerical values, or predicted values on those realizations. They may be useful for selecting superior realizations: for example, picking superior clinics in our clinic example. It turns out (see Chapter 13) that the best predictor (minimum mean squared error) is a conditional mean. Thus if **y** represents the data the best predictor (BP) of a_i is $\mathrm{BP}(a_i) = \mathrm{E}[a_i|\mathbf{y}]$. And, similar to a confidence interval, we will at times also want to calculate a prediction interval for a_i using $\mathrm{BP}(a_i)$.

1.8   COMPUTER SOFTWARE

Nowadays there is a host of computer software packages that will compute some or many of the analysis methods described in the following chapters. Their existence is exceedingly important and useful to today's disciplines of statistics and data analysis. Nevertheless, this is not a book on software, its merits, its demerits, or the mechanics of using it. Software expands and (usually) improves so rapidly that whatever we might write about it today would be at least somewhat outdated by the time the book is read. So this is a statistics, not a software, user's manual. Very occasionally we mention SAS, which is a widely available package that can compute much of what we describe. In doing so, we give few or no details — and hope that our mentioning of only a single package is not construed as anything negative about the many other packages!

1.9   EXERCISES

E 1.1 Suppose a clothing manufacturer has collected data on the number of defective socks it makes. There are six subsidiary companies (factor C) that make knitted socks. At each company, there are five brands (B) of knitting machines with 20 machines of each brand at each company. All machines of all brands are used on the different types of yarn (Y) from which socks are made: cotton, wool, and nylon. At each company, data have been collected from just two machines (M) of each brand for operation by each of four locally resident workers, using each of the yarns. And on each occasion the number of defective socks in each of two replicate samples of 100 socks is recorded.

Which factors do you think should be treated as fixed and which as random? Give reasons for your decisions.

E 1.2 For $\mathrm{E}[y_{ij}|a_i] = \mu + a_i$ and $\mathrm{var}(y_{ij}|a_i) = \sigma^2$ use formulae (1.14) and (1.16) to derive $\mathrm{var}(y_{ij}) = \sigma_a^2 + \sigma^2$ and $\mathrm{cov}(y_{ij}, y_{i\ell}) = \sigma_a^2$.

E 1.3 For $y_{ij} \sim \mathrm{Poisson}(\lambda_i)$, repeat E 1.2; take $\lambda_i = e^{\mu + a_i}$, with $a_i \sim \mathcal{N}(0, \sigma_a^2)$. *Hint:* $\mathrm{E}[e^{a_i}]$ is the moment-generating function of a_i, i.e., $\mathrm{E}[e^{ta_i}]$, with t set equal to 1.

E 1.4 For $y_{ij} \sim \mathcal{N}(\mu_i, \sigma^2)$ with $i = 1, 2$, use (a) the LRT and (b) Wald's procedure to test $H_0: \mu_1 = \mu_2$.

E 1.5 Derive (1.14), starting from

$$\begin{aligned}
\mathrm{var}(y) &= \mathrm{E}\left(y - \mathrm{E}[y]\right)^2 \\
&= \mathrm{E}\left(y - \mathrm{E}[y|u] + \mathrm{E}[y|u] - \mathrm{E}[y]\right)^2 \\
&= \mathrm{E}_u\left[\mathrm{E}\left(y - \mathrm{E}[y|u] + \mathrm{E}[y|u] - \mathrm{E}[y]\right)^2|u\right].
\end{aligned}$$

In doing so, explain why, in the expansion of the squared term, the cross-product is zero.

E 1.6 With $y_t \sim$ i.i.d. $\mathcal{N}(\mu, \sigma^2), t = 1, 2, \ldots, n$ and $\mathbf{y}' = [y_1, y_2, \ldots, y_n]$ we have $\mathbf{y} \sim \mathcal{N}(\mu \mathbf{1}_n, \sigma^2 \mathbf{I}_n)$. Then the distribution function for $\mathbf{y}$ is

$$f(\mathbf{y}) = \exp\left[-\tfrac{1}{2}(\mathbf{y} - \mu\mathbf{1})'(\mathbf{y} - \mu\mathbf{1})/\sigma^2\right]/(2\pi\sigma^2)^{n/2}$$

and the log likelihood is

$$l_1 = -\tfrac{1}{2}(\mathbf{y} - \mu\mathbf{1})'(\mathbf{y} - \mu\mathbf{1})/\sigma^2 - \frac{n}{2}\log\sigma^2 - \frac{n}{2}\log 2\pi.$$

(a) By differentiating l_1 with respect to μ and σ^2 show that the ML estimators are

$$\hat{\mu} = \bar{y}_. = \sum_{t=1}^{n} y_t/n$$

and

$$\tilde{\sigma}_1^2 = \sum_{t=1}^{n} (y_t - \bar{y})^2/n.$$

(b) For REML estimation of σ^2, $\mathbf{K}'$ for $\mathbf{K}'\mathbf{y}$ such that $\mathbf{K}'\mathbf{1} = \mathbf{0}$ can be taken as the first $n-1$ rows of $\mathbf{C}_n = \mathbf{I}_n - \mathbf{J}_n/n$. Thus

$$\mathbf{K}' = [\mathbf{I}_{n-1} \quad \mathbf{0}\mathbf{1}_{n-1}] - \frac{1}{n}\mathbf{J}_{(n-1)\times n}$$

of order $(n-1) \times n$. Then

$$f(\mathbf{K}'\mathbf{y}) = \frac{\exp\left[-\frac{1}{2}\mathbf{y}'\mathbf{K}(\mathbf{K}'\mathbf{K})^{-1}\mathbf{K}'\mathbf{y}/\sigma^2\right]}{(2\pi\sigma^2)^{(n-1)/2} \mid \mathbf{K}'\mathbf{K} \mid^{1/2}}.$$

Thus the log likelihood of $\mathbf{K}'\mathbf{y}$ is

$$l_2 = -\frac{\mathbf{y}'\mathbf{K}(\mathbf{K}'\mathbf{K})^{-1}\mathbf{K}'\mathbf{y}}{2\sigma^2} - \frac{1}{2}\log \mid \mathbf{K}'\mathbf{K} \mid^{1/2} - \frac{n-1}{2}\log \sigma^2.$$

By differentiating this with respect to σ^2, show that the REML estimator of σ^2 is

$$\sigma_{\text{REML}}^2 = \frac{\mathbf{y}'\mathbf{K}(\mathbf{K}'\mathbf{K})^{-1}\mathbf{K}'\mathbf{y}}{n-1}.$$

(c) Show that the REML estimator of σ^2 is $\sum_{t=1}^{n}(y_t-\bar{y})^2/(n-1)$. *Hint:* To do so, show that $\mathbf{K}'\mathbf{K} = \mathbf{I}_{n-1} - \mathbf{J}_{n-1}/n$ and $\mathbf{K}'\mathbf{y} = \left\{ _c y_t - \bar{y} \right\}_{t=1}^{n-1}$.

Chapter 2

ONE-WAY CLASSIFICATIONS

We begin by describing fixed and random effects models for the one-way classification for both normally and Bernoulli (binary) distributed data. Not only do these constitute a convenient starting point for explaining many of the concepts described later in the book, but they are also commonly employed in practice.

For example, in a modification of the comprehension of humor example (Section 1.3b) suppose that we have three cartoons, each of a different type (visual only, linguistic only, and visual-linguistic combined) and each is rated by separate people on a scale from 1 to 9, where 9 represents extremely funny and 1 represents not funny at all. We might consider the responses as approximately normally distributed and be interested in assessing whether the mean rating is the same for the three cartoons. Alternatively or additionally, we might measure a yes/no response of whether the rater "got" the cartoon. The goal is the same: to compare the cartoons, but now, because of the binary nature of the data ("yes" or "no"), it is no longer valid to consider the data as approximately normally distributed. Statistical techniques acknowledging the binary nature of the data would be required.

If the inferential goal were to compare cartoon types, it would be insufficient to consider only a single cartoon of each type. A different modification of the humor example might have only visual cartoons, but would test, say, 15 different cartoons of this type. In this scenario, it is likely that we would regard cartoon as a random effect and the response could be either humor rating (from 1 to 9) or "got it?" (yes/no), or both.

In this chapter we first consider normally distributed data with a fixed or random classification and then consider Bernoulli distributed data. In dealing with the one-way classification throughout the chapter we let m be the number of classes and n_i the number of observations in the ith class. Then, with y_{ij} being the jth observation in the ith class, we have $i = 1, 2, \ldots, m$ and $j = 1, 2, \ldots, n_i$.

2.1   NORMALITY AND FIXED EFFECTS

a.   Model

A model for responses y_{ij} is

$$E[y_{ij}] = \mu_i \text{ with } y_{ij} \sim \text{indep. } \mathcal{N}(\mu_i, \sigma^2), \tag{2.1}$$

where "indep." means that the random variables are mutually independent and $\mathcal{N}(\mu_i, \sigma^2)$ indicates a normal distribution with mean μ_i and variance σ^2. An alternative but equivalent specification to (2.1) is the *overparameterized model:* :

$$E[y_{ij}] = \mu + \alpha_i \text{ with } y_{ij} \sim \text{indep. } \mathcal{N}(\mu + \alpha_i, \sigma^2), \tag{2.2}$$

so-called because the mean of y is a function of more parameters than there are distinct values for the mean.

b.   Estimation by ML

Derivation of maximum likelihood estimators of the parameters requires the likelihood. Using (2.1) it is

$$L = (2\pi\sigma^2)^{-N/2} \exp\left[-\frac{1}{2\sigma^2} \sum_i \sum_j (y_{ij} - \mu_i)^2 \right] \tag{2.3}$$

for $N \equiv \sum_{i=1}^m n_i$. Then

$$l = \log L = -\frac{N}{2}\log(2\pi) - \frac{N}{2}\log(\sigma^2) - \frac{1}{2\sigma^2}\sum_i\sum_j(y_{ij} - \mu_i)^2.$$

The derivatives are

$$\frac{\partial l}{\partial \mu_k} = -\frac{1}{2\sigma^2}\sum_j (y_{kj} - \mu_k)(-2) \qquad \text{and} \tag{2.4}$$

$$\frac{\partial l}{\partial \sigma^2} = -\frac{N}{2\sigma^2} + \frac{1}{2\sigma^4}\sum_i\sum_j(y_{ij} - \mu_i)^2.$$

Setting these equal to zero gives solutions to the ML equations which, in this case, are the ML estimators; they are denoted using "hats":

$$\hat{\mu}_i = \bar{y}_{i.} \tag{2.5}$$

$$\hat{\sigma}^2 = \frac{1}{N}\sum_i\sum_j(y_{ij} - \hat{\mu}_i)^2 = \frac{1}{N}\sum_i\sum_j(y_{ij} - \bar{y}_{i.})^2.$$

These ML solutions are indeed ML estimators because they do not lie outside their corresponding parameter ranges (see Section 2.2b–iii), namely, $-\infty < \mu_i < \infty$ and $0 \le \sigma^2$.

It is easily shown that $\hat{\mu}_i$ is unbiased; but $\hat{\sigma}^2$, the ML estimator of σ^2 in (2.5), is biased since

$$E[\hat{\sigma}^2] = E\left[\frac{1}{N}\sum_i(n_i - 1)s_i^2\right] = \frac{N - m}{N}\sigma^2 \ne \sigma^2, \tag{2.6}$$

where $s_i^2 = 1/(n_i - 1)\sum_j(y_{ij} - \bar{y}_{i.})^2$ is the sample variance for the ith class. A common modification is to use an unbiased estimator:

$$s^2 = \frac{1}{N - m}\sum_i\sum_j(y_{ij} - \bar{y}_{i.})^2.$$

This is also the REML (see Section 1.7a–ii) estimator. Derivation of this and some other estimators (see Sections 2.2b-vi and 3.2c) is left to Chapter 6, wherein a general equation is given for REML estimator of variances. And exercises in that chapter require using that general equation to obtain results for some simple cases such as those here.

The variances of these unbiased estimators are easily derived:

$$\text{var}(\hat{\mu}_i) = \frac{\sigma^2}{n_i} \quad \text{and} \quad \text{var}(s^2) = \frac{2\sigma^4}{N - m}. \tag{2.7}$$

This result for s^2 follows easily from the standard result for the variance of a sample variance for a single sample from a normal distribution:

$$\text{var}(s_i^2) = \frac{2\sigma^4}{n_i - 1}.$$

If we work with model (2.2) the derivatives of the log likelihood are

$$\frac{\partial l}{\partial \mu} = \frac{1}{\sigma^2}\sum_i\sum_j(y_{ij} - \mu - \alpha_i) \quad \text{and} \tag{2.8}$$

$$\frac{\partial l}{\partial \alpha_k} = \frac{1}{\sigma^2}\sum_j(y_{kj} - \mu - \alpha_k).$$

Setting these equal to zero gives the equations

$$\hat{\mu} + \hat{\alpha}_k = \bar{y}_{k\cdot} \tag{2.9}$$
$$\hat{\mu} + \sum n_i \hat{\alpha}_i / N = \bar{y}_{\cdot\cdot} \ .$$

The latter equation is redundant since it equals the sum of each of the former equations multiplied by its n_k/N. Hence there is no unique solution to the equations. But we can get *a* solution by placing a constraint on the α_i. A commonly used constraint, which clearly makes the equations easy to solve, is $\sum n_i \hat{\alpha}_i = 0$, which then yields

$$\hat{\mu} = \bar{y}_{\cdot\cdot} \qquad \text{and} \qquad \hat{\alpha}_k = \bar{y}_{k\cdot} - \bar{y}_{\cdot\cdot} \tag{2.10}$$

However, some people find it distasteful to have a constraint which depends on the sample sizes n_i, as is inherent in (2.10). These may be, to some extent, random variables for a given experiment.

This gives the ML estimators of μ and α_i. For σ^2, (2.2) gives the same estimator as does (2.1).

c.   Generalized likelihood ratio test

A starting point for many statistical investigations is to test the hypothesis

$$H_0 : \mu_i \text{ all equal.}$$

Denoting the common value of μ_i under H_0 as μ, the ML estimator under H_0 is $\hat{\mu}_0 = \bar{y}_{\cdot\cdot}$ and the corresponding ML estimator of σ^2 is

$$\hat{\sigma}_0^2 = \frac{1}{N} \sum_i \sum_j (y_{ij} - \bar{y}_{\cdot\cdot})^2$$
$$= \frac{1}{N} \left[\sum_i \sum_j (y_{ij} - \bar{y}_{i\cdot})^2 + \sum_i \sum_j (\bar{y}_{i\cdot} - \bar{y}_{\cdot\cdot})^2 \right].$$

This gives a generalized likelihood ratio statistic (see Section 1.7b–i) of

$$\Lambda = \frac{(2\pi\hat{\sigma}_0^2)^{-N/2} \exp\left(-N\hat{\sigma}_0^2 / 2\hat{\sigma}_e^2\right)}{(2\pi\hat{\sigma}^2)^{-N/2} \exp\left(-N\hat{\sigma}^2 / 2\hat{\sigma}_e^2\right)}$$
$$= \left(\frac{\hat{\sigma}_0^2}{\hat{\sigma}^2}\right)^{-N/2} . \tag{2.11}$$

Taking logarithms and multiplying by -2 gives

$$-2\log\Lambda \;=\; N\log\frac{\hat{\sigma}_0^2}{\hat{\sigma}^2}$$

$$=\; N\log\left(\frac{\sum_i\sum_j(y_{ij}-\bar{y}_{i\cdot})^2+\sum_i\sum_j(\bar{y}_{i\cdot}-\bar{y}_{\cdot\cdot})^2}{\sum_i\sum_j(y_{ij}-\bar{y}_{i\cdot})^2}\right)$$

$$=\; N\log\left(1+\frac{m-1}{N-m}F\right),\qquad\qquad(2.12)$$

where

$$F=\frac{\sum_i\sum_j(\bar{y}_{i\cdot}-\bar{y}_{\cdot\cdot})^2/(m-1)}{\sum_i\sum_j(y_{ij}-\bar{y}_{i\cdot})^2/(N-m)}$$

is the usual F-statistic from the analysis of variance for a one-way classification. Under $H_0 : \mu_i$ all equal, F has a central $\mathcal{F}$-distribution with numerator degrees of freedom $m-1$ and denominator degrees of freedom $N-m$. We denote this by $F\sim\mathcal{F}_{N-m}^{m-1}$.

Rejecting the null hypothesis when the likelihood ratio, Λ, is small ("the null hypothesis is unlikely") is equivalent to $-2\log\Lambda$ being large or $F\geq\mathcal{F}_{N-m,1-\alpha}^{m-1}$, the $100(1-\alpha)\%$ percentile of the $\mathcal{F}$–distribution (i.e., $P\{F_{N-m}^{m-1}\geq\mathcal{F}_{N-m,1-\alpha}^{m-1}\}=\alpha$) .

Asymptotic theory (Mood et al., 1974) tells us that the large-sample distribution of $-2\log\Lambda$ under H_0 is chi-square with degrees of freedom equal to the difference in the number of parameters in the parameter space and the number under H_0. In this situation there are $m+1$ parameters $\mu_1,\mu_2,\dots,\mu_m$ and σ^2 . Under H_0 there is a single μ-parameter and σ^2, so the difference is $m+1-2=m-1$.

The large-sample test would thus be to reject H_0 when $-2\log\Lambda\geq\chi_{m-1,1-\alpha}^2$, where the percentile is defined as with the $\mathcal{F}$-distribution above. Exact distribution theory based on the $\mathcal{F}$-distribution is to reject H_0 when $-2\log\Lambda\geq N\log\left(1+\frac{m-1}{N-m}\mathcal{F}_{N-m,1-\alpha}^{m-1}\right)$. How do these compare? Figure 2.1 plots the χ^2- and $\mathcal{F}$-based critical values for $\alpha = 0.05$, $m = 2$, 5, and 10 versus N. For total sample sizes $N < 50$ the differences can be appreciable. For the impact of using the chi-square critical values see E 2.2.

d.   Confidence intervals

Confidence intervals for μ_i, $\mu_i-\mu_k$, and σ^2 are easily derived and widely available (Snedecor and Cochran, 1989, Chapters 5 and 6). For completeness we list them here. In doing so we emphasize that these

Figure 2.1: Critical values based on $\mathcal{F}$-distribution (dashed lines) and χ^2-distributions (solid lines) plotted versus N for $m = 10$ (top set of lines), 5 (middle set of lines), and 2 (bottom set of lines).

Critical Value

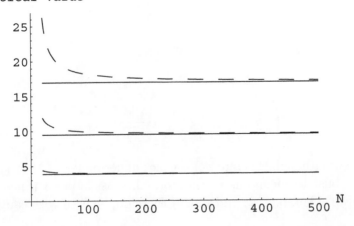

intervals are for fixed effects models only, not for mixed models. For example, with $\mu_i = \mu + \alpha_i$ the interval shown here for $\mu_i - \mu_k = \alpha_i - \alpha_k$ is for the αs being fixed effects. The case of mixed models is considered in Chapter 6.

– i.   *For means*

A confidence interval for μ_i is given by

$$\bar{y}_{i\cdot} \pm t_{N-m,\alpha/2}\frac{s}{\sqrt{n_i}},$$

where $t_{\nu,\alpha}$ denotes the upper $100\alpha\%$ percentile of the $\mathcal{T}$-distribution on ν degrees of freedom, that is,

$$P\{\mathcal{T}_\nu > t_{\nu,\alpha}\} = \alpha.$$

– ii.   *For differences in means*

A confidence interval for $\mu_i - \mu_k$ is given by

$$\bar{y}_{i\cdot} - \bar{y}_{k\cdot} \pm t_{N-m,\alpha/2}s\sqrt{\left(\frac{1}{n_i} + \frac{1}{n_j}\right)}.$$

– iii.   *For linear combinations*

A linear combination of the means is defined as $\sum_i c_i \mu_i$ where the c_i are known constants. A confidence interval for $\sum_i c_i \mu_i$ is given by

$$\sum_i c_i \bar{y}_{i\cdot} \pm t_{N-m,\alpha/2} s \sqrt{\sum_i \frac{c_i^2}{n_i}}.$$

– iv.   *For the variance*

For σ^2 the confidence interval is given by

$$\left(\frac{(N-m)s^2}{\chi^2_{N-m,1-\alpha/2}}, \frac{(N-m)s^2}{\chi^2_{N-m,\alpha/2}} \right).$$

We note that although this interval is exact under model (2.1) or (2.2), it is not robust to violations of the normality assumption (Snedecor and Cochran, 1989, p. 252) and should therefore be used with caution.

e.   Hypothesis tests

Hypothesis tests concerning the means are straightforward. For example, to test

$$H_0 : \sum_i c_i \mu_i = \eta_0,$$

where η_0 is a hypothesized value, we use the t-statistic

$$t = \frac{\sum_i c_i \bar{y}_{i\cdot} - \eta_0}{s \sqrt{\sum_i c_i^2 / n_i}}.$$

If the alternative is $H_A : \sum_i c_i \mu_i \neq \eta_0$, we reject when $|t| > t_{N-m,\alpha/2}$; if the alternative is $H_A : \sum_i c_i \mu_i > \eta_0$, we reject when $t > t_{N-m,\alpha}$; if the alternative is $H_A : \sum_i c_i \mu_i < \eta_0$, we reject when $t < -t_{N-m,\alpha}$.

2.2   NORMALITY, RANDOM EFFECTS AND MLE

a.   Model

In accordance with Section 1.1, when we assume a random effects classification we attribute a distribution to the effects of the levels of the factor.

A model corresponding to (2.1) is

$$
\begin{aligned}
E[y_{ij}] &= \mu_i \\
y_{ij} &\sim \text{indep. } \mathcal{N}(\mu_i, \sigma^2) \\
\mu_i &\sim \text{i.i.d. } \mathcal{N}(\mu, \sigma_\mu^2),
\end{aligned}
\tag{2.13}
$$

where we have used the usual notation, i.i.d., to indicate that the random variables are mutually *independent with identical distributions*. Since μ_i appears in the expected value of y_{ij} but is later assumed to be random, (2.13) is a somewhat sloppy specification of the distributions. More precisely, the conditional distribution of y_{ij} given μ_i (we indicate the conditioning on μ_i with the vertical bar) is normal with mean μ and variance σ^2 and the distribution of μ_i is $\mathcal{N}(\mu, \sigma_\mu^2)$. This is written as follows:

$$
\begin{aligned}
E[y_{ij}|\mu_i] &= \mu_i \\
y_{ij}\,|\mu_i &\sim \text{indep. } \mathcal{N}(\mu_i, \sigma^2) \\
\mu_i &\sim \text{i.i.d. } \mathcal{N}(\mu, \sigma_\mu^2).
\end{aligned}
\tag{2.14}
$$

An equivalent model, corresponding to (2.2), is traditionally written as

$$
\begin{aligned}
E[y_{ij}|a_i] &= \mu + a_i \\
y_{ij}|a_i &\sim \text{indep. } \mathcal{N}(\mu + a_i, \sigma^2) \\
a_i &\sim \text{i.i.d. } \mathcal{N}(0, \sigma_a^2),
\end{aligned}
\tag{2.15}
$$

where the notation $\sigma_a^2 = \sigma_\mu^2$ is now used in place of σ_μ^2.

It is appropriate at this stage to contrast the random effects model with a Bayesian approach. In a Bayesian approach the parameters μ_i would be assumed to have a distribution just as does the random effects model, (2.14) or (2.15). However, the similarity ends there. In a true Bayesian approach the distribution of the μ_i would represent subjective information on the μ_i, not a distribution across tangible populations (e.g., across animals). The Bayesian approach would further hypothesize a distribution for all other unknown parameters (σ^2 in this case). The method of estimating the parameters would also be different.

– i. *Covariances caused by random effects*

A fundamental difference between the fixed and random effects models is that the observations, y_{ij}, in a random effects model are not independent. In fact, the assumption of a random factor can be viewed as a convenient way to specify a variance-covariance structure. Essentially, observations with model equations that contain the same random effect are correlated. Using (1.6) and (1.16) yields

$$
\begin{aligned}
\mathrm{cov}(y_{ij}, y_{il}) &= \mathrm{cov}(E[y_{ij}|a_i], E[y_{il}|a_i]) + E[\mathrm{cov}(y_{ij}, y_{il}|a_i)] \\
&= \mathrm{cov}(\mu + a_i, \mu + a_i) + 0 \\
&= \mathrm{cov}(a_i, a_i) = \sigma_a^2.
\end{aligned}
\tag{2.16}
$$

Also

$$
\begin{aligned}
\mathrm{var}(y_{ij}) &= \mathrm{var}(E[y_{ij}|a_i]) + E[\mathrm{var}(y_{ij}|a_i)] \\
&= \mathrm{var}(\mu + a_i) + E[\sigma^2] \\
&= \sigma_a^2 + \sigma^2.
\end{aligned}
\tag{2.17}
$$

Thus we have an *intraclass correlation* of

$$
\mathrm{corr}(y_{ij}, y_{il}) = \frac{\sigma_a^2}{\sqrt{(\sigma_a^2 + \sigma^2)(\sigma_a^2 + \sigma^2)}} = \frac{\sigma_a^2}{\sigma_a^2 + \sigma^2}.
\tag{2.18}
$$

– ii.   *Likelihood*

Since observations within the same level of a random effect are correlated the likelihood for the random model is more complicated than for the fixed effects model. For $\mathbf{y}_i = [y_{i1}\ y_{i2} \ldots y_{in_i}]'$ the model (2.15) has

$$
\mathbf{y}_i \sim \mathcal{N}(\mu \mathbf{1}_{n_i}, \mathbf{V}_i),
$$

where $\mathbf{V_i} = \sigma^2 \mathbf{I}_{n_i} + \sigma_a^2 \mathbf{J}_{n_i}$; $\mathbf{I}_n$ is the identity matrix of order n, $\mathbf{J}_n$ is an $n \times n$ matrix of all ones, and $\mathbf{1}_n$ is a column vector of all ones of order n. It is straightforward to show (see Section M.1 in Appendix M) that

$$
\mathbf{V}_i^{-1} = \frac{1}{\sigma^2} \mathbf{I}_{n_i} - \frac{\sigma_a^2}{\sigma^2(\sigma^2 + n_i \sigma_a^2)} \mathbf{J}_{n_i}
$$

and $|\mathbf{V}_i| = (\sigma^2 + n_i \sigma_a^2)(\sigma^2)^{n_i - 1}$. From these the likelihood is

$$
L = \prod_{i=1}^{m} (2\pi)^{-n_i/2} |\mathbf{V}_i|^{-1/2} \exp\{-\tfrac{1}{2}(\mathbf{y}_i - \mu \mathbf{1}_{n_i})' \mathbf{V}_i^{-1}(\mathbf{y}_i - \mu \mathbf{1}_{n_i})\},
\tag{2.19}
$$

or

$$
\begin{aligned}
l &= \log L \\
&= -\tfrac{1}{2} N \log 2\pi - \tfrac{1}{2} \sum_i \log(\sigma^2 + n_i \sigma_a^2) - \tfrac{1}{2}(N - m) \log \sigma^2 \\
&\quad - \frac{1}{2\sigma^2} \sum_i \sum_j (y_{ij} - \mu)^2 + \frac{\sigma_a^2}{2\sigma^2} \sum_i \frac{(y_{i\cdot} - n_i \mu)^2}{\sigma^2 + n_i \sigma_a^2}.
\end{aligned}
\tag{2.20}
$$

b. Balanced data

– i. Likelihood

Balanced data have $n_i = n \; \forall \; i$. This greatly simplifies log L so that it becomes

$$l = \log L = -\tfrac{1}{2}N \log 2\pi - \tfrac{1}{2}m(n-1)\log \sigma^2 - \tfrac{1}{2}m[\log(\sigma^2 + n\sigma_a^2)]$$
$$- \frac{\Sigma_i \Sigma_j (y_{ij} - \mu)^2}{2\sigma^2} + \frac{n^2 \sigma_a^2 \Sigma_i (\bar{y}_{i\cdot} - \mu)^2}{2\sigma^2(\sigma^2 + n\sigma_a^2)}. \tag{2.21}$$

The last two terms in (2.21) can be rewritten after a little algebra (Searle et al., 1992, p. 80) to involve

$$\text{SSA} = \Sigma_i n(\bar{y}_{i\cdot} - \bar{y}_{\cdot\cdot})^2 \quad \text{and} \quad \text{SSE} = \Sigma_i \Sigma_j (y_{ij} - \bar{y}_{i\cdot})^2, \tag{2.22}$$

the familiar sums of squares for classes and error, respectively, in the usual analysis of variance of data from a one-way classification. Further simplification comes from defining

$$\lambda = \sigma^2 + n\sigma_a^2, \tag{2.23}$$

so that l is then

$$l = -\tfrac{1}{2}N\log 2\pi - \tfrac{1}{2}m(n-1)\log\sigma^2 - \tfrac{1}{2}m\log\lambda - \frac{\text{SSE}}{2\sigma^2} - \frac{\text{SSA}}{2\lambda} - \frac{mn(\bar{y}_{\cdot\cdot} - \mu)^2}{2\lambda}. \tag{2.24}$$

Introducing λ in place of $\sigma^2 + n\sigma_a^2$ simplifies the ML estimation process. Then ML yields estimators of σ^2 and σ_a^2 through the standard property of ML estimation that the ML estimator of a function of parameters is that same function of ML estimators of the parameters.

– ii. ML equations and their solutions

The maximum likelihood equations are those equations obtained by equating to zero the partial derivatives of log L with respect to μ, σ^2 and λ:

$$l_\mu = \frac{\partial l}{\partial \mu} = \frac{mn(\bar{y}_{\cdot\cdot} - \mu)}{\lambda},$$

$$l_{\sigma^2} = \frac{\partial l}{\partial \sigma^2} = \frac{-m(n-1)}{2\sigma^2} + \frac{\text{SSE}}{2\sigma^4} = \frac{-m(n-1)}{2\sigma^4}\left[\sigma^2 - \frac{\text{SSE}}{m(n-1)}\right],$$

$$l_\lambda = \frac{\partial l}{\partial \lambda} = \frac{-m}{2\lambda} + \frac{\text{SSA}}{2\lambda^2} + \frac{mn(\bar{y}_{\cdot\cdot} - \mu)^2}{2\lambda^2}$$

$$= \frac{-m}{2\lambda^2}\left(\lambda - \frac{\text{SSA}}{m}\right) + \frac{mn(\bar{y}_{\cdot\cdot} - \mu)^2}{2\lambda^2}. \tag{2.25}$$

In equating these partial derivatives to zero we change the parameter symbols μ, σ_e^2 and λ to be $\dot\mu$, $\dot\sigma_e^2$ and $\dot\lambda$ representing solutions to those equations, and get those solutions as

$$\dot\mu = \bar y_{..}, \quad \dot\sigma^2 = \text{MSE}, \quad \dot\lambda = \frac{\text{SSA}}{m} = \left(1 - \frac{1}{m}\right)\text{MSA}$$

and then

$$\dot\sigma_a^2 = \frac{\dot\lambda - \dot\sigma^2}{n} = \frac{(1 - 1/m)\text{MSA - MSE}}{n}, \tag{2.26}$$

where

$$\text{MSA} = \text{SSA}/(m-1) \quad \text{and} \quad \text{MSE} = \text{SSE}/m(n-1).$$

These are the solutions to the ML equations. But they are not necessarily the ML estimators, even though they maximize the likelihood function, L, for variation in μ, σ^2 and σ_a^2. For example, $\dot\sigma^2$ may be negative, an inadmissible value for a variance.

The theory of maximum likelihood tells us that solutions of ML equations do indeed maximize the likelihood function if the matrix of second derivatives (known as the *Hessian*) of the likelihood is negative definite when the parameters in the Hessian are replaced by the solutions. For $\dot\mu$, $\dot\sigma^2$ and $\dot\gamma$ this is left as E 2.4 at the end of this chapter.

– iii.   *ML estimators*

The very definition of ML demands that the likelihood be maximized over the *parameter space*. And in the one-way classification this space is, from the nature of the parameters, $-\infty < \mu < \infty$, $0 < \sigma^2 < \infty$ and $0 \le \sigma_a^2 < \infty$. Fortunately, in the one-way classification $\dot\sigma_a^2$ is the only one of the three ML solutions $\dot\mu$, $\dot\sigma^2$ and $\dot\sigma_a^2$ that is not necessarily in the parameter space.

We consider the solutions $\dot\mu$, $\dot\sigma^2$ and $\dot\sigma_a^2$ in turn. First, $\dot\mu$ does not depend on $\dot\sigma^2$ or $\dot\sigma_a^2$, and since $\dot\mu = \bar y_{..}$ is clearly in the space of μ it is the MLE of μ:

$$\text{MLE}(\mu) = \hat\mu = \dot\mu = \bar y_{..} \,.$$

Also $\dot\sigma^2 = \text{MSE}$ is in the parameter space for σ^2, since MSE is never negative (and we exclude the naive case where $y_{ij} = \bar y_{i.}$ $\forall$ i and j, which would give $\dot\sigma^2 = 0$). But since $\dot\sigma_a^2$ depends on $\text{MSE} = \dot\sigma^2$, we must ensure not just that $\hat\sigma_a^2$ is in the parameter space for σ_a^2 but that the pair of estimators $(\hat\sigma^2, \hat\sigma_a^2)$ is in the 2-space defined by (σ^2, σ_a^2).

As a result, we find that when $\dot{\sigma}_a^2 \leq 0$ then $\dot{\sigma}^2 = \text{MSE}$ is not the MLE of σ^2. Establishing the ML estimators $\hat{\sigma}^2$ and $\hat{\sigma}_a^2$ from $\dot{\sigma}^2$ and $\dot{\sigma}_a^2$ through taking account of the possibility of $\dot{\sigma}_a^2$ being non-positive was first done by Herbach (1959) and is summarized in Searle *et al.* (1992, pp. 81–83). The consequences are that the MLEs of σ^2 and σ_a^2 are as follows:

$$\hat{\sigma}^2 = \begin{cases} \text{MSE} & \text{if } \left(1 - \dfrac{1}{m}\right) \text{MSA} \geq \text{MSE}, \\[2ex] \dfrac{\text{SST}}{mn} & \text{if } \left(1 - \dfrac{1}{m}\right) \text{MSA} < \text{MSE}, \end{cases} \tag{2.27}$$

and

$$\hat{\sigma}_a^2 = \begin{cases} \left[\left(1 - \dfrac{1}{m}\right) \text{MSA} - \text{MSE}\right] \Big/ n & \text{if } \left(1 - \dfrac{1}{m}\right) \text{MSA} \geq \text{MSE}, \\[2ex] 0 & \text{if } \left(1 - \dfrac{1}{m}\right) \text{MSA} < \text{MSE}. \end{cases} \tag{2.28}$$

Although this is certainly the correct way of stating the MLEs, we also state them in a manner that may well be more useful for data analysts. This is because we state the data conditions first:

$$\text{if } \left(1 - \frac{1}{m}\right) \text{MSA} \geq \text{MSE} \quad \text{then} \quad \hat{\sigma}_a^2 = \left[\left(1 - \frac{1}{m}\right) \text{MSA} - \text{MSE}\right] \Big/ n$$

$$\text{and} \quad \hat{\sigma}^2 = \text{MSE}, \tag{2.29}$$

$$\text{if } \left(1 - \frac{1}{m}\right) \text{MSA} < \text{MSE} \quad \text{then} \quad \hat{\sigma}_a^2 = 0 \text{ and } \hat{\sigma}^2 = \frac{\text{SST}}{mn}, \tag{2.30}$$

where $\text{SST} = \Sigma_i \Sigma_j (y_{ij} - \bar{y}_{..})^2$ is the total sum of squares corrected for the mean.

– iv.   *Expected values and bias*

The expected value of $\hat{\mu} = \bar{y}_{..}$ is easy:

$$E[\hat{\mu}] = E[\bar{y}_{..}] = \mu,$$

that is, the ML estimator $\hat{\mu}$ is unbiased. And expected values of the ML *solutions* (not estimators) are easily defined: from (2.26)

$$E[\dot{\sigma}^2] = E[\text{MSE}] = \sigma^2$$

and

$$E[\dot{\sigma}_a^2] = \frac{(1 - 1/m)E[\text{MSA}] - E[\text{MSE}]}{n}$$

$$= \frac{(1 - 1/m)(\sigma^2 + n\sigma_a^2) - \sigma^2}{n}$$

$$= (1 - 1/m)\sigma_a^2 - \sigma^2/mn.$$

But this direct derivation of expected values does not carry over to ML estimators. The reason is that, as seen in (2.27) and (2.28), each of those estimators takes two different forms: for example, $\hat{\sigma}^2$ is MSE if $(1 - 1/m)\text{MSA} \geq \text{MSE}$, but $\hat{\sigma}^2$ is SST/m if $(1 - 1/m)\text{MSA} < \text{MSE}$. Therefore, for

$$p = Pr\{(1 - 1/m)\text{MSA} < \text{MSE}\} \tag{2.31}$$

$$E[\hat{\sigma}^2] = (1 - p)E[\text{MSE}|\dot{\sigma}_a^2 \geq 0] + pE[\text{SST}/m|\dot{\sigma}_a^2 < 0]. \tag{2.32}$$

Similarly, and because $\hat{\sigma}_a^2 \geq 0$,

$$E[\hat{\sigma}_a^2] = (1 - p)E[\dot{\sigma}_a^2|\dot{\sigma}_a^2 \geq 0]. \tag{2.33}$$

These expectations have no closed form. Not only does p depend on the values σ^2 and σ_a^2, because p can be expressed as

$$p = Pr\left\{\mathcal{F}_{m-1}^{m(n-1)} \geq (1 - 1/m)(1 + n\sigma_a^2/\sigma^2)\right\},$$

but also, the expectations in (2.32) and (2.33) are conditional expectations over restricted ranges. Because the expected values of the ML estimators cannot be expressed in closed form, it is also not possible to express the bias in closed form.

− v.   *Asymptotic sampling variances*

With $\hat{\mu} = \bar{y}_{..}$ it is easily shown that

$$\text{var}(\hat{\mu}) = \frac{\sigma^2 + n\sigma_a^2}{mn}.$$

There is a very general result in ML estimation that the large-sample asymptotic dispersion matrix of ML estimators is the inverse of the negative of the expected value of the matrix of second derivatives of the log likelihood, i.e., of l of (2.21). This general result includes the fact for LMs that covariances between MLs of fixed effects and variance components are zero. See, for example, Searle et al. [1992, p. 239, eq. (39)]. Thus with, leading to (2.25)

$$l_{\sigma^2} = \frac{-m(n-1)}{2\sigma^2} + \frac{\text{SSE}}{2\sigma^4}$$

and

$$l_\lambda = \frac{-m}{2\lambda} + \frac{\text{SSA}}{2\lambda^2} + \frac{mn(\bar{y}.. - \mu)^2}{2\lambda^2},$$

we get the second derivatives

$$l_{\sigma^2\sigma^2} = \frac{m(n-1)}{2\sigma^4} - \frac{2\text{SSE}}{2\sigma^6},$$

$$l_{\sigma^2\lambda} = 0$$

and

$$l_{\lambda\lambda} = \frac{m}{2\lambda^2} - \frac{2\text{SSA}}{2\lambda^3} - \frac{2mn(\bar{y}.. - \mu)^2}{2\lambda^3}.$$

Thus

$$-E[l_{\sigma^2\sigma^2}] = \frac{-m(n-1)}{2\sigma^4} + \frac{2m(n-1)\sigma^2}{2\sigma^6} = \frac{m(n-1)}{2\sigma^4}$$

$$-E[l_{\lambda\lambda}] = \frac{-m}{2\lambda^2} + \frac{(m-1)\lambda}{\lambda^3} + \frac{mnm(n\sigma^2 + n^2\sigma_a^2)}{\lambda^3 m^2 n^2} = \frac{m}{2\lambda^2}.$$

Therefore, with $l_{\sigma^2\lambda} = 0$

$$\text{var}\begin{bmatrix} \hat{\sigma}^2 \\ \hat{\lambda} \end{bmatrix} \rightarrow \left(-E\begin{bmatrix} l_{\sigma^2\sigma^2} & l_{\sigma^2\lambda} \\ l_{\sigma^2\lambda} & l_{\lambda\lambda} \end{bmatrix}\right)^{-1} = \begin{bmatrix} \frac{2\sigma^4}{m(n-1)} & 0 \\ 0 & \frac{2\lambda^2}{m} \end{bmatrix}.$$

$$(2.34)$$

Then, with $\hat{\sigma}_a^2 = (\hat{\lambda} - \hat{\sigma}^2)/n$, the large-sample dispersion matrix for the MLEs of the variance components is

$$\text{var}\begin{bmatrix} \hat{\sigma}^2 \\ \hat{\sigma}_a^2 \end{bmatrix} = 2\sigma^4 \begin{bmatrix} \frac{1}{m(n-1)} & \frac{-1}{mn(n-1)} \\ \frac{-1}{mn(n-1)} & \frac{1}{n^2}\left(\frac{\lambda^2/\sigma^4}{m} + \frac{1}{m(n-1)}\right) \end{bmatrix}. \quad (2.35)$$

Note that in (2.34) the large-sample variance of the ML estimator $\hat{\sigma}^2$ is $2\sigma^4/m(n-1)$. This is the same as $\text{var}(\dot{\sigma}^2) = \text{var}(\text{MSE})$. But $\dot{\sigma}^2 = \text{MSE}$ is not the same as $\hat{\sigma}^2$. As in (2.27), $\hat{\sigma}^2$ is MSE when $\hat{\sigma}_a^2 \geq 0$, but $\dot{\sigma}_a^2 < 0$ leads to $\hat{\sigma}^2 = \text{SST}/mn$. However, (2.34) is an asymptotic result, in which $\hat{\sigma}^2$, by virtue of being an ML estimator, is consistent, and so cannot be negative. Hence, in that asymptotic situation, $\hat{\sigma}_a^2 < 0$ never occurs and so $\hat{\sigma}^2$ is never SST/mn. It is always MSE, with variance $2\sigma^4/m(n-1)$ as in (2.34).

In contrast, the exact variance of $\hat{\sigma}^2$ is, using p of (2.31),

$$\text{var}(\hat{\sigma}^2) =$$

$$(1-p)E[(MSE)^2|\hat{\sigma}_a^2 \geq 0] + pE[(\text{SST})^2|\hat{\sigma}_a^2 < 0]/m^2n^2 - (E[\hat{\sigma}^2])^2.$$

Again intractability is apparent, and numerical evaluation has to be used for each particular case. See Yu, Searle and McCulloch (1991).

– vi. *REML estimation*

In contrast to the ML solutions of (2.26) the REML solutions are $\dot{\sigma}^2 = $ MSE (the same as ML) and $\dot{\sigma}_a^2 = $ (MSA − MSE)/n, as can be derived from the general REML equation, (6.67). Details of this are derived as equation (89) in Section 6.6 of Searle et al. (1992).

c. Unbalanced data

– i. *Likelihood*

Following $\lambda = \sigma^2 + n\sigma_a^2$ in (2.23) we now define

$$\lambda_i = \sigma^2 + n_i\sigma_a^2. \tag{2.36}$$

Then the likelihood of (2.21), after writing $y_{ij} - \mu$ as $y_{ij} - \bar{y}_{i\cdot} + \bar{y}_{i\cdot} - \mu$ and simplifying, becomes

$$l = -\tfrac{1}{2}N\log 2\pi - \tfrac{1}{2}\Sigma_i \log \lambda_i - \tfrac{1}{2}(N-m)\log \sigma^2 - \frac{\text{SSE}}{2\sigma^2} - \Sigma_i\frac{n_i(\bar{y}_{i\cdot}-\mu)^2}{2\lambda_i}. \tag{2.37}$$

– ii. *ML equations and their solutions*

With $\partial\lambda_i/\partial\sigma^2 = 1$ and $\partial\lambda_i/\partial\sigma_a^2 = n_i$ we differentiate l of (2.37) to get (using $l_\theta \equiv \partial \log L/\partial\theta$)

$$l_\mu = \Sigma_i\frac{n_i(\bar{y}_{i\cdot} - \mu)}{\lambda_i}, \tag{2.38}$$

$$l_{\sigma^2} = \frac{-(N-m)}{2\sigma^2} - \frac{1}{2}\sum_i\frac{1}{\lambda_i} + \frac{\text{SSE}}{2\sigma^4} + \sum_i\frac{n_i(\bar{y}_{i\cdot} - \mu)^2}{2\lambda_i^2}, \tag{2.39}$$

and

$$l_{\sigma_a^2} = -\frac{1}{2} \sum_i \frac{n_i}{\lambda_i} + \sum_i \frac{n_i^2(\bar{y}_{i\cdot} - \mu)^2}{\lambda_i^2}. \tag{2.40}$$

The ML equations are obtained by equating the above expressions to zero using $\dot{\mu}$, $\dot{\sigma}^2$ and $\lambda_i = \dot{\sigma}_e^2 + n_i\dot{\sigma}_a^2$ as the solutions. Carrying out this procedure with l_μ of (2.38) gives

$$\dot{\mu} = \Sigma_i \frac{n_i\bar{y}_{i\cdot}}{\lambda_i} \Big/ \Sigma_i \frac{n_i}{\lambda_i} = \frac{\Sigma_i \dfrac{n_i\bar{y}_{i\cdot}}{\dot{\sigma}^2 + n_i\dot{\sigma}_a^2}}{\Sigma_i \dfrac{n_i}{\dot{\sigma}^2 + n_i\dot{\sigma}_a^2}} = \frac{\Sigma_i \bar{y}_{i\cdot}/\text{vâr}(\bar{y}_{i\cdot})}{\Sigma_i [1/\text{vâr}(\bar{y}_{i\cdot})]}. \tag{2.41}$$

We see that $\dot{\mu}$ is a weighted average of the $\bar{y}_{i\cdot}$ weighted inversely by an estimate of $\text{var}(\bar{y}_{i\cdot}) = \sigma_a^2 + \sigma^2/n_i$.

Derivation of $\dot{\sigma}_a^2$ and $\dot{\sigma}^2$ comes from equating the right-hand sides of (2.39) and (2.40) to zero, so giving

$$\frac{\text{SSE}}{\dot{\sigma}^4} - \frac{N-m}{\dot{\sigma}^2} + \Sigma_i \frac{n_i(\bar{y}_{i\cdot} - \dot{\mu})^2}{\lambda_i^2} - \Sigma_i \frac{1}{\lambda_i} = 0 \tag{2.42}$$

and

$$\Sigma_i \frac{n_i^2(\bar{y}_{i\cdot} - \dot{\mu})^2}{\lambda_i^2} = \Sigma_i \frac{n_i}{\lambda_i}. \tag{2.43}$$

With $\dot{\lambda} = \dot{\sigma}_a^2 + n_i\dot{\sigma}_a^2$ occurring in the denominators of the terms being summed (over i) in these equations, there is clearly no analytic solution for the estimators; there is when the data are balanced (i.e., $n_i = n$ and $\lambda_i = \lambda \; \forall \; i$).

– iii. *ML estimators*

As with balanced data, solutions $\dot{\mu}$, $\dot{\sigma}^2$ and $\dot{\sigma}_a^2$ are ML estimators only if the triplet $(\dot{\mu}, \dot{\sigma}^2, \dot{\sigma}_a^2)$ is in the 3-space of $(\mu, \sigma^2, \sigma_a^2)$. And in ensuring that this is achieved, the negativity problem raises its head again. For each data set, equations (2.41), (2.42) and (2.43) have to be solved numerically, using some iterative method suited to the numerical solution of non-linear equations. After doing this, the ML estimators are as follows:

when $\dot{\sigma}_a^2 \geq 0$,

$$\hat{\sigma}^2 = \dot{\sigma}^2, \quad \hat{\sigma}_a^2 = \dot{\sigma}_a^2 \quad \text{and} \quad \hat{\mu} = \dot{\mu}; \tag{2.44}$$

when $\dot{\sigma}_a^2 < 0$,

$$\hat{\sigma}^2 = \text{SST}/N, \quad \hat{\sigma}_a^2 = 0 \quad \text{and} \quad \hat{\mu} = \bar{y}_{\cdot\cdot} \,. \tag{2.45}$$

In the latter case, when $\dot{\sigma}_a^2 < 0$, the argument for having $\hat{\sigma}_a^2 = 0$ is essentially the same as with balanced data, whereupon it is left to the reader to show that $\log L$ reduces to being such that on equating its derivatives to zero one obtains $\hat{\sigma}^2 = \text{SST}/N$, as in (2.45) for balanced data and $\hat{\mu} = \bar{y}_{..}$ (see E 2.4). Having been derived by the method of maximum likelihood, the estimators in (2.44) and (2.45) are, as is well known, asymptotically normally distributed.

The question might well be raised as to what to do if the numerical solution of (2.42) and (2.43) yields a negative value for $\dot{\sigma}^2$. Fortunately, it can be shown that $L = e^l \to 0$ as σ^2 tends to zero or to infinity, and so L must have a maximum at a positive value of σ^2 (see E 2.5).

d.   Bias

With balanced data we were able to specify p, the probability of the solution for $\dot{\sigma}^2$ to the ML equations being negative — in (2.31). But with unbalanced data $F = \text{MSA}/\text{MSE}$ does not have a distribution that is proportional to an F, so this probability cannot be easily specified. Moreover, although we know that $\hat{\sigma}^2 = \text{SST}/N$ with probability p, and the expected value of SST is readily derived, the expected value of $\hat{\sigma}^2$ when $\dot{\sigma}_a^2 < 0$ cannot easily be derived. Thus, in general, the bias in the solutions obtained to (2.42) cannot be derived analytically.

e.   Sampling variances

Large-sample variances come from a matrix similar to (2.34), namely the inverse of the negative of the expected value of the Hessian (matrix of second derivatives) of $\log L$ with respect to μ, σ^2 and σ_a^2. Keeping in mind that, by definition, $\sigma_a^2 > 0$ (because if $\sigma_a^2 = 0$ the model and L change), we differentiate the three first differentials of (2.38), (2.39) and (2.40) and take expected values of the resulting second differentials. [Details are shown in Searle et al. (1992) in Sect. 3.7b.] Arraying these expected values in a matrix gives, after inverting that matrix, the matrix of large-sample variance-covariance matrix:

$$
\text{var} \begin{bmatrix} \hat{\mu} \\ \hat{\sigma}^2 \\ \hat{\sigma}_a^2 \end{bmatrix} \approx \begin{bmatrix} \Sigma_i \dfrac{n_i}{\lambda_i} & 0 & 0 \\ 0 & \dfrac{N-m}{2\sigma^4} + \dfrac{1}{2}\Sigma_i \dfrac{1}{\lambda_i^2} & \dfrac{1}{2}\Sigma_i \dfrac{n_i}{\lambda_i^2} \\ 0 & \dfrac{1}{2}\Sigma_i \dfrac{n_i}{\lambda_i^2} & \dfrac{1}{2}\Sigma_i \dfrac{n_i^2}{\lambda_i^2} \end{bmatrix}^{-1} . \qquad (2.46)
$$

Therefore

$$\text{var}(\hat{\mu}) \approx \left(\sum_i \frac{n_i}{\lambda_i} \right)^{-1} = \left(\sum_i \frac{n_i}{\sigma^2 + n_i \sigma_a^2} \right)^{-1}$$

and

$$\text{var} \begin{bmatrix} \hat{\sigma}^2 \\ \hat{\sigma}_a^2 \end{bmatrix} \approx \frac{2}{D} \begin{bmatrix} \sum_i \dfrac{n_i^2}{\lambda_i^2} & -\sum_i \dfrac{n_i}{\lambda_i^2} \\ -\sum_i \dfrac{n_i}{\lambda_i^2} & \dfrac{N-m}{\sum^4} + \sum_i \dfrac{1}{\lambda_i^2} \end{bmatrix}, \tag{2.47}$$

where

$$D = \frac{N-m}{\sigma^4} \sum_i \frac{n_i^2}{\lambda_i^2} + \sum_i \frac{1}{\lambda_i^2} \sum_i \frac{n_i^2}{\lambda_i^2} - \left(\sum_i \frac{n_i}{\lambda_i^2} \right)^2.$$

2.3   NORMALITY, RANDOM EFFECTS AND REML

a.   Balanced data

– i.   *Likelihood*

For the one-way classification with model equation $E[y_{ij}] = \mu + a_i$ the part of the likelihood of y not involving fixed effects is simply that part not involving μ. And for balanced data that is easily derived. From (2.24) we reconstruct L, the likelihood of $\mathbf{y}$ and write it as

$$L(\mu, \sigma^2, \sigma_a^2 | \mathbf{y}) = \frac{\exp\left\{ -\dfrac{1}{2} \left[\dfrac{\text{SSE}}{\sigma^2} + \dfrac{\text{SSA}}{\lambda} + \dfrac{(\bar{y}.. - \mu)^2}{\lambda/mn} \right] \right\}}{(2\pi)^{\frac{1}{2}mn} \sigma^{2[\frac{1}{2}m(n-1)]} \lambda^{\frac{1}{2}m}}.$$

Since $\bar{y}..$ is independent of both SSE and SSA, the preceding expression can be factored as

$$L(\mu, \sigma^2 \, \sigma_a^2 | \mathbf{y}) = L(\mu | \bar{y}..) L(\sigma^2, \sigma_a^2 | \text{SSA}, \text{SSE}),$$

where $L(\mu | \bar{y}..)$ is the likelihood of μ given $\bar{y}..$, namely

$$L(\mu | \bar{y}..) = \frac{\exp\left[\dfrac{-(\bar{y}.. - \mu)^2}{2\lambda/mn} \right]}{(2\pi)^{\frac{1}{2}} (\lambda/mn)^{\frac{1}{2}}}, \tag{2.48}$$

and

$$L(\sigma^2, \sigma_a^2 | \text{SSE}, \text{SSA}) = \frac{\exp\left[-\frac{1}{2}\left(\dfrac{\text{SSE}}{\sigma^2} + \dfrac{\text{SSA}}{\lambda}\right)\right]}{(2\pi)^{\frac{1}{2}(mn-1)}\sigma^{2[\frac{1}{2}m(n-1)]}\lambda^{\frac{1}{2}(m-1)}(mn)^{\frac{1}{2}}} \qquad (2.49)$$

is the likelihood function of σ^2 and σ_a^2 given SSA and SSE. Thus, because μ is not involved in (2.49) that is the likelihood for REML.

– ii.   REML equations and their solutions

The REML equations come from maximizing the logarithm of (2.49). Denoting this by l_R we find

$$l_R = -\tfrac{1}{2}(mn-1)\log 2\pi - \tfrac{1}{2}\log mn - \tfrac{1}{2}m(n-1)\log\sigma^2$$

$$- \tfrac{1}{2}(m-1)\log\lambda - \frac{\text{SSE}}{2\sigma^2} - \frac{\text{SSA}}{2\lambda}. \qquad (2.50)$$

Equating to zero the derivative of l_R with respect to σ^2 and λ gives solutions $\dot\sigma_R^2$ and $\dot\lambda_R$ as $\dot\lambda_R = \text{SSA}/(m-1) = \text{MSA}$ and

$$\dot\sigma_R^2 = \frac{\text{SSE}}{m(n-1)} = \text{MSE}; \quad \text{and} \quad \text{thus } \dot\sigma_{a,R}^2 = \frac{1}{n}(\text{MSA} - \text{MSE}). \quad (2.51)$$

These are the REML solutions.

– iii.   REML estimators

Similar to the situation with ML, the preceding REML solutions are REML estimators only when both are non-negative. $\dot\sigma_R^2$ is never negative, but $\dot\sigma_{a,R}^2$ can be, whereupon we have to maximize l_R subject to $\dot\sigma_{a,R}^2 = 0$, which leads to $\dot\sigma_R^2$ then being $\text{SST}/(mn-1)$. Thus the REML estimators are

when $\dot\sigma_{a,R}^2 > 0$,

$$\hat\sigma_R^2 = \text{MSE} \quad \text{and} \quad \hat\sigma_{a,R}^2 = \frac{1}{n}(\text{MSA} - \text{MSE});$$

when $\dot\sigma_{a,R}^2 \le 0$, $\qquad\qquad\qquad\qquad\qquad\qquad\qquad\qquad (2.52)$

$$\hat\sigma_R^2 = \frac{\text{SST}}{mn-1} \quad \text{and} \quad \hat\sigma_{a,R}^2 = 0.$$

– iv. **Comparison with ML**

Comparing the REML estimators of (2.52) with the ML estimators of (2.27) and (2.28), we see that the condition for a negative solution for σ_a^2 is not quite the same in the two cases. In REML it is MSA < MSE whereas in ML it is $(1 - 1/m)$MSA < MSE; and the positive estimator is similarly slightly different: (MSA – MSE)$/n$ in REML but $[(1-1/m)$MSA - MSE$]/n$ in ML. Also, when there is a negative solution for σ_a^2, the resulting estimator of σ^2 is not the same in the two cases: SST$/(mn-1)$ in REML but SST$/mn$ in ML. Each of these differences has a common feature: that with REML we see SSA being divided by $m-1$ where it is divided by m in ML; and in REML the divisor of SST is $mn-1$ whereas it is mn in ML. In both instances the REML divisor is one less than the ML divisor. In this way REML is taking account of the degree of freedom that gets utilized in estimating μ — even though REML does not explicitly involve the estimation of μ. Nevertheless, it is a general feature of REML estimation of variance components from balanced data that degrees of freedom for fixed effects get taken into account. The simplest example is that of estimating σ^2 from a simple sample of n independent observations $x_1, x_2, \cdots, x_n$, from $\mathcal{N}(\mu, \sigma^2)$. The ML estimator is $\Sigma_i(x_i - \bar{x})^2/n$ whereas the REML estimator is $\Sigma_i(x_i - \bar{x})^2/(n - 1)$.

– v. **Bias**

What has just been said about REML might lead one to surmise that REML estimators are unbiased. They are not. The same need for non-negative estimates arises as with ML estimation. Similar to (2.31) we define, for balanced data

$$
\begin{aligned}
p_R &= \mathrm{P}\{\dot{\sigma}_{a,R}^2 < 0\} = \mathrm{P}\{\text{MSA} < \text{MSE}\} \\
&= \mathrm{P}\{\mathcal{F}_{m-1}^{m(n-1)} > 1 + n\sigma_a^2/\sigma^2\}.
\end{aligned} \tag{2.53}
$$

Then, based on (2.52), the expected value of $\hat{\sigma}_R^2$ is

$$
E(\hat{\sigma}_R^2) = (1 - p_R)E(\text{MSE}|\dot{\sigma}_{a,R}^2 \geq 0) + p_R E(\text{SST}|\dot{\sigma}_{a,R}^2 < 0)/(mn - 1). \tag{2.54}
$$

– vi. *Sampling variances*

Based on (2.50), we can easily find the large-sample dispersion matrix,

$$
\text{var}
\begin{bmatrix}
\hat{\sigma}^2_{a,R} \\
\hat{\lambda}_R
\end{bmatrix}
\approx
\begin{bmatrix}
-E[l_{R,\sigma^2,\sigma^2_a}] & -E[l_{R,\sigma^2,\lambda}] \\
-E[l_{R,\lambda,\sigma^2}] & -E[l_{R,\lambda,\lambda}]
\end{bmatrix}^{-1},
$$

which leads to exactly the same results as in (2.35) except that in the lower right-hand element the term $(\lambda^2/\sigma^4)/m$ is $(\lambda^2/\sigma^4)/(m-1)$.

b.   Unbalanced data

In keeping with (2.20), the likelihood function for unbalanced data is

$$
L(\mu, \sigma^2, \sigma^2_a | \mathbf{y}) = \frac{\exp\left\{ -\left[\dfrac{\Sigma_i \Sigma_j (y_{ij} - \mu)^2}{2\sigma^2} - \Sigma_i \dfrac{n_i^2 \sigma^2_a (\bar{y}_{i\cdot} - \mu)^2}{2\sigma^2(\sigma^2 + n_i \sigma^2_a)} \right] \right\}}{(2\pi)^{\frac{1}{2}N} \sigma^{2\frac{1}{2}(N-m)} \displaystyle\prod_{i=1}^{a} (\sigma^2 + n_i \sigma^2_a)^{\frac{1}{2}}}.
$$

There is no straightforward factoring of this likelihood that permits separating a function of μ in the manner of (2.48) for balanced data. Nevertheless, equations for REML estimators can be established — as a special case of the equations for the general case. This is left until Chapter 6.

2.4   MORE ON RANDOM EFFECTS AND NORMALITY

a.   Tests and confidence intervals

– i.   *For the overall mean,* μ

With balanced data $(n_i \equiv n)$ we can show (E 2.7) that

$$
\frac{\bar{y}_{\cdot\cdot} - \mu}{\sqrt{\text{MSA}/mn}} \sim \mathcal{T}_{m-1},
$$

the t-distribution on $m-1$ degrees of freedom. A test of $H_0 : \mu = \mu_0$ is then to reject H_0 when

$$
\left| \frac{\bar{y}_{\cdot\cdot} - \mu_0}{\sqrt{\text{MSA}/mn}} \right| > t_{m-1,\alpha/2}
$$

and a corresponding confidence interval for μ is

$$
\bar{y}_{\cdot\cdot} \pm t_{m-1,\alpha/2} \sqrt{\text{MSA}/mn} \quad .
$$

– ii. **For σ^2**

Tests and confidence intervals for σ^2 are based on the result that

$$\frac{(N-m)s^2}{\sigma^2} \sim \chi^2_{N-m},$$

for both balanced and unbalanced data. A somewhat unusual application of this would be to form a test of a specified value of σ^2 or, more commonly, to form a confidence interval:

$$\left(\frac{\text{SSE}}{\chi^2_{N-m,1-\alpha/2}}, \frac{\text{SSE}}{\chi^2_{N-m,\alpha/2}} \right).$$

– iii. **For σ_a^2**

For balanced or unbalanced data a likelihood ratio test (see E 2.10) of $H_0 : \sigma_a^2 = 0$ is to reject H_0 when $F = \text{MSA}/\text{MSE} > \mathcal{F}^{m-1}_{N-m,1-\alpha}$. For balanced data ($n_i \equiv n$) a confidence interval for σ_a^2/σ^2 is given by

$$\left(\frac{F/\mathcal{F}^{m-1}_{N-m,1-\alpha/2} - 1}{n}, \frac{F/\mathcal{F}^{m-1}_{N-m,\alpha/2} - 1}{n} \right).$$

For unbalanced data no exact intervals exist for σ_a^2/σ^2 in closed form. Approximate intervals are described in Searle et al. (1992) in Sections 3.6d–vi. Exact intervals for other functions of the variances are discussed in Khuri et al. (1998).

b. Predicting random effects

– i. **A basic result**

Our model assumes that $a_i \sim$ i.i.d. $\mathcal{N}(0, \sigma_a^2)$ where the a_i are unknown. If we wanted to guess a value for a_i in the absence of any data or information, we could do no better than to guess the mean value of a_i, namely zero. However, suppose we have data known as having a correlation of 0.99 with a_i: then we could use that information to get a prediction of a_i better than its zero mean. The information would adjust our prediction away from that mean of zero.

The basic result for doing this is quite general, as follows. Suppose we have two random variables one of which, Y, cannot be observed but which, in particular cases, we wish to predict; and the other, X, can be observed and which is to be used for predicting Y. Then the predictor we use is the minimum mean squared error predictor of Y, based on X; it is $E[Y|X]$, the conditional mean of Y given X (see E 2.8).

Motivation for this result

$$\text{Best Predictor} = E[Y|X] \tag{2.55}$$

is dealt with extensively in Chapter 13, including its derivation, its properties and applications. Here we just list results for the one-way classification [see Searle, 1971, Sect. 2.4f–v].

– ii.   *In a one-way classification*

Returning to the linear model, we wish to predict a_i given the data. The only portion of the data relevant to a_i is the sample mean for class i, $\bar{y}_i.$ using the general result in Section S.1 of Appendix S for a conditional mean of a normal variate,

$$
\begin{aligned}
E[a_i|\mathbf{y}] &= E[a_i|\bar{y}_i.] = E[a_i] + \mathrm{cov}(a_i, \bar{y}_i.)[\mathrm{var}(\bar{y}_i.)]^{-1}(\bar{y}_i. - E[\bar{y}_i.]) \\
&= 0 + \sigma_a^2 \frac{1}{\sigma_a^2 + \sigma^2/n_i}(\bar{y}_i. - \mu) \\
&= \frac{\sigma_a^2}{\sigma_a^2 + \sigma^2/n_i}(\bar{y}_i. - \mu).
\end{aligned}
\tag{2.56}
$$

This is the best predictor of a_i, which we denote by $\mathrm{BP}(a_i)$.

Immediately a serious problem confronts us concerning the use of (2.56). It depends on the parameters μ, σ_a^2, and σ^2 whose values are unknown. The usual solution is to replace them with estimates and get an estimated best predicted value, which we denote as $\tilde{a}_i$, that is,

$$\tilde{a}_i = \widehat{\mathrm{BP}}(a_i) = \hat{E}[a_i|\bar{y}_i.] = \frac{\hat{\sigma}_a^2}{\hat{\sigma}_a^2 + \hat{\sigma}^2/n_i}(\bar{y}_i. - \hat{\mu}).$$

It is instructive to compare the estimated best predictor with the estimator of α_i under the fixed effects model (2.2) using the constraint $\sum_i n_i \alpha_i = 0$. For prediction in the random model, with balanced data, we have

$$\tilde{a}_i = \frac{\hat{\sigma}_a^2}{\hat{\sigma}_a^2 + \hat{\sigma}^2/n}(\bar{y}_i. - \bar{y}..).$$

For estimation in the fixed model we have

$$\hat{\alpha}_i = (\bar{y}_i. - \bar{y}..).$$

Both are based on $\bar{y}_{i\cdot} - \bar{y}_{\cdot\cdot}$ but $\tilde{a}_i$ is "shrunk" compared to $\hat{\alpha}_i$. It is always smaller than $\hat{\alpha}_i$, the degree to which it is smaller depending on the relative size of $\hat{\sigma}_a^2$ and $\hat{\sigma}^2/n$. If σ_a^2 is estimated to be large with respect to the estimate of σ^2/n (either $\hat{\sigma}_a^2$ is large compared to $\hat{\sigma}^2$ and/or the sample size per class is large) then the two values for the class effect are similar. This corresponds to the situation where there is a lot of variation (relatively speaking) between classes and not much is to be gained by assuming that the effects are selected from a common distribution. On the other hand, when σ_a^2 is estimated to be small with respect to σ^2/n, the shrinkage can be extensive and the two values can differ greatly.

2.5   BINARY DATA: FIXED EFFECTS

We return to the ideas of Section 2.1 wherein y_{ij} is distributed independently, and $E[y_{ij}] = \mu_i$, but use a distribution to accommodate binary data. Hence the only possible distribution is the Bernoulli.

a.   Model equation

As previously, we consider m classes indexed by i, with the ith class having n_i observations. A model for the responses y_{ij} which are coded as 1 or 0 would be

$$E[y_{ij}] \;=\; \pi_i, \tag{2.57}$$

$$y_{ij} \;\sim\; \text{indep. Bernoulli}(\pi_i).$$

The more usual notation for the mean of a Bernoulli distribution is p_i which we reserve for Section 2.6 where p_i is random; and here we use π_i for the fixed effects case — all this being in accord with our convention of Greek letters for fixed effects and Roman letters for random effects.

b.   Likelihood

The likelihood for the data is

$$L \;=\; \prod_i \prod_j \pi_i^{y_{ij}} (1 - \pi_i)^{1 - y_{ij}} = \prod_i \pi_i^{y_{i\cdot}} (1 - \pi_i)^{n_i - y_{i\cdot}}$$

$$=\; \prod_i \left(\frac{\pi_i}{1 - \pi_i} \right)^{y_{i\cdot}} (1 - \pi_i)^{n_i}. \tag{2.58}$$

Therefore

$$l = \log L = \sum_i \{ y_{i\cdot} \log[\pi_i/(1 - \pi_i)] + n_i \log[1 - \pi_i] \} . \qquad (2.59)$$

c. ML equations and their solutions

The ML equations come from differentiating l of (2.59) with respect to the π_i to obtain

$$\begin{aligned}
\frac{\partial l}{\partial \pi_k} &= y_{k\cdot} \left(\frac{1}{\pi_k} + \frac{1}{1 - \pi_k} \right) - \frac{n_k}{1 - \pi_k} \\
&= \frac{y_{k\cdot}}{\pi_k} \left(\frac{1}{1 - \pi_k} \right) - \frac{n_k}{1 - \pi_k} . \qquad (2.60)
\end{aligned}$$

Setting this equal to zero gives

$$\hat{\pi}_k = \bar{y}_{k\cdot} = \text{sample proportion of 1s in class } k.$$

d. Likelihood ratio test

The hypothesis $H_0 : \mu_i$ *all equal* is tested using the likelihood ratio statistic

$$\Lambda = \frac{\prod_i (\bar{y}_{\cdot\cdot})^{y_{i\cdot}} (1 - \bar{y}_{\cdot\cdot})^{n_i - y_{i\cdot}}}{\prod_i (\bar{y}_{i\cdot})^{y_{i\cdot}} (1 - \bar{y}_{i\cdot})^{n_i - y_{i\cdot}}}$$

giving

$$\begin{aligned}
\log \Lambda &= \sum_i \left[y_{i\cdot} \log \left(\frac{\bar{y}_{\cdot\cdot}}{\bar{y}_{i\cdot}} \right) + (n_i - y_{i\cdot}) \log \left(\frac{1 - \bar{y}_{\cdot\cdot}}{1 - \bar{y}_{i\cdot}} \right) \right] \\
&= \sum_i \left[n_i \hat{\pi}_i \log \left(\frac{\hat{\pi}}{\hat{\pi}_i} \right) + n_i (1 - \hat{\pi}_i) \log \left(\frac{1 - \hat{\pi}}{1 - \hat{\pi}_i} \right) \right] , \quad (2.61)
\end{aligned}$$

where $\hat{\pi} = \bar{y}_{\cdot\cdot}$ = overall sample proportions of 1s.

The large sample test is given by

$$\text{Reject } H_0 \text{ if } -2 \log \Lambda \geq \chi^2_{m-1,1-\alpha}. \qquad (2.62)$$

e. The usual chi-square test

A test used more commonly in practice than (2.9) is the chi-square test of independence, or (equivalently) the chi-square test for equality of binomial proportions (Snedecor and Cochran, 1989, Sect. 11.7). It

Table 2.1: Successes and Failures in a One-Way Classification

Outcome	Classification Level					Total
	1	2	$\cdots$ $\quad$ i	$\cdots$	m	
Success	$y_1.$	$y_2.$	$\cdots$ $\quad$ $y_i.$ $\quad$ $\cdots$		$y_m.$	$y_{..}$
Failure	$n_1 - y_1.$	$n_2 - y_2.$	$\cdots$ $\quad$ $n_i - y_i.$ $\quad$ $\cdots$		$n_m - y_m.$	$N - y_{..}$
Total	n_1	n_2	$\cdots$ $\quad$ n_i $\quad$ $\cdots$		n_m	N

is best described by starting with Table 2.1. The usual chi-square test is to reject H_0 when

$$\chi^2 = \sum_i \sum_j \frac{(O_{ij} - E_{ij})^2}{E_{ij}} > \chi^2_{m-1,1-\alpha},$$

where O_{ij} and E_{ij} are, respectively, observed and expected values. For Table 2.1, there are but two values for j, 1 and 2, and

$$O_{i1} = y_i. \quad \text{and} \quad O_{i2} = n_i - y_i.$$
$$E_{i1} = n_i \bar{y}_{..} \quad \text{and} \quad E_{i2} = n_i(1 - \bar{y}_{..}),$$

and hence

$$
\begin{aligned}
\chi^2 &= \sum_i \left[\frac{(y_i. - n_i \bar{y}_{..})^2}{n_i \bar{y}_{..}} + \frac{(n_i - y_i. - (n_i - n_i \bar{y}_{..}))^2}{n_i - n_i \bar{y}_{..}} \right] \\
&= \sum_i \left[\frac{n_i(\hat{\pi}_i - \hat{\pi})^2}{\hat{\pi}} + \frac{n_i(-\hat{\pi}_i + \hat{\pi})^2}{1 - \hat{\pi}} \right] \quad\quad (2.63) \\
&= \sum_i \frac{n_i(\hat{\pi}_i - \hat{\pi})^2}{\hat{\pi}(1 - \hat{\pi})}.
\end{aligned}
$$

An interesting question is: How does χ^2 of (2.63) compare to $-2 \log \Lambda$ of (2.62)? To answer this we use a Taylor series expansion of $f(x) = x \log(c/x)$ about c:

$$f(x) \approx f(c) + f'(c)(x - c) + \tfrac{1}{2} f''(c)(x - c)^2,$$

where

$$f'(c) = \frac{\partial}{\partial x} f(x) \Big|_{x=c} \quad \text{and} \quad f''(c) = \frac{\partial^2}{\partial x^2} f(x) \Big|_{x=c}.$$

This gives

$$f(x) \approx -(x - c) - \tfrac{1}{2}(x - c)^2/c.$$

Applying this to $\log \Lambda$, wherein each term is of the form $x \log(c/x)$, gives

$$\log \Lambda \approx -\sum_i \left[n_i(\hat{\pi}_i - \hat{\pi}) + \frac{n_i}{2} \frac{(\hat{\pi}_i - \hat{\pi})^2}{\hat{\pi}} + n_i[1 - \hat{\pi}_i - (1 - \hat{\pi})] \right.$$

$$\left. + \frac{n_i}{2} \frac{[1 - \hat{\pi}_i - (1 - \hat{\pi})]^2}{1 - \hat{\pi}} \right] \tag{2.64}$$

or

$$-2 \log \Lambda \approx \sum_i \left[\frac{n_i(\hat{\pi}_i - \hat{\pi})^2}{\hat{\pi}} + \frac{n_i(-\hat{\pi}_i + \hat{\pi})^2}{1 - \hat{\pi}} \right]$$

$$= \sum_i \frac{n_i(\hat{\pi}_i - \hat{\pi})^2}{\hat{\pi}(1 - \hat{\pi})}$$

$$= \chi^2$$

of (2.63). Thus, when all the $\hat{\pi}_i$ are close to $\hat{\pi}$, the two statistics give similar results.

f.   Large-sample tests and confidence intervals

Large-sample tests for testing $H_0 : \pi_i = \pi_{i0}$, where π_{i0} is a specified value can be based on

$$z = \frac{\hat{\pi}_i - \pi_{i0}}{\sqrt{\frac{\pi_{i0}(1 - \pi_{i0})}{n_i}}} \sim \mathcal{AN}(0, 1), \tag{2.65}$$

where $\mathcal{AN}$ means asymptotically normally distributed.

The test statistic for the hypothesis $H_0 : \pi_i - \pi_k = \pi_{i0} - \pi_{j0}$ would be

$$z = \frac{\hat{\pi}_i - \hat{\pi}_k - (\pi_{i0} - \pi_{k0})}{\sqrt{\frac{\hat{\pi}_i(1 - \hat{\pi}_i)}{n_i} + \frac{\hat{\pi}_k(1 - \hat{\pi}_k)}{n_k}}} \sim \mathcal{AN}(0, 1),$$

where $\pi_{i0} - \pi_{k0}$ is the hypothesized difference under H_0. For example, to test $H_0 : \pi_i = \pi_k$ versus the alternative $H_1 : \pi_i \neq \pi_k$ we set $\pi_{i0} - \pi_{k0}$ equal to zero and reject H_0 if

$$\left| \frac{\hat{\pi}_i - \hat{\pi}_k}{\sqrt{\frac{\hat{\pi}_i(1 - \hat{\pi}_i)}{n_i} + \frac{\hat{\pi}_k(1 - \hat{\pi}_k)}{n_k}}} \right| > z_{\alpha/2}, \tag{2.66}$$

where z_α is the $100\alpha\%$ percentile of the standard normal distribution, that is, if $Z \sim \mathcal{N}(0,1)$, then $P\{Z > z_\alpha\} = \alpha$. Alternatively, we could perform a likelihood ratio test or a χ^2 test using only the data in columns i and k of Table 2.1. These two tests are quite similar (see E 2.6). Note that in (2.66) we use the estimated values of π_i and π_k rather than the values under H_0 as we did in (2.65). This is because hypothesizing a difference between π_i and π_k under H_0 does not tell us the actual values of π_i and π_k under H_0.

Large-sample confidence intervals for the π_i are given by

$$\hat{\pi}_i \pm z_{\alpha/2}\sqrt{\frac{\hat{\pi}_i(1-\hat{\pi}_i)}{n_i}},$$

again using (2.65). The corresponding confidence intervals for $\pi_i - \pi_k$ are given by

$$\hat{\pi}_i - \hat{\pi}_k \pm z_{\alpha/2}\sqrt{\frac{\hat{\pi}_i(1-\hat{\pi}_i)}{n_i} + \frac{\hat{\pi}_k(1-\hat{\pi}_k)}{n_k}}.$$

g.   Exact tests and confidence intervals

The likelihood ratio and χ^2 tests of $H_0 : \pi_1 = \pi_2 = \cdots = \pi_m$ and the normality-based confidence intervals and tests of the preceding section are based on large-sample distributional approximations which can be inaccurate for small samples. The usual rule of thumb (Snedecor and Cochran, 1989, p. 127) is that the approximation is accurate when most of the "expected values" of Table 2.1 that is, $n_i\bar{y}_{..}$ and $n_i(1-\bar{y}_{..})$, are greater than five and can give inaccurate results when expected values are less than one.

Exact tests can be based on the conditional distribution of the table entries given the marginal totals. Under $H_0 : \pi_i$ *all equal* the conditional probability of a sample is given by

$$\prod_{i=1}^{m} \binom{n_i}{y_{i\cdot}} \Bigg/ \binom{N}{y_{..}}. \tag{2.67}$$

A p-value can be calculated via the usual definition: the probability, under H_0, of a result as extreme or more extreme than that observed. This would be done by summing (2.67) over all the possible data configurations (as in Table 2.1) which are "more extreme" than the observed table.

For two populations ($m = 2$) we have a 2×2 table and it is straight-forward to designate *more extreme* tables. That is, once a single entry in the 2×2 table is known, and conditional on the margins, all the remaining entries are fixed. We can thus enumerate the more extreme tables by varying this single entry from the observed table in a direction "away" from H_0. This test is known as *Fisher's exact test*.

For $m > 2$ populations we must choose a definition of "more extreme." For example, a common choice is whether a table of possible data gives a larger value of the χ^2 statistic than that given by the observed table. The problem is then a computational one since the number of possible tables with given margins gets unmanageably large. Special software (e.g., Mehta and Patel, 1992) is usually required.

Exact confidence intervals for π_i can be calculated (Mood et al., 1974, p. 393) as (π_{iL}, π_{iU}) where the π_{iL} and π_{iU} solve

$$\sum_{k=y_{i\cdot}}^{n_i} \binom{n_i}{k} \pi_{iL}^k (1 - \pi_{iL})^{n_i - k} = \alpha/2 \quad \text{and}$$

$$\sum_{k=1}^{y_{i\cdot}} \binom{n_i}{k} \pi_{iU}^k (1 - \pi_{iU})^{n_i - k} = \alpha/2.$$

These are known as the Clopper-Pearson intervals and can be somewhat conservative. Blyth and Still (1983) give less conservative intervals in tabular form and accurate approximate intervals. Santner and Snell (1980) give intervals for $\pi_i - \pi_k$ and π_i / π_k.

h.   Example: Snake strike data

An experiment was conducted at Cornell University to find factors that determine whether a snake would strike at a target or fail to do so. Snakes were placed in a cage with a target that looked something like an artificial mouse and a binary response was recorded over several observation periods as to whether the snake struck at the target within five minutes. A concern was that some snakes would always strike at targets, whereas others would not strike at all, obscuring any effect due to target differences. Table 2.2 show data for six snakes.

We are interested in testing homogeneity across the snakes. The likelihood ratio test of (2.62) gives a statistic of 9.10 to be compared to a χ^2 distribution with five degrees of freedom. The asymptotic p-value of this test is approximately 0.10.

Table 2.2: Number of Occurrences of Strike or No Strike
for Each Snake

Outcome	Snake						Total
	1	2	3	4	5	6	
Strike	2	2	3	0	1	1	9
No strike	2	2	0	2	0	0	6
Total	4	4	3	2	1	1	15

The chi-square statistic of Section 2.5e is equal to (6.67), again to be compared to a χ^2 distribution with five degrees of freedom, giving an asymptotic p-value of 0.25. Exact calculations using the conditional distribution (2.67) give a p-value for the χ^2 statistic of 0.27 and 0.24 for the likelihood ratio statistic.

2.6 BINARY DATA: RANDOM EFFECTS

a. Model equation

The analog of the random effects model for normally distributed data, (2.13), would be

$$E[y_{ij}] = p_i,$$
$$y_{ij} \sim \text{indep. Bernoulli}(p_i), \qquad (2.68)$$
$$p_i \sim \text{i.i.d. } G,$$

where G is a distribution for the p_i and we maintain our convention of using Roman letters for random effects. Normality cannot be assumed for the p_i since they are probabilities and are restricted to the interval $(0,1)$.

b. Beta-binomial model

A logical choice for G is the beta distribution since it is a flexible distribution on $(0,1)$ and leads to mathematically tractable results. If p_i from (2.68) follows a beta distribution with parameters α and β, its density is given by

$$p_i^{\alpha-1}(1-p_i)^{\beta-1}/B(\alpha,\beta), \qquad (2.69)$$

where

$$B(\alpha,\beta) = \int_0^1 x^{\alpha-1}(1-x)^{\beta-1}\,dx \qquad (2.70)$$

is the beta function. It then follows that

$$E[p_i] \;=\; \frac{\alpha}{\alpha + \beta} \qquad \text{and}$$

$$\text{var}(p_i) \;=\; \frac{\alpha\beta}{(\alpha + \beta)^2(\alpha + \beta + 1)}. \tag{2.71}$$

Model (2.68) along with (2.69) we call the *beta-binomial model*:

$$E[y_{ij}|p_i] \;=\; p_i$$

$$y_{ij}|p_i \;\sim\; \text{indep. Bernoulli}(p_i) \tag{2.72}$$

$$p_i \;\sim\; \text{i.i.d. beta}(\alpha, \beta).$$

– i.   *Means, variances, and covariances*

It is straightforward to calculate moments of the y_{ij} under model (2.72). We have

$$\begin{aligned}
E[y_{ij}] \;&=\; \text{E}\,[\text{E}[y_{ij}|p_i]] = \text{E}[p_i] = \frac{\alpha}{\alpha + \beta} & (2.73)\\[6pt]
\text{var}(y_{ij}) \;&=\; \text{E}[\text{var}(y_{ij})|p_i] + \text{var}(\text{E}[y_{ij}|p_i])\\[4pt]
&=\; \text{E}[p_i(1 - p_i)] + \text{var}(p_i)\\[4pt]
&=\; \text{E}[p_i] - \text{E}[p_i^2] + \text{E}[p_i^2] - \text{E}[p_i]^2\\[4pt]
&=\; \text{E}[p_i](1 - \text{E}[p_i])\\[4pt]
&=\; \frac{\alpha}{\alpha + \beta}\frac{\beta}{\alpha + \beta} = \frac{\alpha\beta}{(\alpha + \beta)^2}. & (2.74)
\end{aligned}$$

The result (2.74) also reflects the fact that being binary forces y_{ij} to have a marginal Bernoulli distribution with variance equal to the mean times 1 minus the mean.

We can calculate a covariance similarly:

$$\begin{aligned}
\text{cov}(y_{ij}, y_{il}) \;&=\; \text{cov}\,(\text{E}[y_{ij}|p_i], \text{E}[y_{il}|p_i]) + \text{E}\,[\text{cov}(y_{ij}, y_{il}|p_i)]\\[4pt]
&=\; \text{cov}(p_i, p_i) + 0 & (2.75)\\[4pt]
&=\; \text{var}(p_i) = \frac{\alpha\beta}{(\alpha + \beta)^2(\alpha + \beta + 1)} \quad \text{for } j \neq l.
\end{aligned}$$

The covariance between y_{ij} and y_{kl} is zero for $i \neq k$. Thus we have an intraclass correlation of

$$\rho = \text{corr}(y_{ij}, y_{il}) = \frac{\alpha\beta/[(\alpha+\beta)^2(\alpha+\beta+1)]}{\alpha\beta/(\alpha+\beta)^2}$$

$$= \frac{1}{\alpha+\beta+1} \tag{2.76}$$

where $\text{corr}(\cdot, \cdot)$ denotes correlation. Thus some authors (e.g., Williams, 1975; Griffiths, 1973) suggest a reparameterization of (2.69) in terms of the mean $\mu = \alpha/(\alpha+\beta)$ and the intraclass correlation, ρ, or a related quantity, $\tau = 1/(\alpha+\beta)$. Reparameterized in such a way, the mean is, of course, μ, and the covariance is $\mu(1-\mu)\rho$ or $\mu(1-\mu)\tau/(\tau+1)$.

– ii. *Overdispersion*

If y_{ij} $(j = 1, 2, \ldots, n_i)$ were independent Bernoulli random variables with mean μ then $y_{i\cdot}$ would follow a binomial(n_i, μ) distribution with variance $n_i\mu(1-\mu)$. Under the beta-binomial model

$$\text{var}(y_{i\cdot}) = \sum_j \text{var}(y_{ij}) + \sum_{j \neq l} \text{cov}(y_{ij}, y_{il}) \tag{2.77}$$

$$= n_i\mu(1-\mu) + 2\binom{n_i}{2}\mu(1-\mu)\rho$$

$$= n_i\mu(1-\mu)\left[1 + (n_i-1)\rho\right]. \tag{2.78}$$

As long as $\rho > 0$, which is required by the beta-binomial model ($\alpha > 0, \beta > 0$), the variance will be larger than the binomial variance. This is often termed *overdispersion*.

Examination of the calculations in and below (2.77) reveals that no detail from the beta-binomial model is used. The only assumption made is that the variances and covariances are the same for all j and l. Thus, overdispersion can arise in a variety of contexts with non-independent data.

– iii. *Likelihood*

The likelihood is given by

$$L = \prod_{i=1}^{m} \int_0^1 \prod_{j=1}^{n_i} p_i^{y_{ij}} (1-p_i)^{1-y_{ij}} g(p_i; \alpha, \beta) \, dp_i$$

$$= \prod_{i=1}^{m} \frac{1}{B(\alpha, \beta)} \int_0^1 p_i^{y_i.} (1-p_i)^{n_i - y_i.} p_i^{\alpha-1} (1-p_i)^{\beta-1} \, dp_i$$

$$= \prod_{i=1}^{m} \frac{B(\alpha + y_i., \beta + n_i - y_i.)}{B(\alpha, \beta)}, \tag{2.79}$$

the last equality coming about from the definition of the beta function (2.70).

Using the results that $B(\alpha, \beta) = \Gamma(\alpha)\Gamma(\beta)/\Gamma(\alpha+\beta)$ and that

$$\Gamma(s+h)/\Lambda(s) = (s+h-1)(s+h-2)\cdots(s+1)s = \prod_{t=0}^{h-1}(s+t),$$

we have

$$
\begin{aligned}
l &= \log L \\
&= \sum_{i=1}^{m} \left[\log \frac{\Gamma(\alpha + y_i.)}{\Gamma(\alpha)} + \log \frac{\Gamma(\beta + n_i - y_i.)}{\Gamma(\beta)} - \log \frac{\Gamma(\alpha + \beta + n_i)}{\Gamma(\alpha + \beta)} \right] \\
&= \sum_{i=1}^{m} \left[\sum_{h=0}^{y_i.-1} \log(\alpha + h) + \sum_{h=0}^{n_i - y_i. - 1} \log(\beta + h) - \sum_{h=0}^{n_i - 1} \log(\alpha + \beta + h) \right].
\end{aligned}
$$

$$\tag{2.80}$$

In (2.80) and (2.81) shown below, any sum with an upper limit of -1 is interpreted as zero. Under the parameterization $\tau = 1/(\alpha + \beta)$, (2.73) takes the form (E 2.12)

$$l = \sum_{i=1}^{m} \left[\sum_{h=0}^{y_i.-1} \log(\mu + h\tau) + \sum_{h=0}^{n_i - y_i. - 1} \log(1 - \mu + h\tau) - \sum_{h=0}^{n_i-1} \log(1 + h\tau) \right]. \tag{2.81}$$

– iv.   *Underdispersion*

Since the parameters of the beta distribution in (2.70) are constrained to be positive, the intraclass correlation (2.76) is also positive and it would seem that the beta binomial model allows only positive within cluster correlation. However, Prentice (1986) shows that the product terms of (2.79) will be valid probabilities as long as

$$\rho \geq \max\{-\mu(n - \mu - 1)^{-1}, -(1 - \mu)(n + \mu - 2)^{-1}\}. \tag{2.82}$$

Thus, these extended beta-binomial models can describe some negative intraclass correlation or *underdispersion*, that is, variance less than that given by the binomial model. A useful feature of this model extension is that the test that the intraclass correlation is zero is no longer a test of a null hypothesis value on the boundary of the parameter space.

– v. ML estimation

Closed-form maximizing values of l do not exist, so numerical maximization must be used to find maximum likelihood estimates for any given data set.

– vi. Large-sample variances

ML estimators based on (2.80) or the reparameterized version (2.81) are asymptotically normally distributed with means equal to the true values and variances given by the inverse of the information matrix. In particular for (2.80)

$$
\begin{pmatrix} \hat{\alpha} \\ \hat{\beta} \end{pmatrix} \sim \mathcal{AN}\left(\begin{pmatrix} \alpha \\ \beta \end{pmatrix}, \Sigma_{\hat{\alpha},\hat{\beta}} = \begin{pmatrix} \sigma^{11} & \sigma^{12} \\ \sigma^{21} & \sigma^{22} \end{pmatrix}^{-1} \right),
\tag{2.83}
$$

where

$$
\begin{aligned}
\sigma^{11} &= -E\left[\frac{\partial^2 l}{\partial \alpha^2} \right] \\
&= \sum_{i=1}^{m} \left[\sum_{k=0}^{n_i} P\{y_{i\cdot} = k\} \left(\sum_{h=0}^{k-1} \frac{1}{(\alpha+h)^2} \right) + \sum_{h=0}^{n_i-1} \frac{1}{(\alpha+\beta+h)^2} \right] \\
\sigma^{12} &= -E\left[\frac{\partial^2 l}{\partial \alpha \partial \beta} \right] = \sigma^{21} \\
&= \sum_{i=1}^{m} \sum_{h=0}^{n_i-1} \frac{1}{(\alpha+\beta+h)^2} \\
\sigma^{22} &= -E\left[\frac{\partial^2 l}{\partial \beta^2} \right] \\
&= \sum_{i=1}^{m} \left[\sum_{k=0}^{n_i} P\{y_{i\cdot} = k\} \left(\sum_{h=0}^{n_i-k-1} \frac{1}{(\beta+h)^2} \right) + \sum_{h=0}^{n_i-1} \frac{1}{(\alpha+\beta+h)^2} \right],
\end{aligned}
$$

and $P\{y_i = k\} = \binom{n_i}{k} B(\alpha + k, n_i + \beta - k)/B(\alpha, \beta)$. As below (2.80) any sum with an upper limit of -1 is set equal to zero.

For the μ and τ parameterization,

$$\begin{pmatrix} \hat{\mu} \\ \hat{\tau} \end{pmatrix} \sim \mathcal{AN}\left(\begin{pmatrix} \mu \\ \tau \end{pmatrix}, \Sigma_{\hat{\mu}, \hat{\tau}} = \begin{pmatrix} \sigma^{11} & \sigma^{12} \\ \sigma^{21} & \sigma^{22} \end{pmatrix}^{-1} \right), \qquad (2.84)$$

where

$$\sigma^{11} = \sum_{i=1}^{m} \left[\sum_{k=0}^{n_i} P\{y_i = k\} \left(\sum_{h=0}^{k-1} \frac{1}{(\mu + h\tau)^2} + \sum_{h=0}^{n_i-k-1} \frac{1}{(1 - \mu + h\tau)^2} \right) \right]$$

$$\sigma^{12} = \sum_{i=1}^{m} \left[\sum_{k=0}^{n_i} P\{y_i = k\} \left(\sum_{h=0}^{k-1} \frac{h}{(\mu + h\tau)^2} + \sum_{h=0}^{n_i-k-1} \frac{h}{(1 - \mu + h\tau)^2} \right) \right]$$

$$\sigma^{21} = \sigma^{12}$$

$$\sigma^{22} = \sum_{i=1}^{m} \left[\sum_{k=0}^{n_i} P\{y_i = k\} \left(\sum_{h=0}^{k-1} \frac{h^2}{(\mu + h\tau)^2} + \sum_{h=0}^{n_i-k-1} \frac{h^2}{(1 - \mu + h\tau)^2} \right) \right.$$

$$\left. + \sum_{h=0}^{n_i-1} \frac{h^2}{(1 + h\tau)^2} \right]$$

and

$$P\{y_i = k\} = \binom{n_i}{k} B(\frac{\mu}{\tau} + k, n_i + \frac{1 - \mu}{\tau} - k)/B\left(\frac{\mu}{\tau}, \frac{1 - \mu}{\tau} \right).$$

– vii.   *Large-sample tests and confidence intervals*

Large-sample inferences concerning μ can be based on the asymptotic distribution (2.83) or (2.84). Let $\mathrm{var}(\hat{\mu})$ denote the (1,1) entry of $\Sigma_{\hat{\mu}, \hat{\tau}}$. Then a large-sample confidence interval would be

$$\hat{\mu} \pm z_{\alpha/2} \sqrt{\widehat{\mathrm{var}}(\hat{\mu})}, \qquad (2.85)$$

where $\widehat{\mathrm{var}}(\cdot)$ indicates that the estimated values of μ and τ have been substituted in the variance formula. Large-sample tests could be based on

$$z = \frac{\hat{\mu} - \mu}{\sqrt{\widehat{\mathrm{var}}(\hat{\mu})}} \sim \mathcal{AN}(0, 1). \qquad (2.86)$$

Alternatively, tests and confidence regions can be based on the like-lihood ratio statistic. To test $H_0 : \mu = \mu_0$, we calculate $\hat{\tau}(\mu_0)$ (that is, the value of τ which maximizes the likelihood when μ is fixed at μ_0). The large-sample test is then to reject H_0 if

$$-2\left\{l[\mu_0, \hat{\tau}(\mu_0)] - l[\hat{\mu}, \hat{\tau}]\right\} > \chi^2_{1,1-\alpha}, \tag{2.87}$$

where $l(\mu, \tau) = \log \Lambda(\mu, \tau)$ is the log likelihood as a function of μ and τ.

A confidence interval for μ which corresponds to the test (2.87) is the set of values μ^* given by

$$\left\{\mu^* : -2\left\{l[\mu^*, \hat{\tau}(\mu^*)] - l[\hat{\mu}, \hat{\tau}]\right\} < \chi^2_{1,1-\alpha}\right\} \tag{2.88}$$

that is, the set of values of μ^* for which we can accept the $H_0 : \mu = \mu^*$. This set must be calculated numerically.

Large-sample inferences for τ must be handled carefully. The usual hypothesis of interest is $H_0 : \tau = 0$, which, as long as μ is not zero or one, is equivalent to no correlation, or equivalently, no variation in the p_i across the one-way classification. When $\tau = 0$ and for large samples, the maximum likelihood estimator is exactly zero half the time (for a similar situation see E 2.17). It thus does *not* have a large-sample normal distribution and the usual large-sample theory for $-2 \log \Lambda$ fails.

In this simple case an easy modification is available: Calculate the usual likelihood ratio statistic and make a simple adjustment to the critical value. Under $H_0 : \tau = 0$ the maximum likelihood estimator of μ is $\bar{y}_{..}$. The test is thus to reject H_0 when

$$-2\left[l(\bar{y}_{..}, 0) - l(\hat{\mu}, \hat{\tau})\right] > \chi^2_{1,1-2\alpha}. \tag{2.89}$$

Roughly speaking, this can be thought of as adjusting a test statistic appropriate for a two-sided test (the likelihood ratio test) in order to test a one-sided hypothesis ($H_0 : \tau = 0$ versus $H_1 : \tau > 0$).

In the less usual case when a specified value of $\tau > 0$ is of interest, the large-sample distribution of $\hat{\tau}$ *is* asymptotically normal. Tests and confidence intervals can then be based on the standard normal distri-bution just as with μ in (2.86) or on the likelihood ratio test with the usual critical point, $\chi^2_{1,1-\alpha}$.

– viii. *Prediction*

As before, we wish to calculate the best predicted values as given by

$$BP(p_i) = E[p_i|y] = E[p_i|y_{i.}]. \tag{2.90}$$

Under the beta-binomial model, (2.72), it is straightforward to show
(E 2.13) that the conditional distribution of p_i given y_i. is beta with
parameters $\alpha + y_i$. and $\beta + n_i - y_i$. . The conditional mean is therefore
given by

$$\mathrm{E}[p_i|y_i.] = \frac{\alpha + y_i.}{\alpha + \beta + n_i} = \frac{\mu/\tau + y_i.}{1/\tau + n_i} \tag{2.91}$$

and the estimated best predictor, $\tilde{p}_i$, is given by

$$\begin{aligned}
\tilde{p}_i &= \frac{\hat{\mu}/\hat{\tau} + y_i.}{1/\hat{\tau} + n_i} \\[2mm]
&= \hat{\mu}\left(\frac{1/\hat{\tau}}{1/\hat{\tau} + n_i}\right) + \bar{y}_i.\left(\frac{n_i}{1/\hat{\tau} + n_i}\right) \\[2mm]
&= \hat{\mu}\left(\frac{1/\hat{\tau}}{1/\hat{\tau} + n_i}\right) + \hat{\pi}_i\left(\frac{n_i}{1/\hat{\tau} + n_i}\right). \tag{2.92}
\end{aligned}$$

As with the normal random effects model, the estimated best predictor,
$\tilde{p}_i$, is a weighted average of an overall estimate, $\hat{\mu}$, and the fixed effects
estimate, $\hat{\pi}_i$. It is also a shrinkage estimator, with the individual
predicted values, $\tilde{p}_i$, being closer to the overall estimate, $\hat{\mu}$, than are
the $\hat{\pi}_i$.

c. Logit-normal model

The reason a normal distribution cannot be assumed for p_i in (2.68) is
that p_i is restricted to the interval $(0,1)$. An alternative approach is to
transform p_i using $\mathrm{logit}(p_i) \equiv \log[p_i/(1 - p_i)]$. The range for $\mathrm{logit}(p_i)$
is $(-\infty, \infty)$ as p_i ranges from zero to one and a normal distribution
can be assumed for $\mathrm{logit}(p_i)$. This gives the model

$$\begin{aligned}
\mathrm{E}[y_{ij}|p_i] &= p_i \\[2mm]
y_{ij}|p_i &\sim \text{indep. Bernoulli}(p_i) \\[2mm]
\tilde{l}_i = \mathrm{logit}(p_i) &\sim \text{i.i.d. } \mathcal{N}(\mu, \sigma^2). \tag{2.93}
\end{aligned}$$

− i. Likelihood

The likelihood for this model, similar to (2.79), is

$$L = \prod_{i=1}^{m} \int_{-\infty}^{\infty} \prod_{j=1}^{n_i} p_i^{y_{ij}}(1 - p_i)^{1-y_{ij}} \frac{1}{\sqrt{2\pi\sigma^2}} e^{-\frac{1}{2\sigma^2}(\tilde{l}_i - \mu)^2} d\tilde{l}_i. \tag{2.94}$$

This can be written in a slightly simpler fashion as

$$L = \prod_{i=1}^{m} \int_{-\infty}^{\infty} \left(\frac{p_i}{1-p_i} \right)^{y_{i\cdot}} (1-p_i)^{n_i} \frac{1}{\sqrt{2\pi\sigma^2}} e^{-\frac{1}{2\sigma^2}(\tilde{l}_i - \mu)^2} d\tilde{l}_i. \qquad (2.95)$$

With a change of variables to $z_i = (\tilde{l}_i - \mu)/\sigma$ this becomes

$$L = \prod_{i=1}^{m} \int_{-\infty}^{\infty} \frac{e^{(\mu+\sigma z_i)y_{i\cdot}}}{(1+e^{\mu+\sigma z_i})^{n_i}} \frac{1}{\sqrt{2\pi}} e^{-\frac{1}{2}z_i^2} dz_i. \qquad (2.96)$$

Unfortunately, neither L nor $l = \log L$ can be appreciably simplified and both calculation and maximization of l must be done using numerical methods.

– ii.   *Calculation of the likelihood*

Changing variables again, using $\frac{1}{2}z_i^2 = v_i^2$ in (2.96) allows the log likelihood to be written as

$$l = \sum_{i=1}^{m} \log \int_{-\infty}^{\infty} \frac{e^{(\mu+\sqrt{2}\sigma v_i)y_{i\cdot}}}{1+e^{\mu+\sqrt{2}\sigma v_i}} \frac{1}{\sqrt{\pi}} e^{-v_i^2} dv_i. \qquad (2.97)$$

In this form, each integral in l can be evaluated using Gauss-Hermite quadrature wherein

$$\int_{-\infty}^{\infty} g(x)e^{-x^2} dx \approx \sum_{k=-r}^{r} w_k g(x_k), \qquad (2.98)$$

where r is the order of the numerical integration, the w_k are weights, and the x_k are evaluation points. Generally, using large values of r increases the computation time and increases accuracy. Values of w_k and x_k are given in references on numerical integration, for example, (Abramowitz and Stegun, 1964, Table 25.10). Using this approximation,

$$l \approx \sum_{i=1}^{m} \log \left(\sum_{k=-r}^{r} w_k \frac{e^{(\mu+\sqrt{2}\sigma x_k)y_{i\cdot}}}{\sqrt{\pi}(1+e^{\mu+\sqrt{2}\sigma x_k})} \right). \qquad (2.99)$$

For speed of computation, values of r as small as 2 or 3 have been recommended in practice (Goldstein, 1986; Hedeker and Gibbons, 1994), but this can lead to inaccurate results. If μ is not near zero, the integral can be difficult to evaluate accurately (Liu and Pierce, 1994). Values of r of 10 or greater often give good accuracy.

– iii.   *Means, variances, and covariances*

Under model (2.93) we calculate the mean by writing $p_i = 1/(1 + e^{-\tilde{l}_i})$ with $\tilde{l}_i \sim \mathcal{N}(\mu, \sigma^2)$. Therefore,

$$
\begin{aligned}
E[y_{ij}] &= E\left[E[y_{ij} | p_i]\right] = E[p_i] \\
&= E\left[\frac{1}{1 + e^{-\tilde{l}_i}}\right] = \int_{-\infty}^{\infty} \frac{1}{1 + e^{-\tilde{l}_i}} \frac{1}{\sqrt{2\pi\sigma^2}} e^{-\frac{1}{2\sigma^2}(\tilde{l}_i - \mu)^2} \, d\tilde{l}_i \\
&= \int_{-\infty}^{\infty} \frac{1}{1 + e^{-(\mu + \sigma z_i)}} \frac{1}{\sqrt{2\pi}} e^{-\frac{1}{2}z_i^2} \, dz_i.
\end{aligned}
\tag{2.100}
$$

Again, this cannot be evaluated in closed form, but can be approximated as before using Gauss-Hermite quadrature. Interestingly, there is a

Since y_{ij} is binary, it has a marginal Bernoulli distribution with mean, $E[y_{ij}]$, given by (2.99). Its variance is therefore $E[y_{ij}](1 - E[y_{ij}])$.

From first principles, the covariance of two observations in the same level of the one-way classification is $\text{cov}(y_{ij}, y_{il}) = E[y_{ij}y_{il}] - E[y_{ij}]E[y_{il}]$. The second part of this can be evaluated using (2.99) and the first part calculated as

$$
E[y_{ij}y_{il}] = \int_{-\infty}^{\infty} \left(\frac{1}{1 + e^{-(\mu + \sigma z_i)}}\right)^2 \frac{1}{\sqrt{2\pi}} e^{-\frac{1}{2}z_i^2} \, dz_i.
\tag{2.101}
$$

How does σ relate to the correlation between observations in the same level of the classification? Table 2.3 gives values of the correlation for several values of μ and σ.

Table 2.3: Correlations for the Logit-normal Model

σ	μ						
	-2	-1	-0.5	0	0.5	1	2
0	0.000	0.000	0.000	0.000	0.000	0.000	0.000
1	0.118	0.158	0.169	0.174	0.169	0.158	0.118
3	0.521	0.536	0.539	0.541	0.539	0.536	0.521
5	0.694	0.698	0.699	0.699	0.699	0.698	0.694

– iv.   *Large-sample tests and intervals*

As in Section 2.5f, large-sample inferences concerning μ can be based on the large-sample normal distribution of $\hat{\mu}$ or the asymptotic chi-square distribution of $-2 \log \Lambda$. Derivatives for calculating the observed information matrix must be calculated numerically (Hedeker and Gibbons, 1994), which makes dealing with $-2 \log \Lambda$ more attractive.

A large-sample test of $H_0 : \sigma^2 = 0$ is made by rejecting H_0 if

$$-2 \log \Lambda = -2 \left\{ l[\hat{\mu}(\sigma^2 = 0), 0] - l[\hat{\mu}, \hat{\sigma}^2] \right\} > \chi^2_{1,1-2\alpha}. \qquad (2.102)$$

A disadvantage of this approach is that the MLEs must be calculated under the random effects model, which is computationally difficult. An alternative is to consider score tests as given by, for example, Commenges et al. (1994). Scores tests use as their statistic the derivative of the log likelihood evaluated under the null hypothesis (Cox and Hinkley, 1974, p. 315), which does not require estimation under the model with random effects. For the simple case of model (2.92), the test reduces to the usual Pearson chi-square test of Section 2.5e (see E 2.16).

– v. *Prediction*

For prediction we want $E[p_i | y_{ij}]$ for which the conditional distribution of p_i given $y_{i\cdot}$ is required. Again it is more convenient to consider $p_i = 1/(1 + e^{-(\mu + \sigma z_i)})$ where $z_i \sim \mathcal{N}(0,1)$. We thus need

$$f_{z_i | y_{i\cdot}}(z|y) = \frac{f_{y_{i\cdot}|z_i}(y|z) f_{z_i}(z)}{\int f_{y_{i\cdot}|z_i}(y|z) f_{z_i}(z) dz}. \qquad (2.103)$$

The numerator is given by

$$\left(\frac{1}{1 + e^{-(\mu + \sigma z)}} \right)^y \left(1 - \frac{1}{1 + e^{-(\mu + \sigma z)}} \right)^{n_i - y} \frac{1}{\sqrt{2\pi}} e^{-\frac{1}{2} z^2},$$

so the estimated predicted value can now be calculated as

$$\tilde{p}_i = \hat{E}[p_i | y_{i\cdot}] \qquad (2.104)$$

$$= \frac{\int_{-\infty}^{\infty} \left(1 + e^{-(\hat{\mu} + \hat{\sigma} z)} \right)^{-(y_{i\cdot} + 1)} \left(1 + e^{\hat{\mu} + \hat{\sigma} z} \right)^{-(n_i - y_{i\cdot})} \frac{1}{\sqrt{2\pi}} e^{-\frac{1}{2} z^2} dz}{\int_{-\infty}^{\infty} \left(1 + e^{-(\hat{\mu} + \hat{\sigma} z)} \right)^{-(y_{i\cdot})} \left(1 + e^{\hat{\mu} + \hat{\sigma} z} \right)^{-(n_i - y_{i\cdot})} \frac{1}{\sqrt{2\pi}} e^{-\frac{1}{2} z^2} dz}.$$

As in Section 2.6b numerical evaluation must be used; one possible method is Gauss-Hermite quadrature.

d.   Probit-normal model

A model similar to the logit normal model is the probit-normal model, which is obtained by replacing the logit function in (2.93) by Φ^{-1}, where Φ is the standard normal cumulative distribution function (c.d.f.):

$$\Phi(t) = \int_{-\infty}^{t} \frac{1}{\sqrt{2\pi}} e^{-\frac{1}{2}x^2} dx. \qquad (2.105)$$

Using the probit link not appreciably simplify the calculations, except $E[y_{ij}]$ slightly, which is given by

$$E[y_{ij}] = \Phi\left(\frac{\mu}{\sqrt{1+\sigma^2}}\right) \qquad (2.106)$$

(see E 2.18 and, for a similar result, Section 9.3a–i).

Otherwise derivations and formulae closely follow those for the logit-normal in Section 2.7. For example, the analog of (2.101) is

$$E[y_{ij}y_{il}] = \int_{-\infty}^{\infty} [\Phi(\mu + \sigma z_i)]^2 \frac{1}{\sqrt{2\pi}} e^{-\frac{1}{2}z_i^2} dz_i . \qquad (2.107)$$

2.7   COMPUTING

Even the most mathematically tractable of the models for binary data with random effects, the beta-binomial model, poses numerical difficulties. The more easily generalizable logit- and probit-normal models of the preceding section raise further problems. Though we have indicated a possible approach using Gauss-Hermite quadrature, this quickly becomes intractable, even for problems of moderate size. More is said about these issues in Chapter 14.

2.8   EXERCISES

E 2.1 Show that $N \log\left(1 + \frac{m-1}{N-m} F_{N-m,1-\alpha}^{m-1}\right)$ tends to $\chi^2_{m-1,1-\alpha}$ for large N.

E 2.2 For the F-test of Section 2.1c, with $m = 5$ and $N = 20$, 50, and 100, calculate the significance level achieved if the asymptotic critical value is used instead of the exact critical value.

E 2.3 Derive (2.11).

E 2.4 When $\dot{\sigma}_a^2 < 0$, and hence $\hat{\sigma}_a^2 = 0$:

    (a) Use (2.41) and (2.42) to show that for unbalanced data, $\hat{\mu} = \bar{y}_{..}$ and $\hat{\sigma}^2 = \text{SST}/N$.

    (b) Why is (2.43) not used?

E 2.5 With

$$
\begin{aligned}
\log L \;=\; & -\tfrac{1}{2}N \log 2\pi - \tfrac{1}{2}(N - m) \log \sigma^2 - \tfrac{1}{2}\sum_i \log \lambda_i \\
& - \text{SSE}/2\sigma^2 - \sum [n_i(\bar{y}_{i.} - \mu)^2/2\lambda_i] \qquad (2.108)
\end{aligned}
$$

show that $L \to -\infty$ as $\sigma^2 \to 0$ and as $\sigma^2 \to \infty$, so that L must have a maximum for a positive value of σ^2.

E 2.6 For the hypothesis $H_0 : \pi_i = \pi_j$ compare the χ^2 test, which uses only columns i and j of Table 2.1, with the test given by (2.66).

E 2.7 For the balanced data situation ($n_i \equiv n$) for model (2.15) show that

$$
\frac{\bar{y}_{..} - \mu}{\sqrt{\text{MSA}/mn}} \sim \mathcal{T}_{m-1}.
$$

E 2.8 Show that $E[Y|X]$ is the minimum mean square error predictor of Y. That is, show that $g(X) = E[Y|X]$ minimizes $E\left[(Y - g(X))^2\right]$ among all functions $g(\cdot)$ of X.

E 2.9 Suppose that X and Y are bivariate normal with correlation ρ. Show that if X is k standard deviations above its mean, then the minimum mean square error predictor is that Y will be ρk standard deviations above its mean.

E 2.10   (a) Derive $F = \text{MSA}/\text{MSE}$ as the LRT statistic for $H : \sigma_a^2 = 0$ in the one-way classification, random, normal model, balanced data. The derivation is lengthy. The following steps help.

      (i.) Denote (2.24) as $l(\mu,\, \sigma^2,\, \sigma)$.

      (ii.) Find $l(\dot{\mu},\, \dot{\sigma}^2,\, \dot{\sigma})$.

      (iii.) Find $l(\dot{\mu}_0,\, \dot{\sigma}_0^2)$ under H.

      (iv.) Define $q = (m - 1)F/m(n - 1)$.

      (v.) Show that $\partial(-2\log \Lambda)/\partial q > 0$ if $\dot{\sigma}_a^2 > 0$.

      (vi.) Explain how this leads to F being a test statistic.

    (b) For unbalanced data explain why (ii) cannot be obtained analytically, and so neither can $\log \Lambda$. But find (iii).

    (c) Despite (b), show that F is a test statistic for $H : \sigma_a^2 = 0$.
See Searle, Casella and McCulloch (1992, p. 76).

E 2.11 From (2.60) find $-E\left[\partial^2 l / 2\pi_k^2\right]$ and from that the sampling variance of $\hat{\pi}_k$.

E 2.12 For the beta-binomial model given by (2.72) with the parameterization $\mu = \alpha/(\alpha + \beta)$ and $\tau = 1/(\alpha + \beta)$, derive (2.81) from (2.80).

E 2.13 Prove the penultimate sentence before (2.89).

E 2.14 For the beta-binomial model given by (2.72) show that the conditional distribution of p_i given $y_{i\cdot}$ is again beta, but with parameters $\alpha + y_{i\cdot}$ and $\beta + n_i - y_{i\cdot}$.

E 2.15 Prove (2.91).

E 2.16 Show that the derivative of the log of L from (2.95) evaluated at $\sigma = 0$ is a function of

$$\sum_i n_i (\hat{\pi}_i - \pi)^2,$$

as in (2.63). The score test of $H_0: \sigma = 0$ is based on this statistic. Hence show that when properly standardized and with the MLEs substituted for unknown parameters, this is the same as the Pearson chi-square statistic of Section 2.5e for testing homogeneity in the fixed effects model. *Hint*: To calculate the derivative you will need to use L'Hospital's rule.

E 2.17 For large samples from model (2.93), show that the maximum likelihood estimator of σ is equal to zero with probability $1/2$. Hence show that $-2\log\Lambda$ for testing $H_0 : \sigma = 0$ is zero with probability $1/2$.

E 2.18 For the probit-normal model show that $E[y_{ij}] = \Phi(\mu/\sqrt{1 + \sigma^2})$.

E 2.19 Show for Section 2.2b that the ML solutions maximize the likelihood.

    (a) First derivatives of l are shown at (2.25). Use them to find the three second derivatives with respect to μ.

(b) For the results in (a) and for the second derivatives shown in Section 2.2b–v, replace parameters by solutions of the ML equations.

(c) Assemble results from (b) in a matrix and explain by that matrix is negative definite, and hence the solutions maximize l.

Chapter 3

SINGLE-PREDICTOR REGRESSION

3.1 INTRODUCTION

Chapter 2 deals with the one-way classification for data which are either normally or Bernoulli distributed. For each of these distributions this chapter covers simple regression, that is, regression with a single predictor. For the one-way classification in Chapter 2 we described the class means by using different parameters for each class, for instance, equations (2.1) and (2.2). As such, we made no assumptions about the form of the mean of y as a function of the classification variable. In contrast, for simple linear regression we make the restrictive assumption that y is a linear function of a predictor, x. For example, we will consider a case study in which the mean of y, log radial growth of colonies of *Phytophthora infestans sporangia* inoculated onto potato leaflets, is modeled as a linear function of the temperature at which the colonies were allowed to grow.

In practice, there are many situations where the linearity assumption is not met, in which case we must regard it either as a crude approximation or merely the first step in a more in-depth analysis. For some cases it is adequate to assume that the mean of y is a linear function of x, but over only a short interval of x. For example, a plot of the radial growth data (Figure 3.1) for weeks 2, 3 and 4 for temperatures 15°C through 25°C shows an approximately linear relationship. However, over the entire range of the experiment (down to 10°C), the relationship appears nonlinear.

72

Figure 3.1: Plot of log(area) versus temperature for the lesion data.

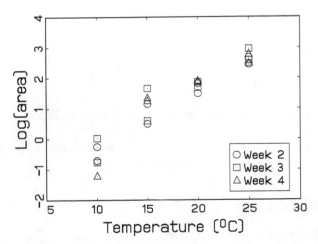

In other cases we must transform y and/or x before the linearity assumption is met even approximately. Alternatively or additionally, we might try more complicated models with multiple predictors designed to encompass more flexible functional forms. We describe such models in subsequent chapters.

A different approach is to model some known function of the mean of y, call it μ, as linear in x. This is called a *generalized linear model*, examples of which are found in Sections 3.7 and 3.8. There we argue that in many instances of Bernoulli-distributed data, it does not make sense to assume a simple linear regression model. Instead we model $\log[\mu/(1 - \mu)]$ as linear in the predictor.

3.2 NORMALITY: SIMPLE LINEAR REGRESSION

a. Model

We begin with normally distributed data, for which the well-known and frequently used simple linear regression model is given by

$$
\begin{aligned}
\mathrm{E}[y_i] &= \mu(x_i) = \alpha + \beta x_i \\
y_i &\sim \text{ indep. } \mathcal{N}[\mu(x_i), \sigma^2] \qquad i = 1, 2, \ldots, N,
\end{aligned} \qquad (3.1)
$$

where the notation $\mu(x)$ is used to indicate that μ is a function of x. The x_i are assumed to be known constants, either fixed as part of the data collection process or regarded as fixed by considering the conditional distribution given the x's. Note that (3.1) encompasses four assumptions:

1. The y_i follow a normal distribution.

2. The y_i are independent.

3. The y_i all have the same variance, σ^2.

4. The mean of y_i is a linear function of the predictor, x_i.

Any or all of these may be regarded simply as adequate approximations or as the beginning of a more serious analysis.

b.   Likelihood

The log likelihood is easily derived:

$$l = \log L = -\frac{N}{2}\log 2\pi - \frac{N}{2}\log\sigma^2 - \frac{1}{2\sigma^2}\sum_i(y_i - \alpha - \beta x_i)^2. \quad (3.2)$$

c.   Maximum likelihood estimators

The maximum likelihood estimators can be found by equating the derivatives of the log likelihood to zero. Those derivatives are

$$\frac{\partial l}{\partial \alpha} = \frac{1}{\sigma^2}\sum_{i=1}^{n}(y_i - \alpha - \beta x_i) \quad (3.3)$$

$$\frac{\partial l}{\partial \beta} = \frac{1}{\sigma^2}\sum_{i=1}^{n}x_i(y_i - \alpha - \beta x_i) \quad (3.4)$$

$$\frac{\partial l}{\partial \sigma^2} = -\frac{N}{2\sigma^2} + \frac{1}{\sigma^4}\sum_{i=1}^{n}(y_i - \alpha - \beta x_i)^2. \quad (3.5)$$

In equating these to zero, we can replace parameters by MLEs (e.g., replace α by $\hat{\alpha}$) because the solutions to the resulting equations are indeed the MLEs. The distinction between solutions and estimators (discussed in Section 1.7a–iii) is not problematic here because α and β can be any real numbers and therefore so can $\hat{\alpha}$ and $\hat{\beta}$, as evident in

(3.6) and (3.7). And, from equation (3.8), $\hat{\sigma}^2$ is non-negative, in accord with the definition of σ^2. We therefore have

$$\sum_i (y_i - \mathrm{E}[y_i]) = 0 \quad \text{or} \tag{3.6}$$

$$\sum_i y_i = n\hat{\alpha} + \hat{\beta} \sum_i x_i$$

from (3.3);

$$\sum_i x_i (y_i - \mathrm{E}[y_i]) = 0 \quad \text{or} \tag{3.7}$$

$$\sum_i x_i y_i = \hat{\alpha} \sum_i x_i + \hat{\beta} \sum_i x_i^2$$

from (3.4); and

$$\hat{\sigma}^2 = \frac{1}{N} \sum_i (y_i - \hat{\alpha} - \hat{\beta}x_i)^2 \tag{3.8}$$

from (3.5). We can straightforwardly solve these equations:

$$\hat{\beta} = \frac{\sum_i x_i y_i - (\sum_i x_i)(\sum_i y_i)/N}{\sum_i x_i^2 - (\sum_i x_i)^2/N}$$

$$\hat{\alpha} = \bar{y} - \hat{\beta}\bar{x} \tag{3.9}$$

$$\hat{\sigma}^2 = \frac{1}{N} \sum_i (y_i - \hat{\alpha} - \hat{\beta}x_i)^2.$$

Again, the general REML equation, (6.67), yields the REML estimator of σ^2. It is exactly $\hat{\sigma}^2$ of (3.9) except for the important replacement of N by $N-2$ to account for the two fixed effects α and β (see Section 1.7a–ii).

d. Distributions of MLEs

Standard derivations (e.g., Weisberg, 1980, p. 4) give the distributions of the MLEs. Defining $S_{xx} = \sum_i (x_i - \bar{x})^2$, the MLEs of α and β are bivariate normal:

$$\begin{pmatrix} \hat{\alpha} \\ \hat{\beta} \end{pmatrix} \sim \mathcal{N}\left[\begin{pmatrix} \alpha \\ \beta \end{pmatrix}, \frac{\sigma^2}{S_{xx}} \begin{pmatrix} \sum x_i^2 & -\bar{x} \\ -\bar{x} & 1 \end{pmatrix} \right]. \tag{3.10}$$

They are independent of $\hat{\sigma}^2$, which is distributed as a multiple of a chi-square distribution:

$$\frac{N\hat{\sigma}^2}{\sigma^2} \sim \chi^2_{N-2}. \tag{3.11}$$

e.   Tests and confidence intervals

Tests and confidence intervals can be derived utilizing the t-distribution. For example, a confidence interval for β is given by

$$\hat{\beta} \ \pm \ t_{N-2,\alpha/2}\sqrt{\widehat{\text{var}}(\hat{\beta})},$$

which is

$$\hat{\beta} \ \pm \ t_{N-2,\alpha/2}\sqrt{\frac{\frac{N}{N-2}\hat{\sigma}^2}{S_{xx}}}. \tag{3.12}$$

Similarly, a test of $H_0 : \beta \leq 0$ versus $H_A : \beta > 0$ would be to reject the H_0 if

$$\hat{\beta}\Big/\sqrt{\frac{\frac{N}{N-2}\hat{\sigma}^2}{S_{xx}}} > t_{N-2,\alpha/2}. \tag{3.13}$$

Tests and confidence intervals for α can be derived in the same manner. A confidence interval for σ^2 can be calculated using (3.11) as

$$\left(\frac{N\hat{\sigma}^2}{\chi^2_{N-2,1-\alpha/2}}, \frac{N\hat{\sigma}^2}{\chi^2_{N-2,\alpha/2}}\right). \tag{3.14}$$

f.   Illustration

For the *Phytophthora* data of Figure 3.1 (using weeks 2 through 4, temperatures 15°C through 25°C only) we consider a linear regression model for the average of the two measurements on a leaflet

$$\text{E}[\text{ALD}_j] \ = \ \mu_j = \alpha + \beta\text{TEMP}_j$$

$$\text{ALD}_j \ \sim \ \text{indep. } \mathcal{N}(\mu_j, \sigma^2),$$

where $\text{ALD}_j = y_j$ is the average log diameter of the lesions on the jth leaflet, and $\text{TEMP}_j = x_j$ is the temperature for the jth leaflet. For these data, the maximum likelihood estimators are $\hat{\alpha} = -1.296, \hat{\beta} =$

0.157, and $\hat{\sigma}^2 = 0.0755$. From these we calculate a 95% confidence interval for β as

$$0.157 \pm t_{N-2,0.025}\sqrt{\frac{\frac{N}{N-2}0.0755}{S_{xx}}} \quad = \quad 0.157 \pm t_{16,0.025}\sqrt{\frac{\frac{18}{16}0.0755}{300}}$$

$$= \quad 0.157 \pm 2.120\sqrt{0.000283}$$

$$= \quad 0.157 \pm 0.0357$$

$$= \quad (0.121, 0.193).$$

We are thus 95% confident that the average log lesion diameter increases by between 0.121 and 0.193 with each increase in temperature of $1°$ C over the range of temperatures from $15°$C through $25°$C.

3.3 NORMALITY: A NONLINEAR MODEL

a. Model

A variation on (3.1) is to assume that the model is nonlinear in its parameters. A simple example is

$$\begin{aligned} E[y_i] &= \mu(x_i) = e^{\alpha + \beta x_i} \\ y_i &\sim \text{ indep. } \mathcal{N}[\mu(x_i), \sigma^2]; \qquad i = 1, 2, \ldots, N. \quad (3.15) \end{aligned}$$

Note that we have changed only the mean of y_i, not assumptions about the shape of its distribution.

b. Likelihood

The log likelihood is basically the same as (3.2):

$$l = \log L = -\frac{N2}{\log}2\pi - \frac{N}{2}\log\sigma^2 - \frac{1}{2\sigma^2}\sum_i [y_i - \mu(x_i)]^2 \quad (3.16)$$

except that $\mu(x_i)$ in (3.16) is $e^{\alpha + \beta x_i}$ instead of $\alpha + \beta x_i$ of (3.1).

c.   Maximum likelihood estimators

From (3.16) we can see that to maximize the likelihood with respect to α and β we must minimize the residual sum of squares: $\sum[y_i - \mu(x_i)]^2$. So maximum likelihood for homoscedastic, normal distribution models is equivalent to least squares, even with nonlinear models. More formally, we can differentiate l of (3.16) to try to maximize it. The derivatives are

$$\frac{\partial l}{\partial \alpha} = \frac{1}{\sigma^2} \sum_{i=1}^{N} (y_i - e^{\alpha + \beta x_i}) e^{\alpha + \beta x_i} \tag{3.17}$$

$$\frac{\partial l}{\partial \beta} = \frac{1}{\sigma^2} \sum_{i=1}^{N} (y_i - e^{\alpha + \beta x_i}) e^{\alpha + \beta x_i} x_i \tag{3.18}$$

$$\frac{\partial l}{\partial \sigma^2} = -\frac{N}{2\sigma^2} + \frac{1}{\sigma^4} \sum_{i=1}^{N} (y_i - e^{\alpha + \beta x_i})^2. \tag{3.19}$$

Setting these equal to zero, with parameters replaced by MLEs (e.g., α replaced by $\hat{\alpha}$ as in (3.9)), gives

$$\sum_i y_i e^{\hat{\alpha} + \hat{\beta} x_i} = \sum_i e^{2(\hat{\alpha} + \hat{\beta} x_i)} \tag{3.20}$$

$$\sum_i y_i e^{\hat{\alpha} + \hat{\beta} x_i} x_i = \sum_i e^{2(\hat{\alpha} + \hat{\beta} x_i)} x_i \tag{3.21}$$

$$\hat{\sigma}^2 = \frac{1}{N} \sum_i (y_i - e^{\hat{\alpha} + \hat{\beta} x_i})^2 \tag{3.22}$$

The first two equations can be solved for $\hat{\alpha}$ and $\hat{\beta}$ which are then substituted in the third equation to find $\hat{\sigma}^2$. To solve (3.20) and (3.21) for $\hat{\alpha}$ and $\hat{\beta}$ reduce the equations to

$$\sum_i y_i e^{\hat{\beta} x_i} = e^{\hat{\alpha}} \sum_i e^{2\hat{\beta} x_i} \tag{3.23}$$

and

$$\sum_i y_i e^{\hat{\beta} x_i} x_i = e^{\hat{\alpha}} \sum_i e^{2\hat{\beta} x_i} x_i \tag{3.24}$$

or

$$\sum_i y_i e^{\hat{\beta} x_i} / \sum_i y_i e^{\hat{\beta} x_i} x_i = \sum_i e^{2\hat{\beta} x_i} / \sum_i e^{2\hat{\beta} x_i} x_i. \tag{3.25}$$

This equation must be solved numerically to find $\hat{\beta}$. We can then obtain

$$\hat{\alpha} = \log \frac{\sum_i y_i e^{\hat{\beta} x_i}}{\sum_i e^{2\hat{\beta} x_i}}$$

$$= \log \frac{\sum_i y_i e^{\hat{\beta} x_i} x_i}{\sum_i e^{2\hat{\beta} x_i} x_i} \tag{3.26}$$

and

$$\hat{\sigma}^2 = \frac{1}{N} \sum_i (y_i - e^{\hat{\alpha} + \hat{\beta} x_i})^2. \tag{3.27}$$

What about restricted maximum likelihood for this model? The usual basis for REML is linear combinations of the data chosen to be free of the fixed effects. With a nonlinear model such as this one, that is not possible.

d.   Distributions of MLEs

The MLEs are nonlinear functions of the data and do not having closed-form expressions and this precludes our working out their exact, small-sample distributions. Simulations and calculations for small sample sizes show that the estimators for the parameters α, β, and σ^2 are biased.

The large-sample distributions of the MLEs (see Section S.4d of Appendix S) are, as usual, tractable:

$$
\begin{pmatrix} \hat{\alpha} \\ \hat{\beta} \\ \hat{\sigma}^2 \end{pmatrix} \sim \mathcal{AN} \left[\begin{pmatrix} \alpha \\ \beta \\ \sigma^2 \end{pmatrix}, \frac{\sigma^2}{\Delta} \begin{pmatrix} \sum \mu_i^2 x_i^2 & -\sum \mu_i^2 x_i & 0 \\ -\sum \mu_i^2 x_i & \sum \mu_i^2 & 0 \\ 0 & 0 & \frac{2\sigma^2 \Delta}{N} \end{pmatrix} \right],
$$
$$\tag{3.28}$$

where $\Delta = \sum \mu_i^2 x_i^2 \sum \mu_i^2 - (\sum \mu_i^2 x_i)^2$, and $\mu_i = \exp(\alpha + \beta x_i)$.

This points out that the nice properties of ML estimators under the linear model (3.1) are quite delicate. A modest change in modeling $E[y_i]$ from $\alpha + \beta x_i$ in (3.1) to $e^{\alpha + \beta x_i}$ in (3.15) causes the estimators not to have closed-form expressions, makes it impossible to work out their small-sample distribution and causes the estimator to be biased (in small samples).

3.4   TRANSFORMING VERSUS LINKING

a.   Transforming

A temptation to resolve the difficulties inherent in (3.15) is to work with the log transform of the data and assume the model

$$E[\log y_i] \;=\; \alpha^* + \beta^* x_i$$
$$\log y_i \;\sim\; \text{indep. } \mathcal{N}(\alpha^* + \beta^* x_i, \sigma^{*2}); \quad i = 1, 2, \ldots, N, \quad (3.29)$$

where we have used superscript asterisks to indicate that the parameters are different from those of (3.15). Since the model for $\log y_i$ is now linear with homoscedastic, normal distributions, we regain the nice properties of ML estimators.

b.   Linking

The second way is to assume in (3.15) that a function of the mean is linear in the parameters. We will call this the link function. In that case we are assuming that $\log E[y_i] = \alpha + \beta x_i$, so we are using a log link. However, this is not the same as in (3.29). First, using (3.15), $E[\log y_i]$ does not exist (since negative values of y_i are possible). More practically, even if y_i had a distribution such that it was positive with probability 1, by Jensen's inequality (Casella and Berger, 1990, p. 182)

$$E[\log y_i] < \log(E[y_i]) = \log[e^{\alpha + \beta x_i}] = \alpha + \beta x_i. \qquad (3.30)$$

Second, (3.15) has y_i homoscedastic on the original scale while (3.29) has y_i homoscedastic after taking the log transformation. Under (3.29) it is easy to show (E 3.3) that the standard deviation of y_i increases proportionally with the mean.

c.   Comparisons

Thus a log transformation of the data is not the same as using the log link, namely that $\log(E[y_i])$ follows a linear model. Choosing between (3.15) and (3.29) would ordinarily be done by checking whether the variance is constant on the original scale or increases with the mean (Ruppert et al., 1989).

3.5   RANDOM INTERCEPTS: BALANCED DATA

From the example of observing lesion growth on potato leaflets described at the start of this chapter, let us concentrate on y_{ij} being the lesion area observed in week i at temperature x_{ij}. Then, as an extension of traditional single-predictor regression, $E[y_{ij}] = \alpha + \beta x_{ij}$, we now consider

$$E[y_{ij}|a_i] = \mu + a_i + \beta x_{ij} \tag{3.31}$$

for $i = 1, 2, \ldots, m$ and (initially for balanced data) for $j = 1, 2, \ldots, n$. The important feature attributed to (3.31) is that the intercepts $\mu + a_i$ are taken as being random; that is, we treat the a_i as random effects, normally distributed with zero mean:

$$a_i \sim \text{ i.i.d. } \mathcal{N}(0, \sigma_a^2). \tag{3.32}$$

In every week, $i = 1, 2, \ldots, m$, the same n temperatures are used so that $x_{ij} = x_j$ for $j = 1, 2, \ldots, n$ for all i. Thus (3.31) becomes

$$E[y_{ij}|a_i] = \mu + a_i + \beta x_j. \tag{3.33}$$

a.   The model

Suppose that for a given i we write down the equation (3.33) for each $j = 1, 2, \ldots, n$. This gives n equations

$$
\begin{aligned}
E[y_{i1}|a_i] &= \mu + a_i + \beta x_1, \\
E[y_{i2}|a_i] &= \mu + a_i + \beta x_2, \\
&\ \vdots \\
E[y_{in}|a_i] &= \mu + a_i + \beta x_n.
\end{aligned}
\tag{3.34}
$$

Now define three column vectors of order n:

$$
\mathbf{y}_i = \begin{bmatrix} y_{i1} \\ y_{i2} \\ \vdots \\ y_{in} \end{bmatrix}, \qquad
\mathbf{1}_n = \begin{bmatrix} 1 \\ 1 \\ \vdots \\ 1 \end{bmatrix} \qquad \text{and} \qquad
\mathbf{x}_0 = \begin{bmatrix} x_1 \\ x_2 \\ \vdots \\ x_n \end{bmatrix}. \tag{3.35}
$$

Thus $\mathbf{y}_i$ is the vector of all n observations in the ith week, $\mathbf{1}_n$ is a vector of n ones, and $\mathbf{x}_0$ is the vector of the n different x-values associated with the n observations each week. Then, with the definitions of (3.35), equations (3.34) can be written succinctly as

$$E[\mathbf{y}_i|a_i] \quad = \quad \mu \mathbf{1}_n + a_i \mathbf{1}_n + \beta \mathbf{x}_0, \tag{3.36}$$

which we rewrite as

$$E[\mathbf{y}_i|a_i] \quad = \quad \mathbf{X}_0 \begin{bmatrix} \mu \\ \beta \end{bmatrix} + \mathbf{Z}_0 a_i \tag{3.37}$$

with

$$\mathbf{X}_0 \quad = \quad [\mathbf{1}_n \ \ \mathbf{x}_0] \qquad \text{and} \qquad \mathbf{Z}_0 = \mathbf{1}_n. \tag{3.38}$$

We now assume normality for $y_{ij}|a_i$ in the form

$$\mathbf{y}_i|a_i \sim \text{ indep. } \mathcal{N}\left(\mu \mathbf{1}_n + a_i \mathbf{1}_n + \beta \mathbf{x}_0, \ \sigma^2 \mathbf{I}_n\right). \tag{3.39}$$

Using (1.14) and (1.17) we have $\text{var}(y_{ij}) = \sigma_a^2 + \sigma^2$ and $\text{cov}(y_{ij}, y_{ij'}) = \sigma_a^2$. Thus on defining

$$\mathbf{V}_0 \quad = \quad \text{var}(\mathbf{y}_i) = \sigma^2 \mathbf{I}_n + \sigma_a^2 \mathbf{J}_n, \tag{3.40}$$

we have

$$\mathbf{V}_0^{-1} \quad = \quad \frac{1}{\sigma^2}\left(\mathbf{I}_n - \frac{\sigma_a^2}{\sigma^2 + n\sigma_a^2}\mathbf{J}_n\right), \tag{3.41}$$

where, as described in Section M.1 of Appendix M, $\mathbf{I}_n$ is an identity matrix and $\mathbf{J}_n$ is a matrix of all ones.

Notation

For purposes of subsequent algebra it turns out to be useful to define

$$\tau \quad = \quad \frac{n\sigma_a^2}{\sigma^2 + n\sigma_a^2} = \frac{\sigma_a^2}{\sigma^2/n + \sigma_a^2} \tag{3.42}$$

with

$$1 - \tau \quad = \quad \frac{\sigma^2}{\sigma^2 + n\sigma_a^2} = (\sigma^2/n\sigma_a^2)\tau. \tag{3.43}$$

This gives

$$\mathbf{V}_0^{-1} = \frac{1}{\sigma^2}(\mathbf{I}_n - \tau \bar{\mathbf{J}}_n) \qquad \text{for} \qquad \bar{\mathbf{J}} = \frac{1}{n}\mathbf{J}_n. \tag{3.44}$$

To encompass all the data, namely $\mathbf{y}_i$ for $i = 1, 2, \ldots, m$, we define

$$\mathbf{y} = \left\{ {}_c \mathbf{y}_i \right\}_{i=1}^m \qquad \text{and} \qquad \mathbf{a} = \left\{ {}_c a_i \right\}_{i=1}^m. \qquad (3.45)$$

Then from (3.37) we get

$$E[\mathbf{y}|\mathbf{a}] \;=\; \mathbf{X} \begin{bmatrix} \mu \\ \beta \end{bmatrix} + \mathbf{Z}\mathbf{a}$$

and

$$E[\mathbf{y}] \;=\; \mathbf{X} \begin{bmatrix} \mu \\ \beta \end{bmatrix} \qquad (3.46)$$

with

$$\mathbf{X} = \mathbf{1}_m \otimes \mathbf{X}_0 \qquad \text{and} \qquad \mathbf{Z} = \mathbf{I}_m \otimes \mathbf{Z}_0, \qquad (3.47)$$

where $\otimes$ represents the direct (or Kronecker) product operation as described in Section M.2 of Appendix M. Similarly, with the $\mathbf{y}_i$-vectors being independent (being data from different weeks) and with every $\mathbf{y}_i$ having the same variance-covariance matrix, namely $\mathbf{V}_0$ of (3.40),

$$\mathbf{V} = \text{var}(\mathbf{y}) = \mathbf{I}_m \otimes \mathbf{V}_0, \qquad (3.48)$$

a block diagonal matrix of m matrices $\mathbf{V}_0$ on the diagonal. And

$$\mathbf{V}^{-1} = \left(\mathbf{I}_m \otimes \mathbf{V}_0^{-1} \right). \qquad (3.49)$$

Thus

$$\mathbf{y} \sim \mathcal{N} \left(\mathbf{X} \begin{bmatrix} \mu \\ \beta \end{bmatrix}, \; \mathbf{V} \right). \qquad (3.50)$$

b. Estimating μ and β

ML estimators (under normality, and assuming $\mathbf{V}$ known) of μ and β come from the general expression given in Section S.6 of Appendix S.

$$\begin{bmatrix} \hat{\mu} \\ \hat{\beta} \end{bmatrix} = (\mathbf{X}'\mathbf{V}^{-1}\mathbf{X})^{-1}\mathbf{X}'\mathbf{V}^{-1}\mathbf{y}. \qquad (3.51)$$

– i. *Estimation*

On substituting for $\mathbf{X}$ and $\mathbf{V}^{-1}$ from (3.47) and (3.49) we find (3.51) reduces to

$$\begin{bmatrix} \hat{\mu} \\ \hat{\beta} \end{bmatrix} = \begin{bmatrix} \dfrac{1}{m} \otimes (\mathbf{X}_0'\mathbf{V}_0^{-1}\mathbf{X}_0)^{-1} \end{bmatrix} \left(\mathbf{1}_m' \otimes \mathbf{X}_0'\mathbf{V}_0^{-1} \right) \mathbf{y}. \qquad (3.52)$$

To simplify this expression note that

$$\mathbf{X}_0'\mathbf{V}_0^{-1} = \begin{bmatrix} \mathbf{1}_n' \\ \mathbf{x}_0' \end{bmatrix} \frac{1}{\sigma^2} (\mathbf{I}_n - \tau\bar{\mathbf{J}}_n) = \begin{bmatrix} (\tau/n\sigma_a^2)\mathbf{1}_n' \\ (\mathbf{x}_0' - \tau\bar{x}.\mathbf{1}_n')/\sigma^2 \end{bmatrix} \qquad (3.53)$$

and

$$\mathbf{X}_0'\mathbf{V}_0^{-1}\mathbf{X}_0 = \mathbf{X}_0'\mathbf{V}_0^{-1}[\mathbf{1}_n \ \ \mathbf{x}_0] = \left(\tau/\sigma_a^2 \right) \begin{bmatrix} 1 & \bar{x}. \\ \bar{x}. & S_{xx}\sigma_a^2/\tau\sigma^2 + \bar{x}^2. \end{bmatrix}, \qquad (3.54)$$

where

$$S_{xx} = \sum_{j=1}^{n}(x_j - \bar{x}.)^2 \quad \text{with} \quad \bar{x}. = \sum_{j=1}^{n} x_j/n. \qquad (3.55)$$

Also for (3.52), using (3.53) leads to

$$\left(\mathbf{1}_m' \otimes \mathbf{X}_0'\mathbf{V}_0^{-1} \right) \mathbf{y} = \left(m\tau/\sigma_a^2 \right) \begin{bmatrix} \bar{y}.. \\ S_{xy}\sigma_a^2/\tau\sigma^2 + \bar{x}.\bar{y}.. \end{bmatrix}, \qquad (3.56)$$

where

$$S_{xy} = \sum_{j}(x_j - \bar{x}.)(\bar{y}._j - \bar{y}..).$$

The inverse of the left-hand-side of (3.54) is easily derived, being

$$\left(\mathbf{X}_0'\mathbf{V}_0^{-1}\mathbf{X}_0 \right)^{-1} = \frac{\sigma^2}{S_{xx}} \begin{bmatrix} S_{xx}\sigma_a^2/\tau\sigma^2 + \bar{x}^2. & -\bar{x}. \\ -\bar{x}. & 1 \end{bmatrix}, \qquad (3.57)$$

and post-multiplying this by (3.56) leads (see E 3.1), from (3.52), to

$$\hat{\mu} = \bar{y}.. - \hat{\beta}\bar{x}.. \qquad (3.58)$$

and

$$\hat{\beta} = \frac{S_{xy}}{S_{xx}} = \frac{\sum_{j=1}^{n} x_j \bar{y}_{\cdot j} - n\bar{x}_{\cdot}\bar{y}_{\cdot\cdot}}{\sum_{j=1}^{n} x_j^2 - n\bar{x}_{\cdot}^2}. \tag{3.59}$$

An immediately noticeable feature of this result is that $\hat{\mu}$ and $\hat{\beta}$ do not depend on the unknown variances σ_a^2 and σ^2. Also, these estimators are exactly the same (for balanced data) as when the a_i-effects are fixed rather than random, or indeed if there are no a_i-effects. Moreover, (3.58) and (3.59) are precisely the results that occur in traditional analysis of covariance as in, for example, equations (6) and (7) of Searle (1987, Chapter 6).

– ii. *Unbiasedness*

On using $\mathrm{E}[a_i] = 0$ of (3.32) the expected values of $\hat{\mu}$ and $\hat{\beta}$ are

$$\mathrm{E}[\hat{\beta}] = \frac{\sum_{j=1}^{n} x_j(\mu + \mathrm{E}[\bar{a}_{\cdot}] + \beta x_j) - n\bar{x}_{\cdot}.(\mu + \mathrm{E}[\bar{a}_{\cdot}] + \beta \bar{x}_{\cdot})}{\sum_{j=1}^{n} x_j^2 - n\bar{x}_{\cdot}^2} = \beta$$

and

$$\mathrm{E}[\hat{\mu}] = (\mu + \mathrm{E}[\bar{a}_{\cdot}] + \beta \bar{x}_{\cdot}) - \beta \bar{x}_{\cdot} = \mu.$$

Thus the estimators $\hat{\mu}$ and $\hat{\beta}$ are unbiased.

– iii. *Sampling distributions*

From (3.58) and (3.59) we see that $\hat{\mu}$ and $\hat{\beta}$ are linear combinations of the normally distributed y_{ij} — see (3.39), and so the estimators themselves are bivariate normally distributed. Their means are μ and β, and using the well-known result that the variance-covariance matrix of estimators $(\mathbf{X}'\mathbf{V}^{-1}\mathbf{X})^{-1}\mathbf{X}'\mathbf{V}^{-1}\mathbf{y}$ of (3.51) is $(\mathbf{X}'\mathbf{V}^{-1}\mathbf{X})^{-1}$, we see that this is $(1/m)(\mathbf{X}_0'\mathbf{V}_0^{-1}\mathbf{X}_0)^{-1}$ as occurs in (3.52). And so from (3.57)

$$\mathrm{var}(\hat{\mu}) = \frac{\sigma_a^2}{m\tau} + \frac{\sigma^2 \bar{x}_{\cdot}^2}{mS_{xx}} = \frac{\sigma_a^2 + \sigma^2/n}{m} + \frac{\sigma^2 \bar{x}_{\cdot}^2}{mS_{xx}}, \tag{3.60}$$

$$\mathrm{var}(\hat{\beta}) = \frac{\sigma^2}{mS_{xx}}, \tag{3.61}$$

and

$$\mathrm{cov}(\hat{\mu}, \hat{\beta}) = -\bar{x}_{\cdot}.\mathrm{var}(\hat{\beta}). \tag{3.62}$$

Except for the occurrence of σ_a^2 in (3.60), these results are essentially the same as with standard, simple regression of Section 3.1. One apparent difference is the presence of m in the denominators, arising from the fact that S_{xx} is defined, in (3.55) as $S_{xx} = \Sigma_j(x_j - \bar{x}.)^2 = \Sigma_j x_j^2 - n\bar{x}_.^2$ and not as

$$\sum_{i=1}^{m}\sum_{j=1}^{n} x_j^2 - mn\bar{x}_.^2 = m\left(\sum_{j=1}^{n} x_j^2 - n\bar{x}_.^2\right) = mS_{xx}.$$

c. Estimating variances

In Section 3.5b the ML estimators of μ and β were derived assuming normality and known $\mathbf{V}$. Interestingly, the resulting estimators, $\hat{\mu}$ and $\hat{\beta}$ of (3.58) and (3.59) do not depend on $\mathbf{V}$; i.e., they do not depend on σ_a^2 and σ^2. However, these variances are often unknown, in which case we will want to estimate them, not only for their own sake but also to use them in, for example, the variances of the estimators in Section 3.5b–iii. And if we are going to estimate σ_a^2 and σ^2 from the same data set as will be used for estimating μ and β (as is often done) it will be advisable to use ML for estimating all four parameters, μ, β, σ_a^2 and σ^2, simultaneously. In doing this, the equations for estimating μ and β will be exactly the same as those already considered except that the ML estimators $\hat{\sigma}_a^2$ and $\hat{\sigma}^2$ will replace σ_a^2 and σ^2 in those equations. However, because σ_a^2 and σ^2 do not occur in those equations the estimators $\hat{\mu}$ and $\hat{\beta}$ when estimating all four parameters will be exactly the same as already derived in (3.58) and (3.59). This means that to estimate σ_a^2 and σ^2 we can maximize, with respect to σ_a^2 and σ^2, the likelihood with μ and β replaced by $\hat{\mu}$ and $\hat{\beta}$. Thus if we write the log likelihood as $l(\mu, \beta, \sigma^2, \sigma_a^2)$ we maximize

$$l^* = \log L(\hat{\mu}, \hat{\beta}, \sigma^2, \sigma_a^2).$$

– i. *When ML solutions are estimators*

Suppose we write $\boldsymbol{\theta} = \mathrm{E}[\mathbf{y}] = \mathbf{X}\begin{bmatrix} \mu \\ \beta \end{bmatrix}$ of (3.50), in which the normality assumed therein gives the likelihood as

$$L = \frac{\exp\{-\frac{1}{2}(\mathbf{y} - \boldsymbol{\theta})'\mathbf{V}^{-1}(\mathbf{y} - \boldsymbol{\theta})\}}{(2\pi)^{\frac{mn}{2}}|\mathbf{V}|^{\frac{1}{2}}}. \tag{3.63}$$

Then, using $\mathbf{V}^{-1}$ of (3.49) and

$$
\begin{aligned}
|\mathbf{V}|^{\frac{1}{2}} &= |\mathbf{V}_0|^{\frac{1}{2}m} = |\sigma^2 \mathbf{I}_n + \sigma^2 \mathbf{J}_n|^{\frac{1}{2}m} = \left[\sigma^{2(n-1)}(\sigma^2 + n\sigma_a^2)\right]^{\frac{1}{2}m} \\
&= \sigma^{2\left[\frac{m(n-1)}{2}\right]}(\sigma^2 + n\sigma_a^2)^{\frac{1}{2}m},
\end{aligned}
\tag{3.64}
$$

it can be shown that

$$
\begin{aligned}
l^* = -\frac{1}{2}\Big[& mn\log 2\pi + m(n-1)\log\sigma^2 + m\log(\sigma^2 + n\sigma_a^2) \\
& + \frac{S_1}{\sigma^2} + \frac{n\sigma_a^2 S_2}{\sigma^2(\sigma^2 + n\sigma_a^2)}\Big],
\end{aligned}
\tag{3.65}
$$

where

$$
S_1 = \sum_i \sum_j (y_{ij} - \hat{\mu} - \hat{\beta} x_j)^2 \quad \text{and} \quad S_2 = \sum_i n(\bar{y}_{i\cdot} - \hat{\mu} - \hat{\beta}\bar{x}_\cdot)^2. \tag{3.66}
$$

After substituting for $\hat{\mu}$ and $\hat{\beta}$ from (3.58) and (3.59)

$$
S_1 = \sum_{i=1}^{m}\sum_{j=1}^{n}[y_{ij} - \bar{y}_{\cdot\cdot} - \hat{\beta}(x_j - \bar{x}_\cdot)]^2 = \text{SST} - \hat{\beta}^2 m S_{xx} \tag{3.67}
$$

and

$$
S_2 = \sum_{i=1}^{m} n\left[\bar{y}_{i\cdot} - \bar{y}_{\cdot\cdot} - \hat{\beta}(\bar{x}_\cdot - \bar{x}_\cdot)\right]^2 = \sum_{i=1}^{m} n(\bar{y}_{i\cdot} - \bar{y}_{\cdot\cdot})^2 = \text{SSA}, \tag{3.68}
$$

where the familiar notation for sums of squares in analysis of variance is introduced:

$$
\text{SSA} = \Sigma_i \Sigma_j (\bar{y}_{i\cdot} - \bar{y}_{\cdot\cdot})^2, \quad \text{with} \quad \text{MSA} = \text{SSA}/(m-1) \tag{3.69}
$$

$$
\text{SSE} = \Sigma_i \Sigma_j (y_{ij} - \bar{y}_{i\cdot})^2, \quad \text{with} \quad \text{MSE} = \text{SSE}/[m(n-1)] \tag{3.70}
$$

$$
\text{SST} = \Sigma_i \Sigma_j (y_{ij} - \bar{y}_{\cdot\cdot})^2 = \text{SSA} + \text{SSE}. \tag{3.71}
$$

We also have occasion later to use notation familiar to analysis of covariance:

$$
\text{SSC} = \hat{\beta}^2 m S_{xx} = \frac{\left[\Sigma_i \Sigma_j (x_j - \bar{x}_\cdot)(y_{ij} - \bar{y}_{i\cdot})\right]^2}{\Sigma_i \Sigma_j (x_j - \bar{x}_\cdot)^2}, \tag{3.72}
$$

the sum of squares due to covariance, and

$$\text{SSR} \quad = \quad \text{SSE} - \text{SSC} = \sum_i \sum_j \left[y_{ij} - \bar{y}.. - \hat{\beta}(x_j - \bar{x}.) \right]^2, \quad (3.73)$$

the sum of squares for residual. This gives (3.67) as

$$S_1 \quad = \quad \text{SSA} + \text{SSR}. \tag{3.74}$$

Now, differentiating l^* with respect to σ_a^2 and σ^2 and equating the results to zero yields ML solutions $\dot{\sigma}_a^2$ and $\dot{\sigma}^2$ (see Section 1.7a–iii, for the overhead dot notation). First, $\partial l^* / \partial \sigma_a^2 = 0$ yields

$$\frac{-mn}{\dot{\sigma}^2 + n\dot{\sigma}_a^2} + \frac{n\,S_2}{(\dot{\sigma}^2 + n\dot{\sigma}_a^2)^2} = 0$$

so giving

$$\dot{\sigma}^2 + n\dot{\sigma}_a^2 = \text{SSA}/m. \tag{3.75}$$

Next, after some considerable algebra, and using (3.75) we find that $\dfrac{\partial l^*}{\partial \sigma^2}$ ultimately yields

$$\hat{\sigma}^2 = \frac{\text{SSE} - m\hat{\beta}^2 S_{xx}}{m(n-1)} = \frac{\text{SSR}}{m(n-1)}. \tag{3.76}$$

This is exactly the same result, for balanced data, as when the a_i-effects are fixed, not random. Indeed it is the standard analysis of covariance result as in, for example, equations (23) and (25) of Chapter 6 of Searle (1987). Then from (3.75)

$$\dot{\sigma}_a^2 = \frac{1}{n} \left(\frac{\text{SSA}}{m} - \hat{\sigma}^2 \right) = \frac{1}{n} \left[\left(1 - \frac{1}{m} \right) \text{MSA} - \text{MSE} \right] + \frac{\hat{\beta}^2 S_{xx}}{n(n-1)}. \tag{3.77}$$

Providing $\dot{\sigma}_a^2$ is not negative it is the ML estimator $\hat{\sigma}_a^2 = \dot{\sigma}_a^2$. In passing we note that (3.77) is the same as $\dot{\sigma}_a^2$ in (2.25) except for the addition of the term in $\hat{\beta}^2$.

– ii.   *When an ML solution is negative*

But it is possible for $\dot{\sigma}_a^2$ to be negative, in which case it is not the ML estimator. And then neither is $\hat{\sigma}^2$ of (3.76) the ML estimator of σ^2. This is so because the general method of maximum likelihood estimation

demands that ML estimators be within the range of their corresponding parameters; and negative $\dot{\sigma}_a^2$ is not within the non-negative range
of σ_a^2, which affects estimation of σ^2. To overcome this difficulty we
must adopt the ML methods for this situation (see Searle et al., 1992,
Section 3.7a–iii) which lead to taking $\hat{\sigma}_a^2 = 0$. This being so we then
have to maximize

$$l^*(\hat{\mu}, \hat{\beta}, \sigma^2, \hat{\sigma}_a^2 = 0) = -\frac{1}{2}\left[mn \log 2\pi + mn \log \sigma^2 + \frac{S_1}{\sigma^2}\right], \qquad (3.78)$$

obtained from (3.65) by replacing σ_a^2 with zero. Equating to zero the
differential of (3.78) with respect to σ^2 gives what will be denoted as
$\hat{\sigma}_0^2$, which is $\hat{\sigma}_0^2 = S_1/mn$. And so using (3.74) for S_1 we have

$$\hat{\sigma}_0^2 = \frac{S_1}{mn} = \frac{\text{SSA} + \text{SSR}}{mn}. \qquad (3.79)$$

d.   Tests of hypotheses — using LRT

The likelihood ratio technique (LRT) of hypothesis testing is described
in general terms in Section 1.7b–i. Suppose $\hat{\theta}$ is the value of θ that
maximizes a likelihood function $L(\theta)$ involving parameters θ. And
denote by $\hat{\theta}_0$ the value of θ which maximizes the likelihood when the
parameters are limited (restricted or defined) by a null hypothesis H_0
pertaining to some of the elements of θ. Then the likelihood ratio is

$$\Lambda = \frac{L(\hat{\theta}_0)}{L(\hat{\theta})}. \qquad (3.80)$$

It leads to a test statistic for the hypothesis H. We illustrate this for
two hypotheses that are often of interest, namely $H_0 : \beta = 0$, and
$H_0: \sigma_a^2 = 0$.
    Rather than using (3.80) in practical application, it is often easier
to use the negative of twice its logarithm:

$$\begin{aligned}
-2 \log \Lambda &= -2 \log L(\hat{\theta}_0) + 2 \log L(\hat{\theta}) \\
&= -2l^*(\hat{\theta}_0) + 2l^*(\hat{\theta}). \qquad (3.81)
\end{aligned}$$

– i.   *Using the maximized log likelihood $l^*(\hat{\theta})$*

The maximized log likelihood is l^* of (3.65) with σ^2 replaced by $\hat{\sigma}^2$ of
(3.76) and, on assuming $\dot{\sigma}_a^2$ is positive, with σ_a^2 replaced by $\hat{\sigma}_a^2 = \dot{\sigma}_a^2$ of

(3.77). Then, with $\hat{\mu}$ and $\hat{\beta}$ used in S_1 and S_2 as in (3.67) and (3.68) we have

$$-2l^*(\hat{\boldsymbol{\theta}}) = mn \log 2\pi + m(n-1) \log \left(\frac{\text{SSR}}{m(n-1)} \right) + m \log \left(\frac{\text{SSA}}{m} \right)$$

$$+ \frac{\text{SSA} + \text{SSR}}{\text{SSR}/[m(n-1)]} - \left[\frac{(n-1)\text{SSA}}{\text{SSR}} - 1 \right] \frac{\text{SSA}}{\text{SSA}/m}, \qquad (3.82)$$

after some simplification of $n\hat{\sigma}_a^2/\hat{\sigma}^2$ for the last term. Then, after collecting all the terms in (3.82) that do not involve SSA and SSR into what we will call $f_1(m,n)$, we get

$$-2l^*(\hat{\boldsymbol{\theta}}) = f_1(m,n) + m(n-1) \log \text{SSR} + m \log \text{SSA}. \qquad (3.83)$$

It is to be noticed that the LRT is defined in terms of maximum likelihood estimators. Yet in going from (3.65) to (3.83) we did, in fact, use $\hat{\sigma}_a^2 = \dot{\sigma}_a^2$, as if $\dot{\sigma}_a^2 > 0$. But we know that when $\dot{\sigma}_a^2 < 0$ we take $\hat{\sigma}_a^2 = 0$; and no account of this has been used in deriving (3.83). The reason for this is that (see E 3.3), for $H_0: \sigma_a^2 = 0$, having $\hat{\sigma}_a^2 = 0$ leads to $\Lambda = 1$, for which value one would never reject H_0.

– ii.   *Testing the hypothesis* $H_0 : \sigma_a^2 = 0$

To derive $-2\log\Lambda$ for $H_0 : \sigma_a^2 = 0$ we need first to estimate μ, β and σ^2 from the likelihood adapted by using 0 for σ_a^2. But with $\hat{\mu}$ and $\hat{\beta}$ of (3.58) and (3.59) not involving σ^2 or σ_a^2, we know they will be the same for $\hat{\boldsymbol{\theta}}_0$. And the estimator of σ^2 for $\hat{\boldsymbol{\theta}}_0$ will be $\hat{\sigma}_0^2$ obtained in (3.79). Therefore, $-2l^*(\hat{\boldsymbol{\theta}}_0)$ can be found by replacing σ^2 by $\hat{\sigma}_0^2$ of (3.79) and σ_a^2 by 0 in (3.65). This gives

$$
\begin{aligned}
-2l^*(\hat{\boldsymbol{\theta}}_0) &= mn \log 2\pi + m(n-1) \log \hat{\sigma}_0^2 + m \log \hat{\sigma}_0^2 + S_1/\hat{\sigma}_0^2 \\
&= mn \log 2\pi + mn \log \left(\frac{\text{SSA} + \text{SSR}}{mn} \right) + \frac{\text{SSA} + \text{SSR}}{(\text{SSA} + \text{SSR})/m} \\
&= f_2(m,n) + mn \log(\text{SSA} + \text{SSR}). \qquad (3.84)
\end{aligned}
$$

Therefore, on ignoring $f_1(m,n)$ and $f_2(m,n)$, we get from (3.81)

$$
\begin{aligned}
-2\log\Lambda &= -2l^*(\hat{\boldsymbol{\theta}}_0) + 2l^*(\hat{\boldsymbol{\theta}}) \\
&= mn \log(\text{SSA} + \text{SSR}) \\
&\qquad - m(n-1) \log \text{SSR} - m \log \text{SSA} \qquad (3.85)
\end{aligned}
$$

$$= mn \log \{SSR(SSA/SSR + 1)\}$$

$$- m(n - 1) \log SSR - m \log SSA.$$

We now write

$$q = \frac{SSA}{SSR} = \frac{(m - 1)MSA}{[m(n - 1) - 1]MSR} = \frac{m - 1}{m(n - 1) - 1}F,$$

where F is the F-statistic in analysis of covariance for testing $H_0 : a_i$ all equal; and if the a_i are all equal then, with probability 1.0, we have $\sigma_a^2 = 0$. And, on using q

$$-2 \log \Lambda = m[n \log(q + 1) - \log q].$$

To investigate the monotonicity of $-2 \log \Lambda$ with respect to q we consider

$$\frac{\partial}{\partial q}[-2 \log \Lambda] = m \left[\frac{n}{q + 1} - \frac{1}{q} \right] = \frac{m}{q(q + 1)}[(n - 1)q - 1]$$

$$> 0 \quad \text{if} \quad q > \frac{1}{n - 1} \qquad\qquad (3.86)$$

$$\Rightarrow \quad SSA > \frac{SSR}{n - 1}$$

$$\Rightarrow \quad \dot{\sigma}_a^2 > 0 \qquad \text{from (3.77);} \qquad (3.87)$$

and this is the condition for $\hat{\sigma}_a^2 = \dot{\sigma}^2$. And, in (3.86) we see that $-2 \log \Lambda$ is monotonic increasing with q, that is, $\log \Lambda$ is monotonic decreasing with increasing q, which is just what we want. This algebra shows that the usual F-test in an analysis of covariance is the LRT for $H_0 : \sigma_a^2 = 0$.

– iii.   *Testing $H_0 : \beta = 0$*

This is easy. First, $l^*(\hat{\boldsymbol{\theta}})$ stays as is, in (3.83). Second, under $H_0 : \beta = 0$ we simply put $\hat{\beta} = 0$ in the estimators $\hat{\mu}$, $\hat{\sigma}^2$ and $\hat{\sigma}_a^2$. Thus $\hat{\mu}$ of (3.58) becomes $\hat{\mu}_0 = \bar{y}_{..}$, S_1 of (3.67) becomes $S_{10} = SST$ and S_2 of (3.68) stays the same, $S_2 = SSA$. Also, with no β in the model, SSC of (3.73) becomes $SSR = SSE$. Therefore from (3.75) and (3.76)

$$\dot{\sigma}_0^2 + n\dot{\sigma}_{a,0}^2 = SSA/m \quad \text{and} \quad \hat{\sigma}_0^2 = \dot{\sigma}_0^2 = SSE/[m(n - 1)].$$

The only effective change in all of this is that $l^*(\hat{\boldsymbol{\theta}}_0)$ will be $l^*(\hat{\boldsymbol{\theta}})$ with SSR replaced by SSE. Doing this in (3.83) gives

$$-2l^*(\hat{\boldsymbol{\theta}}_0) \;=\; f_3(m,n) + m(n-1)\log\text{SSE} + m\log\text{SSA}. \quad (3.88)$$

and so, on subtracting (3.83) from (3.88) and ignoring f_1 and f_3,

$$\begin{aligned}
-2\log\Lambda \;&=\; -m(n-1)\log\text{SSR} + m(n-1)\log\text{SSE} \\
&=\; -m(n-1)\log\frac{\text{SSR}}{\text{SSE}} \\
&=\; -m(n-1)\log\left(1 - \frac{\text{SSC}}{\text{SSE}}\right), \quad (3.89)
\end{aligned}$$

from (3.73) where, in (3.72) SSC is $\hat{\beta}^2 mS_{xx}$, which is what is usually called SS(Regression). Thus (3.89) suggests what we will denote by q_β as the test statistic:

$$q_\beta = \frac{\text{SS(Regression)}}{\text{SSE}} = \frac{\hat{\beta}^2 S_{xx}}{(n-1)\text{MSE}}.$$

In point of fact the usual analysis of variance statistic is

$$\begin{aligned}
F \;&=\; \frac{\text{SS(Regression)}/1}{\text{SS(Residual after fitting } \mu + a_i + \beta x_{ij})/[m(n-1)-1]} \\
&=\; \frac{[m(n-1)-1]\text{SS(Regression)}}{\text{SSE - SS(Regression)}} \\
&=\; \frac{kq_\beta}{1 - q_\beta}, \quad (3.90)
\end{aligned}$$

where SS(Residual) has $k = m(n-1)-1$ degrees of freedom. And from (3.90)

$$q_\beta = \frac{F}{k+F}.$$

As previously in Section (3.5d-ii), we see that, with balanced data, the usual analysis of variance F-test is the LRT of $H_0 : \beta = 0$.

e.   Illustration

We return to the *Phytophthora* data of Figure 3.1 using all six weeks of data but only temperatures 15°C through 25°C. Since conditions vary from week to week, possibly causing lesion sizes to change, and since our goal would probably be to draw conclusions about a hypothetical population of experiments replicated over time, we treat the week-specific intercepts as random. Accordingly, we model the average log diameter (ALD) per leaflet as

$$E[ALD_{ij}|a_i] = \mu_{ij} = \mu + a_i + \beta TEMP_{ij}$$

$$ALD_{ij}|a_i \sim \text{indep. } \mathcal{N}(\mu_{ij}, \sigma^2), \qquad (3.91)$$

$$a_i \sim \text{i.i.d.} \mathcal{N}(0, \sigma_a^2),$$

where $ALD_{ij} = y_{ij}$ is the average log diameter of the lesion, and $TEMP_{ij} = x_{ij}$ is the temperature, both being defined for the jth leaflet during the ith week.

Using SAS PROC MIXED (SAS Institute, 1998) the restricted maximum likelihood estimators are $\hat{\mu} = -1.818$, $\hat{\beta} = 0.170$, $\hat{\sigma}^2 = 0.219$ and $\hat{\sigma}_a^2 = 0.250$. So, for example, the estimate of β tells us that log diameter increases about 0.170 with each increase in temperature of one degree. The value of $\hat{\sigma}_a^2$ indicates the magnitude of the variation of the weekly intercepts: they have a standard deviation of about 0.5. We can also use it to provide an estimate of the correlation of the observations taken in the same week. Using (2.18) the ML estimate of the correlation is

$$\widehat{\text{corr}}(y_{ij}, y_{ij'}) = \frac{\hat{\sigma}_a^2}{\hat{\sigma}_a^2 + \hat{\sigma}^2}$$

$$= \frac{0.250}{0.250 + 0.219} = 0.65,$$

which is appreciably high. Is the correlation statistically significantly different from zero? We can test it by testing $H_0 : \sigma_a^2 = 0$ versus $H_A : \sigma_a^2 > 0$. As in Snedecor and Cochran (1989) we form the F-statistic, $F = MSA/MSR$, which is equal to 7.85. The critical value is $\mathcal{F}_{30,0.95}^5 = 2.54$, so the test easily rejects H_0 with a p-value, $P\{\mathcal{F}_{30}^5 \geq 7.85\}$, which is less than 0.001.

f. Predicting the random intercepts

As a predictor of a we use, as in Section 2.4b, the best predictor, BP:

$$BP(\mathbf{a}) = E[\mathbf{a}|\mathbf{y}].$$

This is valid for all forms of probability distributions of $\mathbf{a}$ and $\mathbf{y}$ (see, e.g., Searle et al., 1992, Sect. 7.2). In the model being considered here the individual vectors $\mathbf{y}_i$ are independent and the only information about a_i is that contained in $\bar{y}_{i.}$. Therefore we consider just

$$\mathrm{BP}(a_i) = \mathrm{E}[a_i | \bar{y}_{i.}]$$

and under the normality conditions of (3.32) and (3.34) this is

$$\mathrm{BP}(a_i) = \mathrm{E}[a_i] + \mathrm{cov}(a_i, \bar{y}_{i.})[\mathrm{var}(\bar{y}_{i.})]^{-1}(\bar{y}_{i.} - \mathrm{E}[\bar{y}_{i.}])$$

$$= \sigma_a^2 \left(\sigma_a^2 + \sigma^2/n\right)^{-1}[\bar{y}_{i.} - (\mu + \beta\bar{x}_{.})]$$

$$= \frac{n\sigma_a^2}{\sigma^2 + n\sigma_a^2}(\bar{y}_{i.} - \mu - \beta\bar{x}_{.}). \qquad (3.92)$$

This is very similar to $\mathrm{BP}(a_i)$ of Section 2.4b–ii.

We now face a problem: how to convert the algebraic expression of (3.92) to a numerical value that can be of practical use? In other words, how can we estimate $\mathrm{BP}(a_i)$? Because it is a ratio of variances multiplying a linear function of μ and β, the derivation of an optimum estimator is undoubtedly difficult. Several alternative possibilities do exist. The easy part is to use $\hat{\mu}$ and $\hat{\beta}$ in place of μ and β. At least for balanced data, $\hat{\mu}$ and $\hat{\beta}$ do not involve σ^2 and σ_a^2, so no matter what we do about those variances, using $\hat{\mu}$ of (3.58) and $\hat{\beta}$ of (3.59) seems appropriate. Thus if σ^2 and σ_a^2 are known we could use, as an estimator of $\mathrm{BP}(a_i)$,

$$\mathrm{BP}^0(a_i) = \frac{n\sigma_a^2}{\sigma^2 + n\sigma_a^2}\left(\bar{y}_{i.} - \hat{\mu} - \hat{\beta}\bar{x}_{.}\right)$$

$$= \frac{n\sigma_a^2}{\sigma^2 + n\sigma_a^2}(\bar{y}_{i.} - \bar{y}_{..}) \qquad (3.93)$$

after substituting for $\hat{\mu}$ and $\hat{\beta}$. The sampling variance would be

$$\mathrm{var}[\mathrm{BP}^0(a_i)] = \frac{(1 - 1/m)\sigma_a^4}{\sigma_a^2 + \sigma^2/n}. \qquad (3.94)$$

If we do not know σ^2 and σ_a^2 we may be prepared to assume that some prior estimates are true values, in which case we could use them in (3.93) and (3.94).

But lacking true (or satisfactory prior) values for σ^2 and σ_a^2 we need to estimate them. Suppose we do this, using ML. Then, bearing in mind that under large-sample theory the ML estimator of a function of parameters represented by $\boldsymbol{\theta}$, say $f(\boldsymbol{\theta})$, is $f(\hat{\boldsymbol{\theta}})$ for $\hat{\boldsymbol{\theta}}$ being the ML estimator of $\boldsymbol{\theta}$, we calculate the ML estimate of $\mathrm{BP}(a_i)$ as

$$\widehat{\mathrm{BP}}(a_i) = \frac{\hat{\sigma}_a^2}{\hat{\sigma}_a^2 + \hat{\sigma}^2/n} \left(\bar{y}_{i\cdot} - \bar{y}_{\cdot\cdot} \right). \tag{3.95}$$

But if we use $\widehat{\mathrm{BP}}(a_i)$ of (3.95), goodness knows how we could derive its variance, especially if, as is often the case, the estimates of all four parameters have been obtained from the same data set. Moreover, the very form of $\widehat{\mathrm{BP}}(a_i)$ creates complications for ascertaining variances. For example, what is the variance of a ratio of estimated variance components, let alone of that ratio multiplied by a mean?

A practical way out of this predicament is to assume the variance components are known, leading to BP^0 of (3.93); derive its variance and in that variance replace σ^2 and σ_a^2 by estimates (or assumed values) thereof. Thus use as the variance (3.94) with σ^2 and $\hat{\sigma}_a^2$ replaced by their estimates:

$$\widehat{\mathrm{var}} \left[\mathrm{BP}^0(a_i) \right] = \frac{(1 - 1/m)(\hat{\sigma}_a^2)^2}{\hat{\sigma}_a^2 + \hat{\sigma}^2/n}.$$

Properties of this are unknown — but at least it is a practical procedure. Of course, if $\hat{\sigma}_a^2 = 0$ every $\widehat{\mathrm{BP}}(a_i)$ is the same, namely zero, consistent with $\hat{\sigma}_a^2 = 0$.

3.6   RANDOM INTERCEPTS: UNBALANCED DATA

The preceding section deals with balanced data, by which we mean that every data vector $\mathbf{y}_i$ for $i = 1, 2, \ldots, m$ contains n observations y_{ij} for $j = 1, 2, \ldots, n$. And corresponding to every $\mathbf{y}_i$ is the same vector $\mathbf{x}_0$ of the n x-values, x_j for $j = 1, 2, \ldots, n$. Now we deal with unbalanced data, which is the situation when for each i (i.e., each week of the example) there may be some y-values missing (i.e., at some temperatures data are missing). Thus $\mathbf{y}_i$ may contain fewer than n observations: we denote the number of observations in $\mathbf{y}_i$ by n_i. Likewise, corresponding to $\mathbf{y}_i$ there will be only n_i x-values. They will, of course, be n_i values from the $\mathbf{x}_0$ vector, but not necessarily the same n_i values even for two values of n_i that are the same. So now, instead of the balanced data

Table 3.1: Illustrative Examples of $\mathbf{x}_i$

Balanced Data	Unbalanced Data			
$\mathbf{x}_0$	$\mathbf{x}_1$ $n_1 = 5$	$\mathbf{x}_2$ $n_2 = 6$	$\mathbf{x}_3$ $n_3 = 4$	$\mathbf{x}_4$ $n_4 = 5$
	Elements of $\mathbf{x}_i$			
2	$x_{11} = 2$	$x_{21} = 2$		
3	$x_{12} = 3$		$x_{31} = 3$	$x_{41} = 3$
4		$x_{22} = 4$		
5		$x_{23} = 5$	$x_{32} = 5$	$x_{42} = 5$
6	$x_{13} = 6$	$x_{24} = 6$		$x_{43} = 6$
7	$x_{14} = 7$	$x_{25} = 7$	$x_{33} = 7$	
8	$x_{15} = 8$		$x_{34} = 8$	$x_{44} = 8$
9		$x_{26} = 9$		$x_{45} = 9$

case of every $\mathbf{y}_i$ of order n being associated with $\mathbf{x}_0$ of order n, each $\mathbf{y}_i$ of order n_i is associated with its own $\mathbf{x}_i$ of order n_i, its elements being n_i values occurring in $\mathbf{x}_0$. Table 3.1 shows some illustrative examples.

Notation

Elements of $\mathbf{y}_i$ are y_{ij} for $j = 1, 2, \ldots, n_i$. The n_i-values associated with elements of $\mathbf{y}_i$ are denoted x_{ij} for $j = 1, 2, \ldots, n_i$. But this use of j is for y_{ij} and x_{ij} being numbered consecutively from $j = 1$ through $j = n_i$ without regard for the value of j when it is used for x_j in $\mathbf{x}_0$ of (3.35) for the balanced data case. For example, in Table 3.1, the entry 6 in $\mathbf{x}_0$ is x_j for $j = 5$. But that same 6 in $\mathbf{x}_1$ is x_{13}, in $\mathbf{x}_2$ it is x_{24} and in $\mathbf{x}_4$ it is $\mathbf{x}_{43}$. And, of course, as these second subscripts indicate, the elements of $\mathbf{x}_i$ are written one after the other in the usual way, so that $\mathbf{x}_i$ is $n_i \times 1$; it is not $n \times 1$ with gaps or zeros as might be suggested by Table 3.1.

This particularly affects one's understanding of average x-values. With balanced data the average x-value was the same for every i, the average of all n elements of $\mathbf{x}_0$: it was denoted as $\bar{x}_i.$. Now we have $\bar{x}_{i.}$, the average of the n_i x-values in $\mathbf{x}_i$, namely $\bar{x}_{i.} = \sum_{j=1}^{n_i} x_{ij}/n_i$. And special care in this regard is needed in reducing results for unbalanced data to those for balanced data. For then, not only does $n_i = n$ and $\mathbf{x}_i = \mathbf{x}_0$ but also $\bar{x}_{i.}$ becomes $\bar{x}.$.

a. The model

With $\mathbf{y}_i$ and $\mathbf{x}_i$ having order n_i we have, comparable to (3.36)

$$E[\mathbf{y}_i|a_i] = \mu\mathbf{1}_{n_i} + a_i\mathbf{1}_{n_i} + \beta\mathbf{x}_i$$

$$= \mathbf{X}_i \begin{bmatrix} \mu \\ \beta \end{bmatrix} + \mathbf{1}_{n_i}a_i$$

for

$$\mathbf{X}_i = [\mathbf{1}_{n_i} \quad \mathbf{x}_i].$$

Through steps similar to those leading up to (3.46) we now have

$$E[\mathbf{y}|\mathbf{a}] = \mathbf{X} \begin{bmatrix} \mu \\ \beta \end{bmatrix} + \mathbf{Z}\mathbf{a}$$

for

$$\mathbf{X} = \left\{ {}_c \ [\mathbf{1}_{n_i} \quad \mathbf{x}_i] \right\}_{i=1}^m \quad \text{and} \quad \mathbf{Z} = \left\{ {}_d \ \mathbf{1}_{n_i} \right\}_{i=1}^m. \tag{3.96}$$

For variance specifications we maintain the normality of a_i in (3.32) and of $\mathbf{y}_i|a_i$ in (3.39) (with n and $\mathbf{x}_0$ replaced by n_i and $\mathbf{x}_i$) and so, akin to (3.40), have

$$\mathbf{V}_i = \text{var}(\mathbf{y}_i) = \sigma^2\mathbf{I}_{n_i} + \sigma_a^2\mathbf{J}_{n_i}.$$

Notation

For notational simplification and clarity in this section we make the changes:

$$\mathbf{I}_i \equiv \mathbf{I}_{n_i} \quad \text{and} \quad \mathbf{J}_i \equiv \mathbf{J}_{n_i},$$

$$\tau_i = \frac{n_i\sigma_a^2}{\sigma^2 + n_i\sigma_a^2} \quad \text{and} \quad 1 - \tau_i = \frac{\sigma^2}{n_i\sigma_a^2}\tau_i,$$

and, for example,

$$\left\{ {}_d \quad \right\} \quad \text{for} \quad \left\{ {}_d \quad \right\}_{i=1}^m.$$

Thus we write

$$\mathbf{V}_i \;=\; \sigma^2 \mathbf{I}_i + \sigma_a^2 \mathbf{J}_i$$

with

$$\mathbf{V}_i^{-1} \;=\; \frac{1}{\sigma^2}\left(\mathbf{I}_i - \frac{\sigma_a^2}{\sigma^2 + n_i \sigma_a^2}\mathbf{J}_i\right) = \frac{1}{\sigma^2}\left(\mathbf{I}_i - \tau_i \bar{\mathbf{J}}_i\right), \qquad (3.97)$$

similar to (3.44). Then

$$\mathbf{V} \;=\; \mathrm{var}(\mathbf{y}) = \mathrm{var}\left(\left\{{}_c\,\mathbf{y}_i\right\}\right) = \left\{{}_d\,\mathbf{V}_i\right\}. \qquad (3.98)$$

b.   Estimating μ and β when variances are known

– i.   ML estimators

As in (3.51) we need $\mathbf{X}'\mathbf{V}^{-1}\mathbf{y}$ and $(\mathbf{X}'\mathbf{V}^{-1}\mathbf{X})^{-1}$ for estimating μ and β. Thus from (3.96), (3.97) and (3.98) we get

$$\mathbf{X}'\mathbf{V}^{-1} \;=\; \left\{{}_r\begin{bmatrix} \mathbf{1}_i' \\ \mathbf{x}_i' \end{bmatrix}\right\}\frac{1}{\sigma^2}\left\{{}_d\,\mathbf{I}_i - \tau_i \bar{\mathbf{J}}_i\right\}$$

$$=\; \left\{{}_r\begin{bmatrix} (\tau_i/n_i\sigma_a^2)\mathbf{1}_i' \\ (\mathbf{x}_i' - \tau_i \bar{x}_{i.}\mathbf{1}_i')/\sigma^2 \end{bmatrix}\right\}.$$

Also

$$\mathbf{X}'\mathbf{V}^{-1}\mathbf{X} \;=\; \mathbf{X}'\mathbf{V}^{-1}\left\{{}_c\,[\mathbf{1}_i \;\; \mathbf{x}_i]\right\}$$

$$=\; \frac{1}{\sigma_a^2}\begin{bmatrix} \Sigma\tau_i & \Sigma\tau_i\bar{x}_{i.} \\ \Sigma\tau_i\bar{x}_{i.} & \frac{\sigma_a^2}{\sigma^2}\mathrm{SSE}_{xx} + \Sigma\tau_i\bar{x}_{i.}^2 \end{bmatrix} \qquad (3.99)$$

after adding and subtracting $\Sigma_i n_i \bar{x}_{i.}^2/\sigma^2$ in the 2,2 element to get

$$\mathrm{SSE}_{xx} = \sum_i\left(\sum_j x_{ij}^2 - n_i\bar{x}_{i.}^2\right) = \sum_{i=1}^m\sum_{j=1}^n (x_{ij} - \bar{x}_{i.})^2.$$

And, with algebra similar to that used for deriving $\mathbf{X'V^{-1}X}$,

$$\mathbf{X'V^{-1}y} = \frac{1}{\sigma_a^2} \left[\begin{array}{c} \Sigma \tau_i \bar{y}_{i\cdot} \\ \frac{\sigma_a^2}{\sigma^2} \mathrm{SSE}_{xy} + \Sigma_i \tau_i \bar{x}_{i\cdot} \bar{y}_{i\cdot} \end{array} \right]. \qquad (3.100)$$

For the moment, write

$$\mathbf{X'V^{-1}X} = \frac{1}{\sigma_a^2} \mathbf{B} \quad \text{for} \quad \mathbf{B} = \left[\begin{array}{cc} p & q \\ q & r \end{array} \right] = \sigma_a^2 \mathbf{X'V^{-1}X}. \qquad (3.101)$$

Then

$$(\mathbf{X'V^{-1}X})^{-1} = \sigma_a^2 \mathbf{B}^{-1} = \frac{\sigma_a^2}{|\mathbf{B}|} \left[\begin{array}{cc} r & -q \\ -q & p \end{array} \right]$$

with, from comparing (3.99) and (3.101),

$$|\mathbf{B}| = pr - q^2 = \sum_i \tau_i \left(\frac{\sigma_a^2}{\sigma^2} \mathrm{SSE}_{xx} + \sum_i \tau_i \bar{x}_{i\cdot}^2 - \frac{(\Sigma \tau_i \bar{x}_{i\cdot})^2}{\Sigma \tau_i} \right).$$

Thus

$$\left[\begin{array}{c} \hat{\mu} \\ \hat{\beta} \end{array} \right] = (\mathbf{X'V^{-1}X})^{-1} \mathbf{X'V^{-1}y}$$

$$= \frac{\sigma_a^2}{|\mathbf{B}|} \left[\begin{array}{cc} \frac{\sigma_a^2}{\sigma^2} \mathrm{SSE}_{xx} + \Sigma \tau_i \bar{x}_{i\cdot}^2 & -\Sigma \tau_i \bar{x}_{i\cdot} \\ -\Sigma \tau_i \bar{x}_{i\cdot} & \Sigma \tau_i \end{array} \right]$$

$$\times \left[\begin{array}{c} \Sigma \tau_i \bar{y}_{i\cdot} / \sigma_a^2 \\ \mathrm{SSE}_{xy} / \sigma^2 + \Sigma \tau_i \bar{x}_{i\cdot} \bar{y}_{i\cdot} / \sigma_a^2 \end{array} \right]. \qquad (3.102)$$

This gives

$$\hat{\beta} = \frac{\sigma_a^2}{|\mathbf{B}|} \left[\frac{-\sum \tau_i \bar{x}_{i\cdot} \sum \tau_i \bar{y}_{i\cdot}}{\sigma_a^2} + \frac{\sum \tau_i \mathrm{SSE}_{xy}}{\sigma^2} + \frac{\sum \tau_i \bar{x}_{i\cdot} \bar{y}_{i\cdot}}{\sigma_a^2} \right] \qquad (3.103)$$

$$= \frac{\sigma_a^2}{|\mathbf{B}|} \sum \tau_i \left[\frac{\mathrm{SSE}_{xy}}{\sigma^2} + \frac{1}{\sigma_a^2} \left(\sum \tau_i \bar{x}_{i\cdot} \bar{y}_{i\cdot} - \frac{\Sigma \tau_i \bar{x}_{i\cdot} \Sigma \tau_i \bar{y}_{i\cdot}}{\Sigma \tau_i} \right) \right].$$

$$= \frac{(\sigma_a^2/\sigma^2)\text{SSE}_{xy} + \text{WSSA}_{xy}}{(\sigma_a^2/\sigma^2)\text{SSE}_{xx} + \text{WSSA}_{xx}}, \tag{3.104}$$

where WSSA is for *weighted sum of squares* in the sense of

$$\text{WSSA}_{xx} = \sum \tau_i \bar{x}_{i\cdot}^2 - \frac{(\Sigma \tau_i \bar{x}_{i\cdot})^2}{\Sigma \tau_i} = \sum \tau_i (\bar{x}_{i\cdot} - \tilde{x}_{\cdot\cdot})^2$$

for $\tilde{x}_{\cdot\cdot}$ of (3.105). WSSA_{xy} is defined similarly. $\hat{\mu}$ can also be derived from (3.102). Tedious algebra (see E 3.7) which includes the use of weighted means

$$\tilde{y}_{\cdot\cdot} = \frac{\Sigma_i \tau_i \bar{y}_{i\cdot}}{\Sigma \tau_i} \quad \text{and} \quad \tilde{x}_{\cdot\cdot} = \frac{\Sigma_i \tau_i \bar{x}_{i\cdot}}{\Sigma_i \tau_i}, \tag{3.105}$$

ultimately leads to

$$\hat{\mu} = \tilde{y}_{\cdot\cdot} - \hat{\beta} \tilde{x}_{\cdot\cdot}. \tag{3.106}$$

Notationally this is similar to previous results but with, of course, using the weighted means and the not-so-simple expression for $\hat{\beta}$ in (3.104).

These expressions for $\hat{\mu}$ and $\hat{\beta}$ are ML estimators provided that σ^2 and σ_a^2 are known. And using those known values together with the x- and y-values of the data gives $\hat{\mu}$ and $\hat{\beta}$ as ML estimates.

– ii.   *Unbiasedness*

Using the same procedure as in Section 3.5b–i, it is not difficult to show for $\text{E}[\hat{\beta}]$ that

$$\text{E}[\text{SSE}_{xy}] = \beta(\text{SSE}_{xx}) \quad \text{and} \quad \text{E}[\text{WSSA}_{xy}] = \beta(\text{WSSA}_{xx})$$

and so $\hat{\beta}$ is unbiased. And then the unbiasedness of $\hat{\mu}$ is easily established (see E 3.8).

– iii.   *Sampling variances*

Some solid algebra (see E 3.10), assuming σ^2 and σ_a^2 are known, yields

$$\text{var}(\hat{\beta}) = \frac{1}{\dfrac{\text{SSE}_{xx}}{\sigma^2} + \dfrac{\text{WSSA}_{xx}}{\sigma_a^2}} \tag{3.107}$$

and

$$\text{var}(\hat{\mu}) \quad = \quad \frac{\sigma_a^2}{\Sigma \tau_i} + \tilde{x}^2 \text{var}(\hat{\beta})$$

$$= \quad \sum_i (\sigma^2/n_i + \sigma_a^2) + \tilde{x}_\cdot^2 \text{var}(\hat{\beta}). \tag{3.108}$$

– iv.   *Predicting a_i*

The only change from Section 3.5e is that everywhere n occurs it is replaced by n_i. Thus

$$\text{BP}(a_i) = \frac{n_i \sigma_a^2}{\sigma^2 + n_i \sigma_a^2} \left(\bar{y}_{i\cdot} - \mu - \beta \bar{x}_{i\cdot} \right),$$

and everything follows from this just as in Section 3.5 but with $\hat{\mu}$ and $\hat{\beta}$ of (3.106) and (3.104) in place of μ and β.

3.7   BERNOULLI - LOGISTIC REGRESSION

Consider the example of Section 3.1: We are interested in the growth of *Phytophthora infestans sporangia* lesions as a function of x, the temperature. Suppose our response, y, is the presence (coded as $y = 1$) or absence (coded as $y = 0$) of certain diameter growth. This would be much easier to judge than measuring the radius of colony growth.

What sort of model can we reasonably hypothesize for y as a function of x? Since y is binary it must follow a Bernoulli distribution. Since $\text{E}[y]$ will be modeled as (and vary as) a function of x and since $\text{var}(y) = \text{E}[y](1 - \text{E}[y])$ the variance cannot be assumed constant. Further, since $\text{E}[y] = \text{P}\{y = 1\}$, the mean must be bounded between zero and one. Therefore $\text{E}[y]$ cannot be assumed to be linear in x unless it is modeled over only a short range of x. Otherwise it would lead to values of $\text{E}[y]$ not in the interval $(0,1)$. Alas, three of the four assumptions (all except independence) of the simple linear regression model of Section 3.2a cannot be used for binary data.

One way to deal with the range restriction inherent in $\text{E}[y]$ is in the same parsimonious fashion as in Section 2.6c. Instead of modeling $\text{E}[y]$ directly we instead model $\text{logit}(\text{E}[y]) = \log \left(\dfrac{\text{E}[y]}{1 - \text{E}[y]} \right)$.

**Figure 3.2: Plot of E[y] for the Logistic Regression Model
(3.109) with $\alpha = 1$ and $\beta = 2$.**

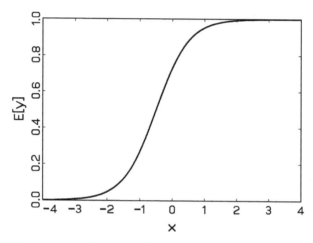

a.   Logistic regression model

The preceding discussion is motivation for the widely used logistic regression model:

$$\text{E}[y_i] \quad = \quad \pi(x_i) = \frac{1}{1 + e^{-(\alpha + \beta x_i)}} \tag{3.109}$$

or equivalently

$$\text{logit}(\text{E}[y_i]) \quad = \quad \text{logit}[\pi(x_i)] = \alpha + \beta x_i$$

$$y_i \quad \sim \quad \text{indep. Bernoulli}[\pi(x_i)].$$

In (3.109) we have again used the notation π for the mean of y since it is a probability.

How is this model different from (3.1)? Clearly we are assuming a different distribution for y_i. Also, by modeling $\text{logit}(\text{E}[y_i])$ as linear in x_i we are, in fact, hypothesizing a nonlinear model for $\text{E}[y_i]$ as given in (3.109). As an example of the form of E[y], Figure 3.2 shows a plot of E[y] as x ranges from -4 to 4 when $\alpha = 1$ and $\beta = 2$.

Several comments about the form of $\text{E}[y_i]$ are in order to understand more fully the logistic regression model. The equation describes a regression line that is always (when viewed over a wide enough range of

x and as long as β is not zero) an S-shaped curve as demonstrated in Figure 3.2. It is increasing if β is greater than zero, decreasing if β is less than zero and flat if β is equal to zero. Thus β governs how quickly the curve increases or decreases whereas α governs its horizontal location. The curve reaches its half height (of 0.5) when $\alpha + \beta x = 0$ or, equivalently, when $x = -\alpha/\beta$.

Another primary interpretation of β is related to the idea of odds. If the probability of an event is π, then the *odds* of the event is defined as $\pi/(1 - \pi)$. For example, if the probability of an event is $1/3$, then its odds are $\frac{1/3}{2/3} = 1/2$ or the odds are 1 to 2; if the probability is $3/4$, then the odds of the event are $\frac{3/4}{1/4} = 3$. Thus, another way to state (3.109) is that the log of the odds of the event $P\{y_i = 1\}$ is $\alpha + \beta x_i$ or that the odds of a success are $e^{\alpha + \beta x_i}$.

A common way to interpret β in the linear regression model (3.1) is to consider how much $E[y]$ changes when x is increased by a single unit. For that model we get the simple result that the difference in expected values is just $[\alpha + \beta(x + 1)] - [\alpha + \beta x] = \beta$.

The equivalent calculation for (3.109) is that

$$\beta = \text{logit}[\pi(x + 1)] - \text{logit}[\pi(x)]$$

$$= \log[\text{odds}(x + 1)] - \log[\text{odds}(x)]$$

$$= \log\left[\frac{\text{odds}(x + 1)}{\text{odds}(x)}\right]$$

or

$$e^\beta = \frac{\text{odds}(x + 1)}{\text{odds}(x)},$$

where $\text{odds}(x) \equiv \pi(x)/[1 - \pi(x)]$. This result is described by saying that β is the *log odds ratio* or that e^β is the *odds ratio*.

While the formulation (3.109) is quite standard and leads to easy interpretations of α and β in the scale of $\text{logit}[\pi(x_i)]$, for which it is linear in x, it does not give a straightforward interpretation on the π scale. For ease of interpretation $E[y_i]$ is sometimes reparameterized in terms of $x_h = -\alpha/\beta$, the halfway point on the x-axis, and γ, the slope of the curve at x_h. In this parameterization

$$E[y_i] = \frac{1}{1 + e^{-4\gamma(x_i - x_h)}}. \tag{3.110}$$

b. Likelihood

Since the y_i are independent and Bernoulli distributed, the likelihood is straightforward to evaluate:

$$L = \prod_{i=1}^{n} [\pi(x_i)]^{y_i} [1 - \pi(x_i)]^{1-y_i}$$

$$= \prod_{i=1}^{n} \{\pi(x_i)/[1 - \pi(x_i)]\}^{y_i} [1 - \pi(x_i)]. \qquad (3.111)$$

Using

$$\pi(x_i)/[1 - \pi(x_i)] = e^{\alpha + \beta x_i}$$

and

$$1 - \pi(x_i) = (1 + e^{\alpha + \beta x_i})^{-1}$$

gives L as

$$L = \prod_{i=1}^{n} e^{y_i(\alpha + \beta x_i)} (1 + e^{\alpha + \beta x_i})^{-1} \qquad (3.112)$$

and the log likelihood as

$$l = \log L = \sum_{i=1}^{n} y_i(\alpha + \beta x_i) - \log(1 + e^{\alpha + \beta x_i}). \qquad (3.113)$$

We immediately see an advantage of assuming the logit of $\pi(x)$ to be linear in x — it yields an extremely simple log likelihood, in (3.113).

c. ML equations

Differentiating (3.113) with respect to α and β gives

$$\frac{\partial l}{\partial \alpha} = \sum_{i=1}^{n} \left[y_i - \frac{e^{\alpha + \beta x_i}}{1 + e^{\alpha + \beta x_i}} \right]$$

$$= \sum_{i=1}^{n} \left[y_i - \frac{1}{1 + e^{-(\alpha + \beta x_i)}} \right]$$

$$= \sum_{i=1}^{n} [y_i - \pi(x_i)] \qquad (3.114)$$

$$\frac{\partial l}{\partial \beta} = \sum_{i=1}^{n} \left[x_i y_i - \frac{x_i e^{\alpha + \beta x_i}}{1 + e^{\alpha + \beta x_i}} \right]$$

$$= \sum_{i=1}^{n} x_i \left[y_i - \pi(x_i) \right]. \qquad (3.115)$$

Noting that $\pi(x_i) = E[y_i]$ we can see that setting (3.114) and (3.115) equal to zero gives exactly the same equations as (3.6) and (3.7) that were derived for the simple linear regression model of Section 3.2. Unfortunately, they are not as easy to solve.

After equating (3.114) and (3.115) to zero we need to solve the (non-linear in $\hat{\alpha}$ and $\hat{\beta}$) equations

$$\sum_{i=1}^{n} y_i = \sum_{i=1}^{n} \frac{1}{1 + e^{-(\hat{\alpha} + \hat{\beta} x_i)}}$$

$$\sum_{i=1}^{n} x_i y_i = \sum_{i=1}^{n} \frac{x_i}{1 + e^{-(\hat{\alpha} + \hat{\beta} x_i)}}. \qquad (3.116)$$

The first equation has a straightforward interpretation: the ML solutions are chosen so that the total predicted number of successes is equal to $\sum_i y_i$, the total observed number of successes. Except in some special cases (e.g E 3.3) (3.116) does not have an explicit solution and must be solved numerically.

The second derivatives of l take a convenient form:

$$\frac{\partial^2 l}{\partial \alpha^2} = -\sum_{i=1}^{n} \frac{\partial \pi(x_i)}{\partial \alpha}$$

$$= -\sum_{i=1}^{n} \frac{e^{-(\alpha + \beta x_i)}}{\left(1 + e^{-(\alpha + \beta x_i)}\right)^2},$$

$$= -\sum_{i=1}^{n} \pi(x_i)[1 - \pi(x_i)] \qquad (3.117)$$

$$\frac{\partial^2 l}{\partial \alpha \, \partial \beta} = -\sum_{i=1}^{n} \frac{x_i e^{-(\alpha + \beta x_i)}}{\left(1 + e^{-(\alpha + \beta x_i)}\right)^2}$$

$$= -\sum_{i=1}^{n} x_i \pi(x_i)[1 - \pi(x_i)] \qquad (3.118)$$

and

$$\frac{\partial^2 l}{\partial \beta^2} = -\sum_{i=1}^{n} x_i^2 \pi(x_i)[1 - \pi(x_i)]. \qquad (3.119)$$

We next write the information matrix in compact notation using $\mathbf{V} = \text{var}(\mathbf{y}) = \text{diag}\{\pi(x_i)[1 - \pi(x_i)]\}$ and $\mathbf{X}' = \begin{bmatrix} 1 & 1 & \cdots & 1 \\ x_1 & x_2 & \cdots & x_n \end{bmatrix}$. Then

$$-E \begin{bmatrix} l_{\alpha\alpha} & l_{\alpha\beta} \\ l_{\beta\alpha} & l_{\beta\beta} \end{bmatrix} = E\left[\mathbf{X}'\mathbf{V}\mathbf{X}\right] = \mathbf{X}'\mathbf{V}\mathbf{X}, \qquad (3.120)$$

which shows that the large-sample variance of $\hat{\alpha}$ and $\hat{\beta}$ is $(\mathbf{X}'\mathbf{V}\mathbf{X})^{-1}$.

This also yields a convenient computing algorithm (see Chapter 14) for finding $\hat{\alpha}$ and $\hat{\beta}$. Since the Hessian is negative definite, the log likelihood is concave. Hence, except in rare cases (see E 3.5), a single local maximum exists (and is the global maximum) and the Newton-Raphson algorithm described below is guaranteed to converge to the MLEs (Santner and Duffy, 1990). The algorithm proceeds as follows:

1. Obtain starting values $\alpha^{(0)}$ and $\beta^{(0)}$. Set $m = 0$.

2. Calculate

$$\begin{pmatrix} \alpha^{(m+1)} \\ \beta^{(m+1)} \end{pmatrix} = \begin{pmatrix} \alpha^{(m)} \\ \beta^{(m)} \end{pmatrix} + (\mathbf{X}'\mathbf{V}^{(m)}\mathbf{X})^{-1}\mathbf{X}'[\mathbf{y} - \boldsymbol{\pi}^{(m)}(\mathbf{x})].$$

3. Check for convergence of $\begin{pmatrix} \alpha^{(m+1)} \\ \beta^{(m+1)} \end{pmatrix}$. If it has converged, stop; otherwise set $m = m + 1$ and return to step 2.

In this algorithm $\boldsymbol{\pi}^{(m)}(\mathbf{x})$ is the notation we use for the vector $\left\{ 1/(1 + e^{-(\alpha^{(m)} + \beta^{(m)} x_i)}) \right\}_{i=1}^{n}$, and $\mathbf{V}^{(m)} = \text{diag}\left\{ \boldsymbol{\pi}^{(m)}(\mathbf{x})[1 - \boldsymbol{\pi}^{(m)}(\mathbf{x})] \right\}$.

d. Large-sample tests and confidence intervals

As in Section 2.6b–v., large-sample tests and confidence intervals can be based on the asymptotic normality ($\mathcal{AN}$) of $\hat{\alpha}$ and $\hat{\beta}$,

$$\begin{pmatrix} \hat{\alpha} \\ \hat{\beta} \end{pmatrix} \sim \mathcal{AN}\left[\begin{pmatrix} \alpha \\ \beta \end{pmatrix}, (\mathbf{X'VX})^{-1}\right],$$

where $\mathbf{X'VX}$ is given in (3.117) through (3.120). For example, to test $H_0 : \beta \leq 0$ versus $H_A : \beta > 0$ we would reject H_0 if

$$\frac{\hat{\beta}}{\sqrt{\widehat{\text{var}}(\hat{\beta})}} > z_\alpha, \tag{3.121}$$

where $\widehat{\text{var}}(\hat{\beta})$ comes from inserting the MLEs into the lower-right-hand entry of $(\mathbf{X'VX})^{-1}$. A large-sample confidence interval for β would be calculated as

$$\hat{\beta} \pm z_{\alpha/2}\sqrt{\widehat{\text{var}}(\hat{\beta})}. \tag{3.122}$$

Similarly, the large-sample confidence interval for the odds ratio, e^β, would be

$$\left(e^{\hat{\beta}-z_{\alpha/2}\sqrt{\widehat{\text{var}}(\hat{\beta})}}, e^{\hat{\beta}+z_{\alpha/2}\sqrt{\widehat{\text{var}}(\hat{\beta})}}\right). \tag{3.123}$$

Alternatively, we can use the likelihood ratio test to test the two-sided hypothesis $H_0 : \beta = 0$ versus $H_A : \beta \neq 0$. Under H_0 the likelihood becomes

$$L = \prod_{i=1}^{n} e^{y_i\alpha}(1 + e^\alpha) \tag{3.124}$$

with maximum $\hat{\alpha}_0 = \log[\bar{y}/(1 - \bar{y})]$. Hence the maximized value of $l = \log L$ under H_0 is $\sum y_i \log \bar{y} + \sum(1 - y_i)\log(1 - \bar{y})$. The likelihood ratio statistic is then

$$-2\log \Lambda = -2\left[\sum y_i \log \bar{y} + \sum(1 - y_i)\log(1 - \bar{y})\right.$$
$$\left. - \sum y_i(\hat{\alpha} + \hat{\beta}x_i) + \sum \log(1 + e^{\hat{\alpha}+\hat{\beta}x_i})\right] \tag{3.125}$$

and the test is to reject H_0 whenever $-2\log \Lambda$ exceeds $\chi^2_{1,1-\alpha}$.

3.8   BERNOULLI - LOGISTIC WITH RANDOM INTERCEPTS

Now consider our example from Section 3.7, but recall that the experiment has been repeated at six different times. We want to analyze all the data together but suspect that the probability of a lesion for a fixed inoculation level varied from time to time. We might hypothesize a model with a common "slope" parameter β but "intercepts" which varied from time to time. Since our goal would probably be to draw conclusions about all the experiments that could be replicated over time, we would want the intercepts to be a random effect, as in Section 3.6e.

a.   Model

A reasonable model for a success for observation j in experiment i would therefore be the following:

$$\mathrm{E}[y_{ij}|a_i] \;=\; \pi(x_{ij}) = \frac{1}{1 + e^{-(\alpha + a_i + \beta x_{ij})}} \qquad (3.126)$$

or, equivalently,

$$\mathrm{logit}[\pi(x_{ij})] \;=\; \alpha + a_i + \beta x_{ij}$$

and

$$y_{ij}|a_i \;\sim\; \text{indep. Bernoulli}[\pi(x_{ij})]$$

$$a_i \;\sim\; \text{i.i.d. } \mathcal{N}(0, \sigma_a^2).$$

Conditional on a_i the y_{ij} follow a logistic regression model with intercepts that vary with the index i (with time in our example). Thus, conditional on a_i, β has the same interpretation as in the usual logistic regression model.

Since the a_i are random effects, there are two important differences between (3.126) and (3.109). First, the y_{ij} do not exactly follow a logistic model marginally. This is because

$$\mathrm{E}[y_{ij}] \;=\; \mathrm{E}\left[\mathrm{E}[y_{ij}|a_i]\right]$$

$$=\; \mathrm{E}[\pi(x_{ij})]$$

$$= \mathrm{E}\left[\frac{1}{1 + e^{-(\alpha + a_i + \beta x_{ij})}}\right]$$

$$= \int_{-\infty}^{\infty} \frac{1}{1 + e^{-(\alpha + a_i + \beta x_{ij})}} \frac{1}{\sqrt{2\pi\sigma_a^2}} e^{-\frac{1}{2\sigma_a^2} a_i^2} \, da_i \quad (3.127)$$

cannot be evaluated in closed form, and, in particular, is not exactly of the logistic form, that is, $1/(1 + e^{\alpha^* + \beta^* x_{ij}})$, for some choice of α^* and β^*. However, it can be well approximated by a marginal logistic model:

$$\mathrm{logit}\,(\mathrm{E}[y_{ij}]) \approx \alpha^* + \beta^* x_{ij},$$

where

$$\alpha^* = \frac{\alpha}{\sqrt{1 + \lambda \sigma_a^2}} \quad \text{and} \quad \beta^* = \frac{\beta}{\sqrt{1 + \lambda \sigma_a^2}}, \quad (3.128)$$

for $\lambda = 256/75\pi$ (see E 3.15).

Second, the y_{ij} and y_{ik} are correlated since they both involve the same random effect a_i. Using (1.16) we obtain

$$\mathrm{cov}(y_{ij}, y_{ik}) = \mathrm{cov}(\mathrm{E}[y_{ij}|a_i], \mathrm{E}[y_{ik}|a_i]) + \mathrm{E}\left[\mathrm{cov}(y_{ij}, y_{ik}|a_i)\right]$$

$$= \mathrm{cov}\left(\frac{1}{1 + e^{-(\alpha + a_i + \beta x_{ij})}}, \frac{1}{1 + e^{-(\alpha + a_i + \beta x_{ik})}}\right) + 0$$

$$= \int_{-\infty}^{\infty} \frac{1}{1 + e^{-(\alpha + a + \beta x_{ij})}} \frac{1}{1 + e^{-(\alpha + a + \beta x_{ik})}} \frac{1}{\sqrt{2\pi\sigma_a^2}} e^{-\frac{1}{2\sigma_a^2} a^2} \, da$$

$$- \int_{-\infty}^{\infty} \frac{1}{1 + e^{-(\alpha + a + \beta x_{ij})}} \frac{1}{\sqrt{2\pi\sigma_a^2}} e^{-\frac{1}{2\sigma_a^2} a^2} \, da$$

$$\times \int_{-\infty}^{\infty} \frac{1}{1 + e^{-(\alpha + a + \beta x_{ik})}} \frac{1}{\sqrt{2\pi\sigma_a^2}} e^{-\frac{1}{2\sigma_a^2} a^2} \, da$$

$$> 0, \text{ as long as } \sigma_a^2 > 0. \quad (3.129)$$

b. Likelihood

The likelihood can be calculated in the usual manner by writing the density conditional on the random effects as in (3.112) and then integrating them out:

$$
L = \prod_{i=1}^{m} \int_{-\infty}^{\infty} \prod_{j=1}^{n_i} e^{(\alpha+a_i+\beta x_{ij})y_{ij}} (1 + e^{\alpha+a_i+\beta x_{ij}})^{-1} \frac{1}{\sqrt{2\pi\sigma_a^2}} e^{-\frac{1}{2\sigma_a^2}a_i^2} \, da_i
$$

$$
= e^{\alpha y_{..} + \beta \sum_{i,j} x_{ij}y_{ij}}
$$

$$
\times \prod_{i=1}^{m} \int_{-\infty}^{\infty} e^{a_i y_{i.}} \prod_{j=1}^{n_i} (1 + e^{\alpha+a_i+\beta x_{ij}})^{-1} \frac{1}{\sqrt{2\pi\sigma_a^2}} e^{-\frac{1}{2\sigma_a^2}a_i^2} \, da_i
$$

and

$$
l = \alpha y_{..} + \beta \sum_{i,j} x_{ij}y_{ij} \tag{3.130}
$$

$$
+ \sum_{i=1}^{m} \log \int_{-\infty}^{\infty} e^{a_i y_{i.}} \prod_{j=1}^{n_i} (1 + e^{\alpha+a_i+\beta x_{ij}})^{-1} \frac{1}{\sqrt{2\pi\sigma_a^2}} e^{-\frac{1}{2\sigma_a^2}a_i^2} \, da_i.
$$

As in Section 2.6c–ii this must be evaluated and maximized numerically. Gauss-Hermite quadrature is again a logical method.

c.   Large-sample tests and confidence intervals

Given the intractibility of the log likelihood in (3.130), calculation of the information matrix for tests or confidence intervals is difficult and numerical at best.

Likelihood ratio tests can be performed by numerically maximizing (3.130) and log likelihoods of reduced models and calculating the negative of twice the difference between their maximized values.

d.   Prediction

Conceptually, the estimated best predictor is given by

$$
\tilde{a}_i = \hat{E}[a_i|\bar{y}_{i.}]. \tag{3.131}
$$

However, as in the preceding section, closed-form expressions do not exist and are numerically difficult to compute.

e. Conditional Inference

If interest focused solely on β of (3.126) as opposed to α, σ_a^2, or the a_i, then another approach is available for inference. Suppose we rewrite model (3.126) as

$$E[y_{ij}] = \pi(x_{ij}) = \frac{1}{1 + e^{-(\alpha_i + \beta x_{ij})}}, \qquad (3.132)$$

(using $\alpha_i = \alpha + a_i$) and make no assumptions about the α_i. Writing the log likelihood gives

$$
\begin{aligned}
l &= \sum_{i,j} y_{ij}(\alpha_i + \beta x_{ij}) - \log(1 + e^{\alpha_i + \beta x_{ij}}) \\
&= \sum_i \alpha_i y_{i\cdot} + \beta \sum_{i,j} y_{ij} x_{ij} - \sum_{i,j} \log(1 + e^{\alpha_i + \beta x_{ij}}), \quad (3.133)
\end{aligned}
$$

from which the sufficient statistics are $y_{1\cdot}, y_{2\cdot}, \ldots, y_{m\cdot}$, and $\sum_{i,j} y_{ij} x_{ij}$.

One reason for using random effects models is that it is known (Neyman and Scott, 1948) that if we let the sample size increase by letting $m \to \infty$ and we try to simultaneously estimate $\alpha_1, \alpha_2, \ldots, \alpha_m$, and β by maximum likelihood then inconsistent MLEs can result.

Standard theory (Lehmann, 1986, Sec. 4.4) is to consider the conditional distribution of the sufficient statistic "associated" with the parameter of interest conditional on all the others. Conditioning on the sufficient statistic removes dependence on the remaining "nuisance" parameters. Such a methodology leads to tests which, for the marginal problem, are uniformly most powerful unbiased.

Applied in our situation, where we are assuming β is the parameter of interest, the conditional methodology leads to consideration of the conditional distribution of $T = \sum_{i,j} y_{ij} x_{ij}$ given $S_1 = y_{1\cdot}, S_2 = y_{2\cdot}, \ldots, S_m = y_{m\cdot}$. We start with their joint distribution.

Since the y_{ij} are discrete, to calculate the probability that $S_1 = s_1$, $S_2 = s_2$, ..., $S_m = s_m$ and $T = t$, we merely sum over the y_{ij} that give $y_{1\cdot} = s_1$, $y_{2\cdot} = s_2$, ..., $y_{m\cdot} = s_m$, and $\sum_{i,j} y_{ij} x_{ij} = t$:

$$P\{S_1 = s_1, S_2 = s_2, \ldots, S_m = s_m, T = t\} = \sum_R \frac{e^{\sum \alpha_i y_{i\cdot} + \beta \sum y_{ij} x_{ij}}}{\prod_1^m \prod_1^n (1 + e^{\alpha_i + \beta x_{ij}})}, \qquad (3.134)$$

where $R = \{y_{ij} : y_{1\cdot} = s_1, \ldots, y_{m\cdot} = s_m, \sum y_{ij} x_{ij} = t\}$. This is equal to

$$C(s_1, \ldots, s_m, t) \frac{e^{\sum \alpha_i s_i + \beta t}}{\prod_1^m \prod_1^n (1 + e^{\alpha_i + \beta x_{ij}})}, \qquad (3.135)$$

where $C(s_1, \ldots, s_m, t)$ is the number of combinations of the y_{ij} that are in R. To find the marginal distribution of $S_1, S_2, \ldots, S_m$ we sum out T:

$$f_{\mathbf{S}}(\mathbf{s}) = \sum_z C(s_1, \ldots, s_m, z) \frac{e^{\sum \alpha_i s_i + \beta z}}{\prod_1^m \prod_1^n (1 + e^{\alpha_i + \beta x_{ij}})}. \tag{3.136}$$

Then

$$
\begin{aligned}
f_{T|\mathbf{S}}(t|\mathbf{s}) &= \frac{C(s_1, \ldots, s_m, t) e^{\sum \alpha_i s_i + \beta t}}{\sum_z C(s_1, \ldots, s_m, z) e^{\sum \alpha_i s_i + \beta z}} \\[2mm]
&= \frac{C(s_1, \ldots, s_m, t) e^{\beta t}}{\sum_z C(s_1, \ldots, s_m, z) e^{\beta z}},
\end{aligned}
\tag{3.137}
$$

which is independent of the α_i, as promised by sufficiency. This can be used to form tests or calculate estimates.

For example, to test $H_0 : \beta \leq 0$ versus $H_A : \beta > 0$, we use the null hypothesis distribution of $T|\mathbf{S}$, namely

$$P_{H_0}\{T = t | \mathbf{S} = \mathbf{s}\} = \frac{C(s_1, \ldots, s_m, t)}{\sum_z C(s_1, \ldots, s_m, z)}, \tag{3.138}$$

which depends only on the combinatorial coefficients and on no unknown parameters. The p-value corresponding to an observed value, $t_0 = \sum y_{ij} x_{ij}$, is

$$p = P_{H_0}\{T \geq t_0 | \mathbf{S} = \mathbf{s}\} = \frac{\sum_{t \geq t_0} C(s_1, \ldots, s_m, t)}{\sum_z C(s_1, \ldots, s_m, z)}. \tag{3.139}$$

Clearly we need to know the values of $C(s_1, \ldots, s_m, t)$ to perform the calculation. Conceptually this is a simple counting task, but from a practical point of view the task can get tedious or, for larger problems, insurmountable. Specialized software (e.g., Mehta and Patel, 1992) has been written to efficiently compute these quantities for small and moderate-sized problems.

A conditional MLE for β can be calculated by maximizing (3.137).

3.9   EXERCISES

E 3.1 Showing all intermediate steps, derive from details given in Section 3.5b–i, the $\hat{\mu}$ and $\hat{\beta}$ of (3.58) and (3.59).

E 3.2 Derive (3.65), and from that derive (3.75) and (3.76).

E 3.3 Explain why the LRT for $H : \sigma_a^2 = 0$ is 1.0 when $\hat{\sigma}_a^2 = 0$.

E 3.4 Derive (3.94) and develop expressions for the covariance of $\mathrm{BP}^0(a_i)$ and $\mathrm{BP}^0(a_k)$; and for the variance of $\mathrm{BP}^0(a_i) - a_i$.

E 3.5 Show that the covariances of $\hat{\mu}$ with $\mathrm{BP}^0(a_i)$, of $\hat{\beta}$ with $\bar{y}_{i.}$, and of $\hat{\beta}$ with $\bar{y}_{..}$ are all zero.

E 3.6 Derive (3.99) and (3.100).

E 3.7 From (3.102) derive $\hat{\mu}$ of (3.106). *Note:* This is quite lengthy.

E 3.8 Show that $\hat{\beta}$ and $\hat{\mu}$ of (3.104) and (3.106) are unbiased.

E 3.9 Simplify $\hat{\beta}$ and $\hat{\mu}$ of (3.104) and (3.106) for $n_i = n \, \forall \, i$.

E 3.10 Derive (3.107) and (3.108).

E 3.11 Under model (3.29) show that the standard deviation of y is proportional to its mean.

E 3.12 For x_i taking on only the values 0 and 1, find the MLEs of α and β from (3.116).

E 3.13 Suppose there is a value d such that $y_i = 1$ for all observations with $x_i > d$ and $y_i = 0$ for all observations with $x_i < d$. Show that the log likelihood (3.113) is an increasing function of β for an appropriately chosen value of α and hence that a finite MLE does not exist.

E 3.14 Under $\beta = 0$ for model (3.109), show that $\hat{\alpha}_0$, the estimate of α, is $\log[\bar{y}/(1 - \bar{y})]$.

E 3.15 Using the fact that $\dfrac{1}{1 + e^{-t}} \approx \Phi\left(t\dfrac{16\sqrt{3}}{15\pi}\right)$ (Johnson and Kotz, 1970, p. 6) and the results of (2.106), derive α^* and β^* of (3.128).

Chapter 4

LINEAR MODELS (LMs)

This chapter provides a thumbnail discussion of linear models (LMs), one of the most widely treated branches of statistics, both in theory and in practice, embracing, as it does, regression, analysis of variance, and analysis of covariance. These topics are dealt with in varying degrees of detail in a myriad of books and papers, so it is not our intention to have this book or this chapter replicate them to any great extent. We simply assume at this point that the reader is familiar with the basic ideas dealt with in Chapters 1 through 3.

The prime object of the chapter is to describe the general ideas of LMs and the analysis of data based thereon. In doing this we establish notation and concepts for use in the succeeding chapters on linear mixed models (LMMs), generalized linear models (GLMs) and some nonlinear models. We begin with an introductory example.

Consider a portion of the experiment on the growth of potato lesions described in Chapter 3. That experiment extended over several weeks; we consider just one week, the first, say. In that week there are 16 observations, consisting of the extent of lesions on each of four leaves, at four different temperatures. Let y_{ij} be the average log diameter of the lesions on leaf j at temperature i, for $j = 1, 2, \ldots, 4$ and $i = 1, 2, \ldots, 4$. Then for μ being an overall mean, and τ_i the effect of temperature on lesion number we could take, for each i

$$\mathrm{E}[y_{ij}] = \mu + \tau_i, \tag{4.1}$$

for $j = 1, 2, \ldots, 4$. For

$$E[\mathbf{y}_i] = E \begin{bmatrix} y_{i1} \\ y_{i2} \\ y_{i3} \\ y_{i4} \end{bmatrix} = \begin{bmatrix} \mu + \tau_i \\ \mu + \tau_i \\ \mu + \tau_i \\ \mu + \tau_i \end{bmatrix} = (\mu + \tau_i)\mathbf{1}_4$$

we can define

$$\mathbf{y} = \left\{ {}_c \mathbf{y}_i \right\}_{i=1}^{4} \tag{4.2}$$

and then write

$$E[\mathbf{y}] = \begin{bmatrix} \mathbf{1}_4 & \mathbf{1}_4 & \mathbf{0} & \mathbf{0} & \mathbf{0} \\ \mathbf{1}_4 & \mathbf{0} & \mathbf{1}_4 & \mathbf{0} & \mathbf{0} \\ \mathbf{1}_4 & \mathbf{0} & \mathbf{0} & \mathbf{1}_4 & \mathbf{0} \\ \mathbf{1}_4 & \mathbf{0} & \mathbf{0} & \mathbf{0} & \mathbf{1}_4 \end{bmatrix} \begin{bmatrix} \mu \\ \tau_1 \\ \tau_2 \\ \tau_3 \\ \tau_4 \end{bmatrix} \tag{4.3}$$

$$= \begin{bmatrix} \mathbf{1}_{16} & \left\{ {}_d \mathbf{1}_4 \right\}_{i=1}^{4} \end{bmatrix} \boldsymbol{\beta} \quad \text{for} \quad \boldsymbol{\beta} = \begin{bmatrix} \mu \\ \tau_1 \\ \tau_2 \\ \tau_3 \\ \tau_4 \end{bmatrix} \tag{4.4}$$

$$= \mathbf{X}\boldsymbol{\beta} \quad \text{for} \quad \mathbf{X} = \begin{bmatrix} \mathbf{1}_{16} & \left\{ {}_d \mathbf{1}_4 \right\}_{i=1}^{4} \end{bmatrix}. \tag{4.5}$$

Throughout all this the μ and τs are taken as fixed effects which we wish to estimate. That is what is now considered. But we go no further with this example, using (4.5) simply as a base from which to describe a general model.

4.1   A GENERAL MODEL

We think of dealing with N items of data, arrayed as a vector $\mathbf{y}$ of order $N \times 1$, and we take the basic (vector) equation of a model to be

$$E[\mathbf{y}] = \boldsymbol{\mu}, \tag{4.6}$$

where, for example, $\boldsymbol{\mu}$ may have the form $\boldsymbol{\mu} = \mathbf{X}\boldsymbol{\beta}$ as in (4.5). We turn to this form in Section 4.2.

With $\mathbf{y}$ being data, we think of its elements as being realized values of some random variable which would traditionally be denoted as $\mathbf{Y}$.

For simplicity we abandon this notational distinction and use **y** both as a realization of a random vector and as the random vector itself. In this latter sense we attribute to **y** a variance-covariance matrix **V** and write

$$\text{var}(\mathbf{y}) = \mathbf{V}. \tag{4.7}$$

Then, to combine (4.6) and (4.7) we write

$$\mathbf{y} \sim (\boldsymbol{\mu}, \mathbf{V}), \tag{4.8}$$

meaning that **y** has mean $\boldsymbol{\mu}$ and variance-covariance matrix **V**.

Statement (4.8) is very general. It does no more than assign a symbol $\boldsymbol{\mu}$ to the mean **y** and another symbol **V** to the matrix of var(**y**). As they stand, $\boldsymbol{\mu}$ and **V** are nothing more than symbols. $\boldsymbol{\mu}$ has N elements and $\mathbf{V} = \mathbf{V}'$ has $N(N+1)/2$ unique elements. But there are only N data values; so without describing (modeling) $\boldsymbol{\mu}$ and **V** in terms of less than N parameters, $\boldsymbol{\mu}$ and **V** cannot be estimated. Thus we have to specify $\boldsymbol{\mu}$ and **V** in terms of underlying parameters appropriate to the nature of the data being studied. And this is just what we do for the different forms of models, LMs, LMMs, GLMs, GLMMs and some nonlinear models. We start with LMs.

4.2   A LINEAR MODEL FOR FIXED EFFECTS

Special forms of $\boldsymbol{\mu}$ and **V** for the traditional linear model for fixed effects are

$$\boldsymbol{\mu} = \mathbf{X}\boldsymbol{\beta} \tag{4.9}$$

and

$$\mathbf{V} = \sigma^2 \mathbf{I}_N. \tag{4.10}$$

$\boldsymbol{\beta}$ in (4.9) is a $p \times 1$ vector of unknown fixed effects and **X** is a known matrix, of order $N \times p$. Thus (4.8) and (4.9) give

$$\mathrm{E}[\mathbf{y}] = \boldsymbol{\mu} = \mathbf{X}\boldsymbol{\beta}, \tag{4.11}$$

a vector of linear combinations of fixed effects. And from (4.7) and (4.10)

$$\text{var}(\mathbf{y}) = \mathbf{V} = \sigma^2 \mathbf{I}, \tag{4.12}$$

which means that the variance of every element of **y** is taken as being the same, namely σ^2, and the covariance between every pair of elements is taken as being zero. In summary, we therefore have for LMs

$$\mathbf{y} \sim (\mathbf{X}\boldsymbol{\beta}, \sigma^2 \mathbf{I}). \tag{4.13}$$

The equation $E[\mathbf{y}] = \mathbf{X}\boldsymbol{\beta}$ of (4.11) is called the *model equation* and **X** is the *model matrix*. In many situations **X** will have elements which are all 0 or 1, in which case **X** is known as an *incidence matrix*. But it is perfectly permissible for **X** to also have columns of observed or measured variables, such as predictor variables in regression (e.g., Chapter 3) or concomitant variables in analysis of covariance.

Notice that although (4.11) and (4.12) specify the mean and variance-covariance structure for **y** of (4.13), the actual form of the distribution is not specified in (4.13).

4.3 MAXIMUM LIKELIHOOD UNDER NORMALITY

Although estimating the $\boldsymbol{\beta}$ of (4.13) is often done by ordinary least squares (OLSE) or generalized least squares (GLSE), neither of which demand having an underlying distribution for **y**, we follow the general approach of this book and use maximum likelihood (ML). This has the particular merit of simultaneously providing an estimator not only for $\boldsymbol{\beta}$ but also one for σ^2 of (4.13).

Starting with the assumption that **y** follows a multivariate normal distribution,

$$\mathbf{y} \sim \mathcal{N}(\mathbf{X}\boldsymbol{\beta}, \sigma^2 \mathbf{I})$$

the likelihood function is

$$L = L(\boldsymbol{\beta}, \sigma^2 | \mathbf{y}) = \frac{\exp[-\frac{1}{2}(\mathbf{y} - \mathbf{X}\boldsymbol{\beta})'(\mathbf{I}/\sigma^2)(\mathbf{y} - \mathbf{X}\boldsymbol{\beta}]}{(2\pi\sigma^2)^{\frac{1}{2}N}} \tag{4.14}$$

and so the log likelihood is

$$l = \log L = -\tfrac{1}{2}N\log(2\pi) - \tfrac{1}{2}N\log\sigma^2 - \tfrac{1}{2}(\mathbf{y} - \mathbf{X}\boldsymbol{\beta})'(\mathbf{y} - \mathbf{X}\boldsymbol{\beta})/\sigma^2. \tag{4.15}$$

Denoting $\partial l / \partial \boldsymbol{\beta}$ by $l_{\boldsymbol{\beta}}$ and $\partial l / \partial \sigma^2$ by l_{σ^2}, it is easily found that

$$l_{\boldsymbol{\beta}} = \frac{\mathbf{X}'(\mathbf{y} - E[\mathbf{y}])}{\sigma^2} = \frac{\mathbf{X}'\mathbf{y} - \mathbf{X}'\mathbf{X}\boldsymbol{\beta}}{\sigma^2} \tag{4.16}$$

and

$$l_{\sigma^2} = \frac{(\mathbf{y} - \mathbf{X}\beta)'(\mathbf{y} - \mathbf{X}\beta)}{2\sigma^2} - \frac{N}{2\sigma^2}. \qquad (4.17)$$

Equating l_β to zero, and in doing so denoting β by $\hat{\beta}$ gives $\mathbf{X}'\hat{\mathrm{E}}[\mathbf{y}] = \mathbf{X}'\mathbf{y}$ or

$$\mathbf{X}'\mathbf{X}\hat{\beta} = \mathbf{X}'\mathbf{y}. \qquad (4.18)$$

Equations (4.18) are known as *normal equations*, from which

$$\hat{\beta} = (\mathbf{X}'\mathbf{X})^{-1}\mathbf{X}'\mathbf{y}, \quad \text{if} \quad (\mathbf{X}'\mathbf{X})^{-1} \text{ exists.} \qquad (4.19)$$

And from equating l_{σ^2} to zero, with $\hat{\sigma}^2$ in place of σ^2 we get

$$\hat{\sigma}^2 = (\mathbf{y} - \mathbf{X}\hat{\beta})'(\mathbf{y} - \mathbf{X}\hat{\beta})/N. \qquad (4.20)$$

These are the ML estimators; they are not just solutions, because $\hat{\beta}$ lies in the same range as β, namely $-\infty < \beta < \infty$, and $\hat{\sigma}^2$ is non-negative, as is σ^2.

The first thing to notice about $\hat{\beta}$ of (4.19) is that it exists only if $(\mathbf{X}'\mathbf{X})^{-1}$ exists: and this requires $\mathbf{X}_{N \times p}$ to have full column rank p. This is very restrictive, because in many situations $\mathbf{X}$ does not have rank p.

4.4   SUFFICIENT STATISTICS

Working from (4.14) we can rewrite the density of $\mathbf{y}$ as

$$f_{\mathbf{y}}(\mathbf{y}) = (2\pi\sigma^2)^{-N/2} \exp\left\{ -\frac{1}{2\sigma^2}(\mathbf{y} - \mathbf{X}\beta)'(\mathbf{y} - \mathbf{X}\beta) \right\} \qquad (4.21)$$

$$= (2\pi\sigma^2)^{-N/2} \exp\left\{ \frac{1}{\sigma^2}\beta'\mathbf{X}'\mathbf{y} - \frac{1}{2\sigma^2}\mathbf{y}'\mathbf{y} - \frac{1}{2\sigma^2}\beta'\mathbf{X}'\mathbf{X}\beta \right\}.$$

To identify the sufficient statistics we define $\theta' = (\beta', \sigma^2)'$ as the parameter vector and define the following functions to match with the definition in Section S.3 of Appendix S

$$d(\theta) = (2\pi\sigma^2)^{-N/2} \exp\left(-\frac{1}{2\sigma^2}\beta'\mathbf{X}'\mathbf{X}\beta \right),$$

$$h(\mathbf{y}) = 1,$$

$$(T_1(\mathbf{y}), T_2(\mathbf{y}), \dots, T_p(\mathbf{y}))' \;=\; \mathbf{X}'\mathbf{y},$$

$$T_{p+1}(\mathbf{y}) \;=\; \mathbf{y}'\mathbf{y},$$

$$(\nu_1(\boldsymbol{\theta}), \dots, \nu_p(\boldsymbol{\theta}))' \;=\; \boldsymbol{\beta}'/\sigma^2, \text{ and}$$

$$\nu_{p+1}(\boldsymbol{\theta}) \;=\; -\frac{1}{2\sigma^2}.$$

This results in the distribution of $\mathbf{y}$ being in the exponential family and hence the sufficient statistic is $(\mathbf{y}'\mathbf{X} \;\; \mathbf{y}'\mathbf{y})'$. As expected, the maximum likelihood estimators are functions of the sufficient statistics.

4.5 MANY APPARENT ESTIMATORS

a. General result

When $(\mathbf{X}'\mathbf{X})^{-1}$ does not exist there is no longer just one solution $\hat{\boldsymbol{\beta}}$ given by (4.19). Instead there is an infinite number of solutions to (4.18) of the form

$$\boldsymbol{\beta}^0 = (\mathbf{X}'\mathbf{X})^{-}\mathbf{X}'\mathbf{y} \tag{4.22}$$

for $(\mathbf{X}'\mathbf{X})^{-}$ being any matrix satisfying

$$\mathbf{X}'\mathbf{X}(\mathbf{X}'\mathbf{X})^{-}\mathbf{X}'\mathbf{X}. \tag{4.23}$$

For notational convenience we use $\mathbf{G}$ for $(\mathbf{X}'\mathbf{X})^{-}$:

$$\mathbf{G} \equiv (\mathbf{X}'\mathbf{X})^{-}. \tag{4.24}$$

By virtue of the nature of (4.23), we call $(\mathbf{X}'\mathbf{X})^{-}$ a *generalized inverse* of $\mathbf{X}'\mathbf{X}$. For $\mathbf{X}$ of full column rank, $(\mathbf{X}'\mathbf{X})^{-}$ of (4.24) is $(\mathbf{X}'\mathbf{X})^{-1}$, which exists, and we use $\hat{\boldsymbol{\beta}} = (\mathbf{X}'\mathbf{X})^{-1}\mathbf{X}'\mathbf{y}$ as the estimator of $\boldsymbol{\beta}$. But when $\mathbf{X}$ has less than full column rank, $(\mathbf{X}'\mathbf{X})^{-1}$ does not exist and there are many matrices $(\mathbf{X}'\mathbf{X})^{-}$ satisfying (4.24). Thus there are many solutions $\boldsymbol{\beta}^0$ available from (4.22). That is why the symbol $\boldsymbol{\beta}^0$ is used in (4.22), as emphasis for distinguishing the many solutions $\boldsymbol{\beta}^0$ when $\mathbf{X}$ is less than full column rank from the solitary $\hat{\boldsymbol{\beta}}$ when $\mathbf{X}$ is of full column rank. **Note:** From this point onward we assume $\mathbf{X}$ has less than full column rank, unless otherwise stated.

b.   Mean and variance

Not only are there numerous values of β^0 for any given $\mathbf{X}$, but none of them is unbiased for β. With $\mathrm{E}[\mathbf{y}] = \mathbf{X}\beta$ of (4.11) it is easily seen that the expected value of β^0 is $\mathbf{GX'X}\beta$:

$$\mathrm{E}[\beta^0] = \mathrm{E}[\mathbf{GX'y}] = \mathbf{GX'}\mathrm{E}[\mathbf{y}] = \mathbf{GX'X}\beta. \tag{4.25}$$

In general this is not β; it is β if $\mathbf{GX'X}$ equals $\mathbf{I}$, but this occurs only when $\mathbf{X'X}$ is non-singular, in which case $\mathbf{G}$ is $(\mathbf{X'X})^{-1}$.

The variance of β^0 is

$$\mathrm{var}(\beta^0) = \mathrm{var}(\mathbf{GX'y}) = \mathbf{GX'}\sigma^2\mathbf{IXG'} = \mathbf{GX'XG'}\sigma^2. \tag{4.26}$$

This does not simplify to $\mathbf{G}\sigma^2$, which one might expect on the basis of it being $(\mathbf{X'X})^{-1}$ when that exists.

In view of there being numerous solutions β^0, with none of them unbiased for β, one well might wonder what use there is for any β^0. Fortunately, there is an important invariance property pertaining to $\mathbf{X}\beta^0$ which provides widespread applicability and utility.

c.   Invariance properties

The infinity of values β^0, together with the dependence on $\mathbf{G}$ of $\mathrm{E}[\beta^0]$ and $\mathrm{var}(\beta^0)$ in (4.25) and (4.26), clearly negates using any β^0 as an estimator of β. But three standard properties of $\mathbf{G}$ (Section M.4 of Appendix M) do provide useful results that are invariant to $\mathbf{G}$ (for given $\mathbf{X}$, of course). These results are

$$\mathbf{G'} \text{ is a generalized inverse of } \mathbf{X'X}; \tag{4.27}$$

$$\mathbf{XGX'} = \mathbf{XG'X'} \text{ is invariant to } \mathbf{G}; \tag{4.28}$$

and

$$\mathbf{X} = \mathbf{XGX'X}. \tag{4.29}$$

These results lead to the following useful properties involving $\mathbf{X}\beta^0$. First, the predicted mean of $\mathbf{y}$ is

$$\widehat{\mathrm{E}[\mathbf{y}]} = \hat{\mu} = \mathbf{X}\beta^0 = \mathbf{XGX'y} \tag{4.30}$$

and is invariant to $\mathbf{G}$. Thus for every $\boldsymbol{\beta}^0 = \mathbf{G}\mathbf{X}'\mathbf{y}$, no matter what $\mathbf{G}$ is used, the value of $\mathbf{X}\boldsymbol{\beta}^0$ does not depend on $\mathbf{G}$. Second, in place of (4.25),

$$\mathrm{E}[\hat{\boldsymbol{\mu}}] = \mathrm{E}[\mathbf{X}\boldsymbol{\beta}^0] = \mathbf{X}\mathbf{G}\mathbf{X}'\mathbf{X}\boldsymbol{\beta} = \mathbf{X}\boldsymbol{\beta} \tag{4.31}$$

from (4.29); and third,

$$\mathrm{var}(\hat{\boldsymbol{\mu}}) = \mathrm{var}(\mathbf{X}\boldsymbol{\beta}^0) = \mathbf{X}\mathbf{G}\mathbf{X}'\mathbf{I}\sigma^2\mathbf{X}\mathbf{G}'\mathbf{X}' = \mathbf{X}\mathbf{G}\mathbf{X}'\sigma^2 \tag{4.32}$$

is also invariant to $\mathbf{G}$.

d. Distributions

For

$$\mathbf{y} \sim \mathcal{N}(\mathbf{X}\boldsymbol{\beta},\ \sigma^2\mathbf{I})$$

(4.25) and (4.26) give

$$\boldsymbol{\beta}^0 \sim \mathcal{N}(\mathbf{G}\mathbf{X}'\mathbf{X}\boldsymbol{\beta},\ \mathbf{G}\mathbf{X}'\mathbf{X}\mathbf{G}'\sigma^2), \tag{4.33}$$

whilst from (4.31) and (4.32)

$$\mathbf{X}\boldsymbol{\beta}^0 \sim \mathcal{N}(\mathbf{X}\boldsymbol{\beta},\ \mathbf{X}\mathbf{G}\mathbf{X}'\sigma^2). \tag{4.34}$$

4.6 ESTIMABLE FUNCTIONS

a. Introduction

When we are interested in estimating functions of $\boldsymbol{\beta}$ we must distinguish between those which are functions just of $\mathbf{X}\boldsymbol{\beta}$ and those which are not. Because the parameter $\boldsymbol{\beta}$ affects the distribution of $\mathbf{y}$ only through $\mathbf{X}\boldsymbol{\beta}$, it is only functions of $\mathbf{X}\boldsymbol{\beta}$ which can be estimated satisfactorily. In that context consider a linear combination of elements of $\mathbf{X}\boldsymbol{\beta}$, say $\mathbf{t}'\mathbf{X}\boldsymbol{\beta}$. An unbiased, though perhaps not efficient, estimator of $\mathbf{t}'\mathbf{X}\boldsymbol{\beta}$ is $\mathbf{t}'\mathbf{y}$. Therefore we can estimate $\mathbf{q}'\boldsymbol{\beta}$ whenever $\mathbf{q}'$ is of the form $\mathbf{t}'\mathbf{X}$. But when $\mathbf{q}'$ cannot be written in the form $\mathbf{t}'\mathbf{X}$ it is not possible to estimate $\mathbf{q}'\boldsymbol{\beta}$ unbiasedly. Thus for estimating linear functions of $\boldsymbol{\beta}$ the only ones we can consider are those of the form $\mathbf{q}'\boldsymbol{\beta} = \mathbf{t}'\mathbf{X}\boldsymbol{\beta}$. Such functions are said to be *estimable functions* — functions of $\boldsymbol{\beta}$. Their characteristics are now described.

b. Definition

A linear combination of elements of β is $\mathbf{q}'\beta$ for some row vector $\mathbf{q}'$. It is called an *estimable function*, and is said to be *estimable*, under the following circumstances:

$$\mathbf{q}'\beta \text{ is estimable} \quad \text{iff} \quad \mathbf{q}'\beta = \mathbf{t}'\mathbf{X}\beta \;\; \forall \;\; \beta; \qquad (4.35)$$

that is,

$$\mathbf{q}'\beta \text{ is estimable} \quad \text{iff} \quad \mathbf{q}' = \mathbf{t}'\mathbf{X} \text{ for some } \mathbf{t}'. \qquad (4.36)$$

c. Properties

Three important properties of estimable functions are as follows:

(i) $E[y_k]$ is estimable for any element y_k of $\mathbf{y}$. This is so because for

$$\mathbf{t}' \;\; = \;\; (\text{row of zeros except } k\text{'th element being 1})$$

we have

$$E[y_k] \;\; = \;\; \mu_k = (k\text{'th row of } \mathbf{X})\beta = (\mathbf{t}'\mathbf{X})\beta,$$

which satisfies (4.35). Thus the expected value of each observation is estimable.

(ii) Linear combinations of estimable functions are estimable. Suppose $\mathbf{q}_1'\beta = \mathbf{t}_1'\mathbf{X}\beta$ and $\mathbf{q}_2'\beta = \mathbf{t}_2'\mathbf{X}\beta$ are estimable functions. Combining them using scalars c_1 and c_2 gives

$$c_1\mathbf{q}_1\beta + c_2\mathbf{q}_2\beta = c_1\mathbf{t}_1'\mathbf{X}\beta + c_2\mathbf{t}_2'\mathbf{X}\beta = (c_1\mathbf{t}_1' + c_2\mathbf{t}_2')\mathbf{X}\beta = \mathbf{t}_*'\mathbf{X}\beta$$

for $\mathbf{t}_*' = c_1\mathbf{t}_1' + c_2\mathbf{t}_2'$, thus demonstrating estimability.

(iii) Putting properties (i) and (ii) together enables the establishment of functions (linear combinations of elements of β) that are estimable without having to ascertain the corresponding vectors $\mathbf{t}'$. This is illustrated in the example of Section 4.7.

d. Estimation

With $\mathbf{q}' = \mathbf{t}'\mathbf{X}$ of (4.36) and $\mathbf{X}\boldsymbol{\beta}^0 = \mathbf{X}\mathbf{G}\mathbf{X}'\mathbf{y}$ of (4.30) being invariant to $\mathbf{G}$, we have the ML estimator of estimable $\mathbf{q}'\boldsymbol{\beta}$ as

$$\widehat{\mathbf{q}'\boldsymbol{\beta}} = \mathbf{t}'\mathbf{X}\boldsymbol{\beta}^0 = \mathbf{q}'\boldsymbol{\beta}^0 = \mathbf{t}'\mathbf{X}\mathbf{G}\mathbf{X}'\mathbf{y}, \qquad (4.37)$$

invariant to $\mathbf{G}$. A notable feature of this is the second equality, that the estimator of estimable $\mathbf{q}'\boldsymbol{\beta}$ is $\mathbf{q}'\boldsymbol{\beta}^0$, the same linear combination of elements of $\boldsymbol{\beta}^0$ as is the estimable function $\mathbf{q}'\boldsymbol{\beta}$ of $\boldsymbol{\beta}$. Furthermore, under normality of $\mathbf{y}$, using (4.37), we have

$$\widehat{\mathbf{q}'\boldsymbol{\beta}} = \mathbf{q}'\boldsymbol{\beta}^0 \sim \mathcal{N}(\mathbf{q}'\boldsymbol{\beta}, \mathbf{q}'\mathbf{G}\mathbf{q}\sigma^2). \qquad (4.38)$$

Since the ML estimator of $\mathbf{q}'\boldsymbol{\beta}$ is unbiased and based on the sufficient statistic (Section 4.4) it is a uniform minimum variance unbiased (UMVU) estimator.

The invariance of $\mathbf{q}'\boldsymbol{\beta}^0$ and of its variance to different $\mathbf{G}$ when $\mathbf{q}'\boldsymbol{\beta}$ is an estimable function are two eminently practical features of an estimable function. They totally avoid the impracticality of $\boldsymbol{\beta}^0$ as an estimator of $\boldsymbol{\beta}$ through it and its variance being functions of $\mathbf{G}$ for which there is an infinite number of values.

4.7 A NUMERICAL EXAMPLE

For the sole purpose of numerically illustrating some of the preceding results, suppose for the one-way classification of Chapter 2 we have the following data, for three classes with 3, 2 and 1 observations.

$i = 1$	$i = 2$	$i = 3$
72	48	36
36	12	
12		
120	60	36

Then for

$$\mathbf{y} = \begin{bmatrix} y_{11} \\ y_{12} \\ y_{13} \\ y_{21} \\ y_{22} \\ y_{31} \end{bmatrix} = \begin{bmatrix} 72 \\ 36 \\ 12 \\ 48 \\ 12 \\ 36 \end{bmatrix},$$

$$\mathbf{X\beta} = \begin{bmatrix} \mu + \alpha_1 \\ \mu + \alpha_1 \\ \mu + \alpha_1 \\ \mu \quad +\alpha_2 \\ \mu \quad +\alpha_2 \\ \mu \qquad\quad +\alpha_3 \end{bmatrix} = \begin{bmatrix} 1 & 1 & 0 & 0 \\ 1 & 1 & 0 & 0 \\ 1 & 1 & 0 & 0 \\ 1 & 0 & 1 & 0 \\ 1 & 0 & 1 & 0 \\ 1 & 0 & 0 & 1 \end{bmatrix} \begin{bmatrix} \mu \\ \alpha_1 \\ \alpha_2 \\ \alpha_3 \end{bmatrix} . \qquad (4.39)$$

The resulting normal equations, $\mathbf{X'X\beta^0} = \mathbf{X'y}$ of (4.18), are therefore

$$\begin{bmatrix} 6 & 3 & 2 & 1 \\ 3 & 3 & 0 & 0 \\ 2 & 0 & 2 & 0 \\ 1 & 0 & 0 & 1 \end{bmatrix} \begin{bmatrix} \mu^0 \\ \alpha_1^0 \\ \alpha_2^0 \\ \alpha_3^0 \end{bmatrix} = \begin{bmatrix} 216 \\ 120 \\ 60 \\ 36 \end{bmatrix} . \qquad (4.40)$$

Four different matrices $\mathbf{G} = (\mathbf{X'X})^-$ are

$$\mathbf{G}_1 = \begin{bmatrix} 0 & 0 & 0 & 0 \\ 0 & \frac{1}{3} & 0 & 0 \\ 0 & 0 & \frac{1}{2} & 0 \\ 0 & 0 & 0 & \frac{1}{1} \end{bmatrix}, \quad \mathbf{G}_2 = \begin{bmatrix} 1 & -1 & -1 & 0 \\ -1 & 1\frac{1}{3} & 1 & 0 \\ -1 & 1 & 1\frac{1}{2} & 0 \\ 0 & 0 & 0 & 0 \end{bmatrix},$$

$$\mathbf{G}_3 = \frac{1}{18}\begin{bmatrix} 0 & 2 & 3 & 6 \\ 0 & 4 & -3 & -6 \\ 0 & -2 & 6 & -6 \\ 0 & -2 & -3 & 12 \end{bmatrix} \text{ and } \mathbf{G}_4 = \frac{1}{54}\begin{bmatrix} 17 & -11 & -8 & 1 \\ -11 & 23 & 2 & -7 \\ -8 & 2 & 26 & -10 \\ 1 & -7 & -10 & 35 \end{bmatrix} .$$

Post-multiplying these by $\mathbf{X'y} = [216 \quad 120 \quad 60 \quad 36]'$ gives solutions $\mathbf{\beta^0} = [\mu^0 \ \alpha_1^0 \ \alpha_2^0 \ \alpha_3^0]'$ as

$$\mathbf{\beta}_1^0 = \begin{bmatrix} 0 \\ 40 \\ 30 \\ 36 \end{bmatrix}, \quad \mathbf{\beta}_2^0 = \begin{bmatrix} 36 \\ 4 \\ -6 \\ 0 \end{bmatrix} \text{ and } \mathbf{\beta}_3^0 = \mathbf{\beta}_4^0 = \begin{bmatrix} 35\frac{1}{3} \\ 4\frac{2}{3} \\ -5\frac{1}{3} \\ \frac{2}{3} \end{bmatrix} . \qquad (4.41)$$

Note that $\mathbf{\beta}_2^0$ has $\alpha_3^0 = 0$ (i.e., the last effect has solution zero), a characteristic seen in SAS GLM and Proc MIXED outputs. And $\mathbf{\beta}_3^0 = \mathbf{\beta}_4^0$ has $\alpha_1^0 + \alpha_2^0 + \alpha_3^0 = 0$, a feature of some other computing software.

It is also interesting to note that two different **G**-matrices can give the same $\beta^0 = \mathbf{GX'y}$.

Demonstrating (4.28) we find that $\mathbf{XGX'}$ for each of the four **G**s is

$$\mathbf{XGX'} = \begin{bmatrix} \bar{\mathbf{J}}_3 & \mathbf{0} & \mathbf{0} \\ \mathbf{0} & \bar{\mathbf{J}}_2 & \mathbf{0} \\ \mathbf{0} & \mathbf{0} & \bar{\mathbf{J}}_1 \end{bmatrix},$$

where $\bar{\mathbf{J}}_n$ is $n \times n$ with every element $1/n$. It is then easily verified that $\mathbf{XGX'X} = \mathbf{X}$ of (4.29). Next, from (4.30) it is easily seen that $(\mathbf{X}\beta^0)' = [40 \quad 40 \quad 40 \quad 30 \quad 30 \quad 36]$ for each β^0.

By property (i) of Section 4.6c, having $\mathrm{E}[y_{ij}] = \mu + \alpha_i$ means that $\mu + \alpha_i$ is estimable; and from each β^0 we find that the MLE of $\mu + \alpha_1$ is 40. For example, $\mu^0 + \alpha_1^0$ from β_1^0 is $0 + 40 = 40$, from β_2^0 it is $36 + 4 = 40$ and from β_3^0 it is $35\frac{1}{3} + 4\frac{2}{3} = 40$. Also by property (ii) $\alpha_1 - \alpha_2$ is estimable because $\alpha_1 - \alpha_2 = \mu + \alpha_1 - (\mu + \alpha_2)$; and each β^0 gives $\alpha_1^0 - \alpha_2^0 = 10$. These calculations demonstrate for estimable $\mathbf{q}'\beta^0$ the invariance of $\mathbf{q}'\beta^0$ to β^0.

Writing estimable $\mu + \alpha_1$ as $\mathbf{q}'\beta$ for $\mathbf{q}' = [1 \quad 1 \quad 0 \quad 0]$ it will then be found, using (4.38), that

$$\mathrm{var}(\widehat{\mathbf{q}'\beta}) = \mathrm{var}(\mathbf{q}'\beta^0) = \mathbf{q}'\mathbf{Gq}\sigma^2 = \tfrac{1}{3}\sigma^2$$

no matter which **G** is used.

Remark: Derivation of the four **G**-matrices is as follows: $\mathbf{G}_1$ and $\mathbf{G}_2$ are based on the regular inverse of the lower right and upper left (respectively) 3×3 submatrices. $\mathbf{G}_3$ uses the formula for $\mathbf{G}_r$ in Searle (1987, p. 307) and $\mathbf{G}_4$ uses $(\mathbf{X'X} + \mathbf{H'H})^{-1}$ discussed in Searle (1971, p. 23) and more thoroughly in Searle (1999).

4.8 ESTIMATING RESIDUAL VARIANCE

a. Estimation

Equation (4.20) shows the ML estimator for σ^2 of $\mathbf{y} \sim (\mathbf{X}\beta, \sigma^2\mathbf{I})$ when $\mathbf{X'X}$ is nonsingular. In that equation replacing $\hat{\beta}$ by β^0 gives the ML estimator $\hat{\sigma}_{\mathrm{ML}}^2$ when $\mathbf{X'X}$ is singular as follows:

$$\hat{\sigma}_{\mathrm{ML}}^2 = \frac{(\mathbf{y} - \mathbf{X}\beta^0)'(\mathbf{y} - \mathbf{X}\beta^0)}{N} = \frac{\mathrm{SSE}}{N}. \tag{4.42}$$

The numerator of $\hat{\sigma}^2_{\text{ML}}$ is denoted by SSE, the residual (or error) sum of squares in the context of analysis of variance. For convenience we retain that label but without reference to analysis of variance.

It is of interest to see if $\hat{\sigma}^2_{\text{ML}}$ is unbiased for σ^2. To do this we use the result in Section S.1b of Appendix S for the expected value of a quadratic form. Then after simplifying SSE to be

$$\text{SSE} = (\mathbf{y} - \mathbf{X}\boldsymbol{\beta}^0)'(\mathbf{y} - \mathbf{X}\boldsymbol{\beta}^0) = \mathbf{y}'(\mathbf{I} - \mathbf{X}\mathbf{G}\mathbf{X}')\mathbf{y} \tag{4.43}$$

we find, for

$$r_{\mathbf{X}} = \text{rank of } \mathbf{X} \tag{4.44}$$

that

$$\text{E}[\text{SSE}] = \text{E}[\mathbf{y}'(\mathbf{I} - \mathbf{X}\mathbf{G}\mathbf{X}')\mathbf{y}] = (N - r_{\mathbf{X}})\sigma^2. \tag{4.45}$$

Note that this result does not rely on any distributional form for $\mathbf{y}$, only on $\mathbf{y}$ having mean $\mathbf{X}\boldsymbol{\beta}$ and variance $\sigma^2\mathbf{I}$. Then (4.45) gives

$$\text{E}[\hat{\sigma}^2_{\text{ML}}] = \frac{\text{E}[\text{SSE}]}{N} = \left(1 - \frac{r_{\mathbf{X}}}{N}\right)\sigma^2.$$

Thus the ML estimator of σ^2 is biased downward.

On the other hand, dividing SSE by $(N - r_{\mathbf{X}})$ gives an unbiased estimator

$$\hat{\sigma}^2 = \frac{\text{SSE}}{N - r_{\mathbf{X}}} \qquad \text{with} \qquad \text{E}[\hat{\sigma}^2] = \sigma^2. \tag{4.46}$$

This is, of course, the usual estimator used in analysis of variance, and being based on the sufficient statistic is a UMVU estimator.

b. Distribution of estimators

SSE is a quadratic form; and in Section S.2c of Appendix S is the following important theorem concerning the distribution of quadratic forms.

> **Theorem.** When $\mathbf{y} \sim \mathcal{N}(\boldsymbol{\mu}, \mathbf{V})$ with $\mathbf{V}$ nonsingular then $\mathbf{y}'\mathbf{A}\mathbf{y}$ is distributed as a non-central χ^2 with degrees of freedom $\nu = \text{rank}(\mathbf{A}\mathbf{V})$ and non-centrality parameter $\frac{1}{2}\boldsymbol{\mu}'\mathbf{A}\boldsymbol{\mu}$, if and only if $\mathbf{A}\mathbf{V}$ is idempotent.

In applying this theorem there is also the extension that whenever $\boldsymbol{\mu}'\mathbf{A}\boldsymbol{\mu} = 0$, the distribution becomes a usual (central) χ^2 distribution on ν degrees of freedom, which is denoted by χ^2_ν.

In applying the preceding theorem to SSE of (4.43) it will be found (E 4.12) that

$$\frac{\text{SSE}}{\sigma^2} \sim \chi^2_{N-r_{\mathbf{x}}} . \tag{4.47}$$

Therefore

$$\hat{\sigma}^2 = \frac{\text{SSE}}{N - r_{\mathbf{x}}} \sim \left(\frac{\sigma^2}{N - r_{\mathbf{x}}}\right) \chi^2_{N-r_{\mathbf{x}}}, \tag{4.48}$$

meaning by this that the distribution of $\hat{\sigma}^2$ is a scalar multiple of $\chi^2_{N-r_{\mathbf{x}}}$, that scalar being $\sigma^2/(N - r_{\mathbf{x}})$. And from (4.48)

$$\text{var}(\hat{\sigma}^2) = \frac{2\sigma^4}{N - r_{\mathbf{x}}}, \tag{4.49}$$

and from (4.42)

$$\hat{\sigma}^2_{\text{ML}} \sim \left(\frac{\sigma^2}{N}\right) \chi^2_{N-r_{\mathbf{x}}} \tag{4.50}$$

and so

$$\text{var}\left(\hat{\sigma}^2_{\text{ML}}\right) = \frac{2\sigma^4(N - r_{\mathbf{x}})}{N^2} = \frac{2\sigma^4}{N}\left(1 - \frac{r_{\mathbf{x}}}{N}\right). \tag{4.51}$$

These results, (4.48) and (4.50), rely on the normality of $\mathbf{y}$. An alternative expression comes from the general result (applicable to all ML estimators, regardless of the distribution of the data variable) that ML estimators have asymptotic normal distributions with variance structure given by the inverse of the information matrix. In the case of $\hat{\sigma}^2_{\text{ML}}$ this yields an asymptotic variance of $\hat{\sigma}^2_{\text{ML}}$ of $2\sigma^4/N$. And this is, of course, close in value to (4.51) when $r_{\mathbf{x}}/N$ is small.

4.9 THE ONE- AND TWO-WAY CLASSIFICATIONS

Section 4.7 numerically illustrates some of the basic properties surrounding the estimation of β from $E[\mathbf{y}] = \mathbf{X}\beta$. Here, for the one-way classification, we describe some of its general results. For the two-way classification we merely give a hint as to possible complications. Both of the one-way and two-way classifications are dealt with in great detail in a variety of books (e.g., Searle, 1987, 1997).

a. The one-way classification

The example of Section 4.7 is that of a one-way classification with unbalanced data, that is, not all the same number of observations in the subclasses. Equations (4.40) are an example of the normal equations which for the model equation

$$E[y_{ij}] = \mu + \alpha_i \qquad \text{for} \qquad i = 1, 2, \ldots, a$$

always take the form

$$\begin{bmatrix} N & \{_r n_i\} \\ \{_c n_i\} & \{_d n_i\} \end{bmatrix} \begin{bmatrix} \mu \\ \alpha \end{bmatrix} = \begin{bmatrix} y_{..} \\ \{_c y_{i.}\} \end{bmatrix} \qquad \text{for} \qquad i = 1, 2, \ldots, a.$$

And the easiest solution, exemplified by β_1^0 of (4.41), using

$$G = \begin{bmatrix} 0 & 0 \\ 0 & \{_d 1/n_i\} \end{bmatrix}$$

gives

$$\mu^0 = 0 \qquad \text{and} \qquad \alpha_i^0 = \bar{y}_{i.} \ .$$

Thus with $E(y_{ij}) = \mu + \alpha_i$ being estimable its ML estimator is

$$\widehat{\mu + \alpha_i} = \mu^0 + \alpha_i^0 = 0 + \bar{y}_{i.} = \bar{y}_{i.} \ . \tag{4.52}$$

Also, with $\alpha_i - \alpha_k = \mu + \alpha_i - (\mu + \alpha_k)$ being estimable, its estimator is

$$\widehat{\alpha_i - \alpha_k} = \alpha_1^0 - \alpha_k^0 = \bar{y}_{i.} - \bar{y}_{k.};$$

and the sampling variance of $\alpha_i^0 - \alpha_k^0$ from (4.38) is $(1/n_i + 1/n_k)\sigma^2$.

An alternative model, simpler than $E[y_{ij}] = \mu + \alpha_i$, is $E[y_{ij}] = \mu_i$, often called the *cell means* model. In using it, X of (4.39) would be changed to exclude its first column, and $X'X$ and G would have their first row and column excluded. This change in the model is effectively equating $\mu + \alpha_i$ and μ_i. Hence $\hat{\mu}_i = \widehat{\mu + \alpha_i} = \bar{y}_{i.}$ from (4.52).

b.   The two-way classification

The example in Section 1.3b, where y_{ijk} represented the rating of the ith cartoon type by the kth person in the jth group, suggests using

$$E[y_{ijk}] = \mu + \alpha_i + \beta_j. \tag{4.53}$$

Estimation details for that model are available in many places (e.g., Searle 1971, 1987): Particularly for unbalanced data, those details are extensive and need not occupy us here. We confine attention to estimability. Suppose we are interested in estimating the mean rating for the ith cartoon type. A reasonable estimator for this might be $\bar{y}_{i\cdot\cdot}$, similar to (4.52). If we define $\mu_i = \mu + \alpha_i$ so that

$$E[y_{ijk}] = \mu_i + \beta_j, \tag{4.54}$$

then

$$E[\bar{y}_{i\cdot\cdot}] = E[\Sigma_j\Sigma_k(\mu_i + \beta_j)/n_{i\cdot}] = \mu_i + \Sigma_j n_{ij}\beta_j/n_{i\cdot},$$

where n_{ij} is the number of observations at the intersection of row i and row j. Thus $\bar{y}_{i\cdot\cdot}$ is not an unbiased estimator of μ_i as might have been expected. This illustrates how careful one must be in drawing what seems like an "obvious" conclusion about what it is that some estimators are estimating. In many cases they are not estimating what one might think is "obvious". However, if one follows the $\mathbf{X}'\mathbf{X}\boldsymbol{\beta}^0 = \mathbf{X}'\mathbf{y}$ estimation procedure one finds that $\alpha_i - \alpha_k$ and $\beta_j - \beta_l$ are estimable. On the other hand, when an interaction effect is added to (4.53) so that

$$E[y_{ijk}] = \mu_{ij} \equiv \mu + \alpha_i + \beta_j + \gamma_{ij}, \tag{4.55}$$

then $\hat{\mu}_{ij} = \bar{y}_{ij\cdot}$ is an estimator of $\mu + \alpha_i + \beta_j + \gamma_{ij}$ (providing $n_{ij} \neq 0$). But then it is impossible to estimate $\alpha_i - \alpha_k$ because the interaction terms can never be gotten rid of.

4.10 TESTING LINEAR HYPOTHESES

The general formulation of a linear hypothesis concerning $\boldsymbol{\beta}$ is

$$H : \mathbf{K}'\boldsymbol{\beta} = \mathbf{m}. \tag{4.56}$$

$\mathbf{K}'$ must satisfy three conditions:

1. $\mathbf{K}' = \mathbf{T}'\mathbf{X}$ for some $\mathbf{T}'$, so that $\mathbf{K}'\boldsymbol{\beta}$ is estimable, with unbiased estimator $\mathbf{K}'\boldsymbol{\beta}^0$ invariant to $\boldsymbol{\beta}^0$.

2. $\mathbf{K}'$ must have full row rank, so that $\mathbf{K}'\boldsymbol{\beta}$ contains no redundant elements.

3. $\mathbf{K}'$ must have no more than $r_{\mathbf{X}}$ rows, i.e., the row rank of $\mathbf{K}'$ cannot exceed the rank of $\mathbf{X}$.

We show two ways of deriving a test.

a.   Likelihood ratio test

When $\boldsymbol{\theta}$ represents the vector of parameters in a model, we use $L(\boldsymbol{\theta})$ as the likelihood. Then for $\hat{\boldsymbol{\theta}}$ being the ML estimator of $\boldsymbol{\theta}$, and $\hat{\boldsymbol{\theta}}_0$ the ML estimator under the hypothesis, the likelihood ratio is $L(\hat{\boldsymbol{\theta}}_0)/L(\hat{\boldsymbol{\theta}})$, as discussed in Section 2.5b–iii. With this notation, and $\boldsymbol{\theta}' = [\boldsymbol{\beta}' \; \sigma^2]$, we have $L(\boldsymbol{\theta})$ given in (4.14). For $\hat{\boldsymbol{\theta}}$ (4.22) and (4.42) give the values of $\boldsymbol{\beta}$ and σ^2 that maximize $L(\boldsymbol{\beta} \; \sigma^2)$ as $\boldsymbol{\beta}^0 = \mathbf{GX'y}$ and $\hat{\sigma}^2_{ML} = (\mathbf{y} - \mathbf{X}\boldsymbol{\beta}^0)'(\mathbf{y} - \mathbf{X}\boldsymbol{\beta}^0)/N$. Using these in (4.14) in place of $\boldsymbol{\beta}$ and σ^2 gives

$$L(\hat{\boldsymbol{\theta}}) = L(\boldsymbol{\beta}^0 \; \hat{\sigma}^2_{ML}) = \left[\frac{N/e}{2\pi(\mathbf{y} - \mathbf{X}\boldsymbol{\beta}^0)'(\mathbf{y} - \mathbf{X}\boldsymbol{\beta}^0)} \right]^{\frac{1}{2}N}. \qquad (4.57)$$

Deriving $L(\hat{\boldsymbol{\theta}}_0)$ requires maximizing $L(\boldsymbol{\beta}, \; \sigma^2)$ subject to $\mathbf{K'}\boldsymbol{\beta} = \mathbf{m}$. The results of doing this (E 4.13) are that for $\hat{\boldsymbol{\theta}}_0$

$$\boldsymbol{\beta}^0_0 \;=\; \boldsymbol{\beta}^0 - \mathbf{GK}(\mathbf{K'GK})^{-1}(\mathbf{K'}\boldsymbol{\beta}^0 - \mathbf{m}) \qquad (4.58)$$

and

$$\hat{\sigma}^2_0 \;=\; (\mathbf{y} - \mathbf{X}\boldsymbol{\beta}^0_0)'(\mathbf{y} - \mathbf{X}\boldsymbol{\beta}^0_0)/N. \qquad (4.59)$$

Substituting these values in place of $\boldsymbol{\beta}$ and σ^2 in (4.14) gives, after more algebra,

$$L(\hat{\boldsymbol{\theta}}_0) \;=\; L(\boldsymbol{\beta}^0_0 \; \hat{\sigma}^2_0) = \left[\frac{N/e}{2\pi(\mathrm{SSE} + Q)} \right]^{\frac{1}{2}N} \qquad (4.60)$$

for

$$Q \;=\; (\mathbf{K'}\boldsymbol{\beta}^0 - \mathbf{m})'(\mathbf{K'GK})^{-1}(\mathbf{K'}\boldsymbol{\beta}^0 - \mathbf{m}). \qquad (4.61)$$

Then the likelihood ratio reduces to

$$\Lambda = \frac{L(\hat{\boldsymbol{\theta}}_0)}{L(\hat{\boldsymbol{\theta}})} = \left[\frac{1}{1 + Q/\mathrm{SSE}} \right]^{\frac{1}{2}N}. \qquad (4.62)$$

Clearly this ratio, (4.62), is a single-valued function of Q/SSE, decreasing monotonically when Q/SSE increases. Therefore Q/SSE can be used as a test statistic in place of (4.62). Moreover, by the same reasoning, instead of Q/SSE one can use (with degrees of freedom being abbreviated *df*)

$$\frac{Q}{df \text{ for } Q} \bigg/ \frac{\text{SSE}}{df \text{ for SSE}} \qquad (4.63)$$

as the test statistic.

Then, using the theorem of Section 4.7b one can show (E 4.18) that Q has a non-central χ^2 distribution and SSE has a central χ^2 distribution. Furthermore, Q and SSE are independent, as may be established by using the following theorem.

> **Theorem.** For $\mathbf{y} \sim \mathcal{N}(\boldsymbol{\mu}, \mathbf{V})$ with $\mathbf{V}$ nonsingular, the quadratic forms $\mathbf{y}'\mathbf{A}\mathbf{y}$ and $\mathbf{y}'\mathbf{B}\mathbf{y}$ are independent if and only if $\mathbf{AVB} = \mathbf{0}$.

These χ^2 and independence properties of Q and SSE result in our being able to use (4.63) as an F-statistic for testing $H : \mathbf{K}'\boldsymbol{\beta} = \mathbf{m}$; its degrees of freedom are $r(\mathbf{K}')$ and $N - r_{\mathbf{X}}$.

b. Wald test

An alternative to the likelihood ratio test is to use the Wald test. This is based on the fact that maximum likelihood estimators are asymptotically normal; more specifically that $\mathbf{K}'\boldsymbol{\beta}^0$ is asymptotically normal with asymptotic variance given by $\mathbf{K}'\mathbf{GK}\sigma^2$. In then follows that under $H : \mathbf{K}'\boldsymbol{\beta} = \mathbf{m}$

$$(\mathbf{K}'\boldsymbol{\beta}^0 - \mathbf{m})'(\mathbf{K}'\mathbf{GK}'\sigma^2)^{-1}(\mathbf{K}'\boldsymbol{\beta}^0 - \mathbf{m}) \sim \chi^2 \qquad (4.64)$$

with degrees of freedom equal to $r(\mathbf{K}')$. In this particular case, however, the exact distribution can be worked out so there is no point in using the asymptotic distribution. In the next section we use the exact distributions to work out tests and confidence intervals.

4.11   *t*-TESTS AND CONFIDENCE INTERVALS

When $\mathbf{K}'$ of the hypothesis $H : \mathbf{K}'\boldsymbol{\beta} = \mathbf{m}$ has just a single row $\mathbf{k}'$, then Q of (4.61) becomes

$$\frac{(\mathbf{k}'\boldsymbol{\beta}^0 - \mathbf{m})^2}{\mathbf{k}'\mathbf{Gk}[\text{SSE}/(N - r_{\mathbf{X}})]} = \frac{(\mathbf{k}'\boldsymbol{\beta}^0 - \mathbf{m})^2}{\mathbf{k}'\mathbf{Gk}\hat{\sigma}^2}$$

to be compared to the $\mathcal{F}$-distribution on 1 and $N-r$ degrees of freedom, where we use the notation $r = r_{\mathbf{X}} = \text{rank}(\mathbf{X})$ for this section. Now recall that when a variable is distributed as the t-distribution on n degrees of freedom its square is distributed as $\mathcal{F}_n^1$. Therefore,

$$\frac{\mathbf{k}'\boldsymbol{\beta}^0 - \mathbf{m}}{\hat{\sigma}\sqrt{\mathbf{k}'\mathbf{Gk}}} \sim t_{N-r}$$

provides a test of $H : \mathbf{k}'\boldsymbol{\beta}^0 = \mathbf{m}$. This t-test is also useful for one-sided alternatives, which is not so for the F-test.

Suppose $\mathbf{q}'\boldsymbol{\beta}$ is estimable; then from (4.38) we have the $100(1-\alpha)\%$ confidence interval on $\mathbf{q}'\boldsymbol{\beta}$ as

$$\mathbf{q}'\boldsymbol{\beta}^0 \pm \hat{\sigma} t_{N-r,\frac{\alpha}{2}} \sqrt{\mathbf{q}'\mathbf{Gq}},$$

where $t_{N-r,\frac{\alpha}{2}}$ is defined by the probability statement $\Pr(t \geq t_{N-r,\frac{\alpha}{2}}) = \alpha/2$ for t having the t-distribution with $N-r$ degrees of freedom. When $N-r$ is large, ($N-r \geq 100$, say) $z_{\frac{\alpha}{2}}$ may be used in place of $t_{N-r,\frac{\alpha}{2}}$, where $z_{\frac{\alpha}{2}}$ is defined by

$$(2\pi)^{-\frac{1}{2}} \int_{z_{\frac{\alpha}{2}}}^{\infty} e^{-\frac{1}{2}x^2} dx = \alpha/2.$$

From $\text{SSE}/\sigma^2 \sim \chi_{N-r}^2$ of (4.51) a confidence interval on σ^2 is

$$\frac{\text{SSE}}{\chi_{N-r,U}^2} \leq \sigma^2 \leq \frac{\text{SSE}}{\chi_{N-r,L}^2},$$

where the denominators are defined by

$$P\{\chi_{k,L}^2 \leq \chi_k^2 \leq \chi_{k,U}^2\} = 1 - \alpha.$$

4.12 UNIQUE ESTIMATION USING RESTRICTIONS

In the case of models of the form such as $\mathrm{E}[y_{ij}] = \mu + \alpha_i$ for $i = 1, 2, 3$, readers will undoubtedly have encountered constraints on the $\hat{\alpha}_i$s of the form

$$\hat{\alpha}_1 + \hat{\alpha}_2 + \hat{\alpha}_3 = 0 \quad \text{or} \quad \hat{\alpha}_3 = 0. \tag{4.65}$$

The second of these (and extensions thereof) is especially familiar to users of SAS GLM software where its use is standard practice.

Each equation in (4.64) is what we call a *linear constraint* on the solution. Careful use of such constraints can eliminate having many solutions β^0 to the normal equations, and instead can yield just a single solution, one that satisfies the constraint(s) imposed. That being so, we denote such a solution as $\hat{\beta}$. A brief outline for deriving $\hat{\beta}$ follows. Some of the details are available in Searle (1971, Section 1.5b) and a complete description is given in Searle (1999).

In order to have a unique solution of the ML equations means that for $\mathbf{X}_{N \times p}$ of rank $r = p - m$ we need to have $\mathbf{H}$ in $\mathbf{H}\hat{\beta} = \mathbf{c}$ being of order $m \times p$ of full row rank m. Then ML leads to minimizing $(\mathbf{y} - \mathbf{X}\hat{\beta})'(\mathbf{y} - \mathbf{X}\hat{\beta}) + 2\theta'(\mathbf{H}\hat{\beta} - \mathbf{c})$ which yields equations

$$\begin{bmatrix} \mathbf{X'X} & \mathbf{H'} \\ \mathbf{H} & \mathbf{0} \end{bmatrix} \begin{bmatrix} \hat{\beta} \\ \theta \end{bmatrix} = \begin{bmatrix} \mathbf{X'y} \\ \mathbf{c} \end{bmatrix}. \tag{4.66}$$

After considerable algebra (Searle, 1999) (4.66) yields

$$\hat{\beta} = (\mathbf{X'X} + \mathbf{H'H})^{-1}(\mathbf{X'y} + \mathbf{H'c}). \tag{4.67}$$

Because we have already established that ML generally yields a β as $\mathbf{GX'y}$ for some $\mathbf{G}$ [and $(\mathbf{X'X} + \mathbf{H'H})^{-1}$ is a $\mathbf{G}$] we cannot call (4.67) an ML estimator because of the $\mathbf{H'c}$ term therein. But (4.67) is an ML estimator under the constraints $\mathbf{H}\hat{\beta} = \mathbf{c}$. And if $\mathbf{c} = \mathbf{0}$, which is often the case, then $\hat{\beta} = (\mathbf{X'X} + \mathbf{H'H})^{-1}\mathbf{X'y}$ is an ML estimator.

Instead of having $\mathbf{H}\hat{\beta} = \mathbf{c}$ as constraints on solutions, suppose that we have $\mathbf{H}\beta = \mathbf{c}$ as restrictions on parameters ("restrictions" rather than "constraints" to distinguish parameters from solutions). Then on partitioning $\mathbf{H}$ as $[\mathbf{H}_1 \ \mathbf{H}_2]$ with $\mathbf{H}_1^{-1}$ existing (after perhaps permuting columns of $\mathbf{H}$ to permit this), we can rewrite $\mathbf{H}\beta = \mathbf{c}$ as

$$\beta_1 = \mathbf{H}_1^{-1}\mathbf{c} - \mathbf{H}_1^{-1}\mathbf{H}_2\beta_2. \tag{4.68}$$

This can then be substituted into $\mathrm{E}[\mathbf{y}] = \mathbf{X}\beta = \mathbf{X}_1\beta_1 + \mathbf{X}_2\beta_2$, from which the ML estimator of β_2 can be obtained as

$$\hat{\beta}_2 = (\mathbf{S'S})^{-1}\mathbf{S'}(\mathbf{y} - \mathbf{X}_1\mathbf{H}_1^{-1}\mathbf{c}), \tag{4.69}$$

where $\mathbf{S} = \mathbf{X}_2 - \mathbf{X}_1\mathbf{H}_1^{-1}\mathbf{H}_2$. Replacing β_2 in (4.68) with $\hat{\beta}_2$ of (4.69) then gives $\hat{\beta}_1$, and the resulting $\hat{\beta} = \begin{bmatrix} \hat{\beta}_1 \\ \hat{\beta}_2 \end{bmatrix}$ is identical to (4.67) (Searle, 1999).

4.13   EXERCISES

E 4.1 Write the linear model equation $E[y_{ij}] = \mu + \alpha_i + \beta_j$ for $i = 1, 2, \ldots, m$ and $j = 1, 2, \ldots, n$ in matrix form by identifying $\mathbf{X}$ and $\boldsymbol{\beta}$. For $\mathbf{X}$ use Kronecker product notation (Appendix M).

E 4.2 For *each* of the two different forms of simple linear regressions, namely

$$E[y_i] = \alpha + \gamma x_i \qquad \text{and} \qquad E[y_i] = \mu + \gamma(x_i - \bar{x})$$

for $i = 1, 2, \ldots, n$:

    (a) For each model write $E[\mathbf{y}] = \mathbf{X}\boldsymbol{\beta}$, specifying $\mathbf{X}$ and $\boldsymbol{\beta}$.

    (b) Obtain $\mathbf{X}'\mathbf{X}$, $(\mathbf{X}'\mathbf{X})^{-1}$ and $\hat{\boldsymbol{\beta}}$.

    (c) Obtain $\text{var}(\hat{\boldsymbol{\beta}})$. Do you see any advantage of one model equation over the other?

    (d) Obtain the variance of the estimated value of the mean of y_i at a new value of x, denoted as x^*.

E 4.3 For $\mathbf{X}^-$ being a generalized inverse of $\mathbf{X}$, and for any $\mathbf{z}$ of appropriate order, show for $\boldsymbol{\beta}^0$ of (4.22) that

$$\boldsymbol{\beta}_\mathbf{z} = \boldsymbol{\beta}^0 + (\mathbf{I} - \mathbf{X}^-\mathbf{X})\mathbf{z}$$

is a solution of the normal equations $\mathbf{X}'\mathbf{X}\boldsymbol{\beta} = \mathbf{X}'\mathbf{y}$.

E 4.4 From Appendix M, quote the two calculus results used in deriving (4.16), and explain how they are used.

E 4.5 For $\mathbf{X} = \begin{bmatrix} \mathbf{1}_N & \left\{ {}_d \mathbf{1}_{n_i} \right\}_{i=1}^{5} \end{bmatrix}$ show that $\mathbf{XGX}'$ has the same form as the example of Section 4.7.

E 4.6 For $E[\mathbf{y}] = \mathbf{X}\boldsymbol{\beta}$ with $\mathbf{X}$ of full rank, show that all linear combinations of $\boldsymbol{\beta}$ are estimable.

E 4.7 For the example of Section 4.6, and $\eta = \alpha_1 + 2.7\alpha_2 - 3.7\alpha_3$:

    (a) Explain why η is estimable.

    (b) What is the ML estimate of η?

E 4.8 For $\mathbf{y} \sim \mathcal{N}(\mathbf{X}\boldsymbol{\beta}, \sigma^2\mathbf{I})$ derive (4.38) for estimable $\mathbf{q}'\boldsymbol{\beta}$.

E 4.9 For the one-way classification $E[y_{ij}] = \mu + \alpha_i$

(a) Show that $\Sigma_i \lambda_i \alpha_i$ is estimable if and only if $\Sigma_i \lambda_i = 0$.

(b) For $i = 1, 2, 3$ and $j = 1, 2$ derive the generalized inverses of $\mathbf{X'X}$, which give

$$\beta_1^0 = \begin{bmatrix} \bar{y}_{..} \\ \bar{y}_{1.} - \bar{y}_{..} \\ \bar{y}_{2.} - \bar{y}_{..} \\ \bar{y}_{3.} - \bar{y}_{..} \end{bmatrix} \quad \text{and} \quad \beta_2^0 = \begin{bmatrix} \bar{y}_{3.} \\ \bar{y}_{1.} - \bar{y}_{3.} \\ \bar{y}_{2.} - \bar{y}_{3.} \\ 0 \end{bmatrix}.$$

(c) Verify that your answers in part (b) are indeed generalized inverses of $\mathbf{X'X}$.

E 4.10 Show that

$$\frac{1}{6} \begin{bmatrix} 1 & 16 & 9 & -6 \\ -1 & -14 & -9 & 6 \\ -1 & -16 & -6 & 6 \\ -1 & -16 & -9 & 12 \end{bmatrix}$$

(a) is non-singular;

(b) is a generalized inverse of $\mathbf{X'X}$ in (4.40);

(c) gives a solution of (4.40) very different from the solutions in (4.41), but yields the same estimators of $\mu + \alpha_i$.

E 4.11 Show that the mean squared error of $\hat{\sigma}^2$ is $\text{var}(\hat{\sigma}^2)$ but that of $\hat{\sigma}_{ML}^2$ is $\sigma^4[2N - 1 + (r - 1)^2]/N$.

E 4.12 Derive (4.47).

E 4.13 Derive (4.58).

E 4.14 Derive (4.60).

E 4.15 Derive (4.62).

E 4.16 Simplify (4.63) for $r(\mathbf{K'}) = 1$, and relate it to t of Section 4.10.

E 4.17 Maximize $L(\beta, \sigma^2)$ subject to $\mathbf{K'}\beta = \mathbf{m}$ to yield (4.58) and (4.59).

E 4.18 Show that Q and SSE have the distribution described below (4.63).

Chapter 5

GENERALIZED LINEAR MODELS (GLMs)

5.1 INTRODUCTION

Models for the analysis of non-normal data using nonlinear models have a long history. The use of probit regression for a binary response is a classic example. The word *probit* was traced by David (1995) as far back as Bliss (1934). Finney (1952) attributes the actual origin of probit regression to psychologists in the late 1800s.

In an early example of probit regression, Bliss (1934) describes an experiment in which nicotine is applied to aphids and the proportion killed is recorded (How is that for an early antismoking message?). As an appendix to a paper Bliss wrote a year later (Bliss, 1935), Fisher (1935) outlines the use of maximum likelihood to obtain estimates of the probit model.

However it was years before likelihood estimation for probit models caught on. Finney (1952), in an appendix entitled "Mathematical basis of the probit method" gives some of the rational for maximum likelihood and motivates a computational method that he spends six pages describing in a different appendix.

More specifically, if we let p_i denote the probability of a success for the ith observation, the model is given by

$$y_i \;\sim\; \text{indep. Bernoulli}(p_i)$$

$$p_i \;=\; \Phi\left(\mathbf{x}_i'\boldsymbol{\beta}\right), \tag{5.1}$$

where $\mathbf{x}_i'$ denotes the ith row of a matrix of predictors and $\Phi(\cdot)$ is the standard normal c.d.f. Considering the scalar functions applied elementwise to the vectors, we can rewrite (5.1) as

$$\mathbf{y} \sim \text{indep. Bernoulli}(\mathbf{p})$$

$$\mathbf{p} = \Phi(\mathbf{X}\boldsymbol{\beta}) \qquad (5.2)$$

or equivalently

$$\Phi^{-1}(\mathbf{p}) = \mathbf{X}\boldsymbol{\beta},$$

where $\mathbf{X}$ is the model matrix. The use of the inverse standard normal c.d.f., or probit, to transform the mean of $\mathbf{y}$ to the linear predictor is attractive on two counts. First, it expands the range of $\mathbf{p}$ from $[0,1]$ to the whole real line, making it more reasonable to assume a model of the form $\mathbf{X}\boldsymbol{\beta}$. Second, in many problems, the sigmoidal form of $\mathbf{p}$ as a function of the covariates is often observed in practice.

Finney suggested calculating an estimate of $\boldsymbol{\beta}$ via an iteratively weighted least squares algorithm. He recommended using *working probits* which he defined (ignoring the shift of five units historically used to keep all the calculations positive) as

$$t_i = \mathbf{x}_i'\boldsymbol{\beta} + \frac{y_i - \Phi(\mathbf{x}_i'\boldsymbol{\beta})}{\phi(\mathbf{x}_i'\boldsymbol{\beta})}, \qquad (5.3)$$

where $\phi(\cdot)$ is the standard normal probability density function (p.d.f.). The working probits for a current value of $\boldsymbol{\beta}$ were regressed on the predictors using weights given by $\dfrac{\phi(p_i)^2}{\Phi(p_i)[1 - \Phi(p_i)]}$ (see E 5.1) in order to get the new value of $\boldsymbol{\beta}$. This algorithm was iterated until convergence (or at least until the computer — a person! — got tired of performing the calculations).

Nelder and Wedderburn (1972) recognized that the working probits could be generalized in a straightforward way to unify an entire collection of maximum likelihood problems. This *generalized linear model* (GLM) could handle probit or logistic regression, Poisson regression, log-linear models for contingency tables, variance components estimation from ANOVA mean squares and many other problems in the same way.

They replaced $\Phi^{-1}(\cdot)$ with a general *link* function, $g(\cdot)$, which transforms (or links) the mean of y_i to the linear predictor. With $g_\mu(\mu)$ representing $\partial g(\mu)/\partial\mu$, they then defined a *working variate* via

$$
\begin{aligned}
t_i &\equiv g(\mu_i) + g_\mu(\mu_i)(y_i - \mu_i) \\
&= \mathbf{x}_i'\boldsymbol{\beta} + g_\mu(\mu_i)(y_i - \mu_i).
\end{aligned}
\tag{5.4}
$$

Since the second term on the right-hand side of (5.4) has expectation zero it can be regarded as an *error term* so that t_i follows a linear model, albeit with unequal variances which depend on the unknown $\boldsymbol{\beta}$. This suggests using (5.4) just like (5.3): regress $\mathbf{t}$ on $\mathbf{X}$ using a weighted linear regression (more details are given in Section 5.4e) and iterate until the estimates of $\boldsymbol{\beta}$ stabilize.

More important, it made possible a style of thinking which freed the data analyst from necessarily looking for a transformation which simultaneously achieved linearity in the predictors and normality of the distribution (as in Box and Cox, 1962).

What advantages does this have? First, it unifies what appear to be very different methodologies, which helps us to understand, use and (for those of us in the business) teach the techniques. Second, since the right-hand side of the model equation is a linear model after applying the link, many of the standard ways of thinking about linear models carry over to GLMs.

5.2   STRUCTURE OF THE MODEL

Building a generalized linear model involves three decisions:

1. What is the distribution of the data (for fixed values of the predictors and possibly after a transformation)?

2. What function of the mean will be modeled as linear in the predictors?

3. What will the predictors be?

a.   Distribution of y

Typically the vector $\mathbf{y}$ is assumed to consist of independent measurements from a distribution with density from the exponential family or similar to the exponential family:

$$y_i \sim \text{indep. } f_{Y_i}(y_i)$$

$$f_{Y_i}(y_i) = \exp\{[y_i\gamma_i - b(\gamma_i)]/\tau^2 - c(y_i, \tau)\}, \qquad (5.5)$$

where, for convenience, we have written the distribution in what is called *canonical form*. For example, for the probit model, the data would be independent Bernoulli so that $f_{Y_i}(y_i)$ would be $p_i^{y_i}(1-p_i)^{1-y_i}$, where p_i is the probability of a success and $\gamma_i = \log[p_i/(1 - p_i)]$. Most commonly used distributions can be written in the form (5.5) (see E 5.2).

b. Link function

We typically want to relate the parameters of the distribution to various predictors. We do so by modeling a transformation of the mean, μ_i, which would be some function of γ_i, as a linear model in the predictors:

$$E[y_i] = \mu_i$$

$$g(\mu_i) = \mathbf{x}'_i\boldsymbol{\beta}, \qquad (5.6)$$

where $g(\cdot)$ is a known function, called the *link function* (since it links together the mean of y_i and the linear form of predictors), $\mathbf{x}'_i$ is the ith row of the model matrix, and $\boldsymbol{\beta}$ is the parameter vector in the linear predictor. In the probit example $g(\mu) = \Phi^{-1}(\mu)$ and $\mu = 1/(1 + \exp[-\gamma])$.

c. Predictors

In practice, of course, one must make decisions as to which predictors to include on the right-hand side of (5.6) and in what form to include them. For example, in the classic paper of Bliss (1934) the suggested predictor of survival is log nicotine dose as opposed to nicotine itself.

A key point in using GLMs is that many of the considerations in modeling are the same as for LMMs since the right-hand sides of the model equations for the mean are the same. For example, issues of how to represent predictors and interactions, whether and how to model non-linear relationships and (as we will see in Chapter 7) the incorporation of random factors.

d.   Linear models

This generalized class of models subsumes the linear model of Chapter 4 as a special case. The normal distribution can be written in the form (5.5) by defining:

$$\gamma_i \;=\; \mu_i$$

$$b(\gamma_i) \;=\; \tfrac{1}{2}\mu_i^2$$

$$\tau^2 \;=\; \sigma^2 \tag{5.7}$$

$$c(y_i, \tau) \;=\; \tfrac{1}{2}\log 2\pi\sigma^2 + \tfrac{1}{2}y_i^2/\sigma^2.$$

With $g(\mu_i) = \mu_i$ and $\mu_i = \mathbf{x}_i'\boldsymbol{\beta}$ we generate the linear model of Section 4.3.

5.3   TRANSFORMING VERSUS LINKING

In its earliest incarnations, probit analysis was little more than a transformation technique. It was realized that the frequent sigmoidal shape in plots of observed proportions of successes plotted against a predictor x could be made into a straight line by applying a transformation corresponding to the inverse of the normal c.d.f. However, one of the main ideas of GLMs is to get away from the idea of transforming the data. The strategy, then, is to apply a link function to the mean of the response and fit the resulting model by the method of maximum likelihood.

5.4   ESTIMATION BY MAXIMUM LIKELIHOOD

a.   Likelihood

The log likelihood for (5.5) is given by

$$l = \sum_{i=1}^{n}[y_i\gamma_i - b(\gamma_i)]/\tau^2 - \sum_{i=1}^{n}c(y_i, \tau). \tag{5.8}$$

b. Some useful identities

Before we derive the maximum likelihood equations it is useful to establish some identities. These flow from the results

$$\mathrm{E}\left[\frac{\partial \log f_{Y_i}(y_i)}{\partial \gamma_i}\right] = 0, \tag{5.9}$$

and

$$\mathrm{var}\left(\frac{\partial \log f_{Y_i}(y_i)}{\partial \gamma_i}\right) = -\mathrm{E}\left[\frac{\partial^2 \log f_{Y_i}(y_i)}{\partial \gamma_i^2}\right] \tag{5.10}$$

which hold under regularity conditions (Casella and Berger, 1990, p. 308). Using (5.5) in (5.9) gives

$$\mathrm{E}\left[\left\{y_i - \frac{\partial b(\gamma_i)}{\partial \gamma_i}\right\}\Big/ \tau^2\right] = 0 \tag{5.11}$$

or

$$\mathrm{E}[y_i] = \mu_i = \frac{\partial b(\gamma_i)}{\partial \gamma_i}. \tag{5.12}$$

And using (5.5) in (5.10) we obtain

$$\mathrm{var}\left(\left\{y_i - \frac{\partial b(\gamma_i)}{\partial \gamma_i}\right\}\Big/ \tau^2\right) = -\mathrm{E}\left[-\frac{1}{\tau^2}\frac{\partial^2 b(\gamma_i)}{\partial \gamma_i^2}\right], \tag{5.13}$$

which, using (5.12) gives

$$\mathrm{var}\left(\frac{y_i - \mu_i}{\tau^2}\right) = \frac{1}{\tau^2}\frac{\partial^2 b(\gamma_i)}{\partial \gamma_i^2}$$

or

$$\mathrm{var}(y_i) = \tau^2 \frac{\partial^2 b(\gamma_i)}{\partial \gamma_i^2} \tag{5.14}$$

$$\equiv \tau^2 v(\mu_i),$$

wherein we define $v(\mu_i)$ as $\partial^2 b(\gamma_i)/\partial \gamma_i^2$. Note that $v(\mu_i)$ is often called the *variance function*, since it indicates how the variance of y_i depends on the mean of y_i. Two other useful identities are

$$\frac{\partial \gamma_i}{\partial \mu_i} = \left(\frac{\partial \mu_i}{\partial \gamma_i}\right)^{-1} = \left(\frac{\partial^2 b(\gamma_i)}{\partial \gamma_i^2}\right)^{-1} = \frac{1}{v(\mu_i)} \tag{5.15}$$

and, using the chain rule and (5.6),

$$\frac{\partial \mu_i}{\partial \beta} = \frac{\partial \mu_i}{\partial g(\mu_i)} \frac{\partial g(\mu_i)}{\partial \beta} = \left(\frac{\partial g(\mu_i)}{\partial \mu_i}\right)^{-1} \frac{\partial \mathbf{x}_i' \beta}{\partial \beta}$$

$$= \left(\frac{\partial g(\mu_i)}{\partial \mu_i}\right)^{-1} \mathbf{x}_i' \ . \tag{5.16}$$

As an illustration of these results, consider the linear model in Section 5.1d. With subscripts denoting derivatives we have $b_\gamma(\gamma_i)$ equal to μ_i, the mean, and $b_{\gamma\gamma}(\gamma_i) = 1$ so that, from (5.14), $\text{var}(y_i) = \tau^2 b_{\gamma\gamma}(\gamma_i) = \sigma^2$, as expected. Also, $\partial \gamma_i / \partial \mu_i = \partial \mu_i / \partial \mu_i = 1 = v(\mu_i)^{-1}$, verifying (5.15) and, with $g_\mu(\mu_i) = 1$, $\partial \mu_i / \partial \beta = \mathbf{x}_i'$ as in (5.16). Note that the normal distribution has an unusual feature among distributions given by (5.5): its variance is a constant and not a function of the mean.

c.   Likelihood equations

We are now in a position to derive the maximum likelihood equations for β. From (5.8) we have

$$\frac{\partial l}{\partial \beta} = \frac{1}{\tau^2} \sum \left[y_i \frac{\partial \gamma_i}{\partial \beta} - \frac{\partial b(\gamma_i)}{\partial \gamma_i} \frac{\partial \gamma_i}{\partial \beta} \right]$$

$$= \frac{1}{\tau^2} \sum (y_i - \mu_i) \frac{\partial \gamma_i}{\partial \beta} \qquad \text{using (5.12)}$$

$$= \frac{1}{\tau^2} \sum (y_i - \mu_i) \frac{\partial \gamma_i}{\partial \mu_i} \frac{\partial \mu_i}{\partial \beta} \qquad \text{using the chain rule}$$

$$= \frac{1}{\tau^2} \sum \frac{(y_i - \mu_i)}{v(\mu_i) g_\mu(\mu_i)} \mathbf{x}_i' \qquad \text{using (5.15) and (5.16)}$$

$$= \frac{1}{\tau^2} \sum (y_i - \mu_i) w_i g_\mu(\mu_i) \mathbf{x}_i', \tag{5.17}$$

upon defining $w_i = [v(\mu_i) g_\mu^2(\mu_i)]^{-1}$.

We can write this in matrix notation as

$$\frac{\partial l}{\partial \beta} = \frac{1}{\tau^2} \mathbf{X}' \mathbf{W} \mathbf{\Delta} (\mathbf{y} - \mathbf{\mu}), \tag{5.18}$$

with $\mathbf{W} = \left\{ {}_d w_i \right\}$ and $\boldsymbol{\Delta} = \left\{ {}_d g_\mu(\mu_i) \right\}$.

The ML equations are thus given by

$$\mathbf{X}'\mathbf{W}\boldsymbol{\Delta}\mathbf{y} = \mathbf{X}'\mathbf{W}\boldsymbol{\Delta}\boldsymbol{\mu}, \qquad (5.19)$$

where $\mathbf{W}, \boldsymbol{\Delta}$ and $\boldsymbol{\mu}$ involve the unknown $\boldsymbol{\beta}$. Typically these are non-linear functions of $\boldsymbol{\beta}$ and so (5.19) cannot be solved analytically.

For example, for the probit model of (5.2), the log likelihood and its derivative are

$$l = \sum \left(y_i \left\{ \log \Phi(\mathbf{x}_i'\boldsymbol{\beta}) - \log[1 - \Phi(\mathbf{x}_i'\boldsymbol{\beta})] \right\} + \log[1 - \Phi(\mathbf{x}_i'\boldsymbol{\beta})] \right) \quad (5.20)$$

and

$$
\begin{aligned}
\frac{\partial l}{\partial \boldsymbol{\beta}} &= \sum \left[y_i \left(\frac{\phi(\mathbf{x}_i'\boldsymbol{\beta})}{\Phi(\mathbf{x}_i'\boldsymbol{\beta})}\mathbf{x}_i' + \frac{\phi(\mathbf{x}_i'\boldsymbol{\beta})}{1 - \Phi(\mathbf{x}_i'\boldsymbol{\beta})}\mathbf{x}_i' \right) - \frac{\phi(\mathbf{x}_i'\boldsymbol{\beta})}{1 - \Phi(\mathbf{x}_i'\boldsymbol{\beta})}\mathbf{x}_i' \right] \\
&= \sum \frac{[y_i - \Phi(\mathbf{x}_i'\boldsymbol{\beta})]\phi(\mathbf{x}_i'\boldsymbol{\beta})}{\Phi(\mathbf{x}_i'\boldsymbol{\beta})[1 - \Phi(\mathbf{x}_i'\boldsymbol{\beta})]}\mathbf{x}_i'l \\
&= \sum \frac{(y_i - \mu_i)\phi(\mathbf{x}_i'\boldsymbol{\beta})}{\mu_i(1 - \mu_i)}\mathbf{x}_i'. \qquad (5.21)
\end{aligned}
$$

Identifying $b(\gamma_i)$ of (5.5) as $\log(1+e^{\gamma_i})$ so that $b_\gamma(\gamma_i) = (1+e^{-\gamma_i})^{-1} = \mu_i$ and $b_{\gamma\gamma}(\gamma_i) = \mu_i(1 - \mu_i)$, it is straightforward (see E 5.5) to show that (5.21) is of the form of (5.18).

For solving the ML equations or for deriving the large-sample variance of $\hat{\boldsymbol{\beta}}$, it is useful to have the expected value of the second derivative of the log likelihood. Differentiating (5.18) and using the chain rule we obtain:

$$\frac{\partial^2 l}{\partial \boldsymbol{\beta} \, \partial \boldsymbol{\beta}'} = -\frac{1}{\tau^2}\mathbf{X}'\mathbf{W}\boldsymbol{\Delta}\frac{\partial \boldsymbol{\mu}}{\partial \boldsymbol{\beta}'} + \frac{1}{\tau^2}\mathbf{X}'\frac{\partial \mathbf{W}\boldsymbol{\Delta}}{\partial \boldsymbol{\beta}'}(\mathbf{y} - \boldsymbol{\mu}) \qquad (5.22)$$

so that

$$
\begin{aligned}
-\mathrm{E}\left[\frac{\partial^2 l}{\partial \boldsymbol{\beta} \, \partial \boldsymbol{\beta}'} \right] &= \frac{1}{\tau^2}\mathbf{X}'\mathbf{W}\boldsymbol{\Delta}\frac{\partial \boldsymbol{\mu}}{\partial \boldsymbol{\beta}'} + 0 \\
&= \frac{1}{\tau^2}\mathbf{X}'\mathbf{W}\boldsymbol{\Delta}\boldsymbol{\Delta}^{-1}\mathbf{X} \quad \text{using (5.16)} \\
&= \frac{1}{\tau^2}\mathbf{X}'\mathbf{W}\mathbf{X}, \qquad (5.23)
\end{aligned}
$$

where, again, $\mathbf{W} = \left\{ {}_d w_i \right\} = \left\{ {}_d [v(\mu_i)g_\mu^2(\mu_i)]^{-1} \right\}$.

d.   Large-sample variances

To derive the large-sample variance of $\hat{\boldsymbol{\beta}}$ we first note that

$$-\mathrm{E}\left[\frac{\partial^2 l}{\partial\boldsymbol{\beta}\,\partial\tau^2}\right] \;=\; -\mathrm{E}\left[\frac{\partial}{\partial\tau^2}\frac{1}{\tau^2}\mathbf{X}'\mathbf{W}\boldsymbol{\Delta}(\mathbf{y}-\boldsymbol{\mu})\right]$$

$$=\; \frac{1}{\tau^4}\mathbf{X}'\mathbf{W}\boldsymbol{\Delta}\,\mathrm{E}\,[\mathbf{y}-\boldsymbol{\mu}] \tag{5.24}$$

$$=\; \mathbf{0},$$

so that estimation of τ^2 does not affect the large-sample variance of $\hat{\boldsymbol{\beta}}$. The usual large-sample arguments (see Section S.4c of Appendix S), along with (5.23) and (5.24), show that (see E 5.7)

$$\mathrm{var}_\infty(\hat{\boldsymbol{\beta}}) = \tau^2(\mathbf{X}'\mathbf{W}\mathbf{X})^{-1}, \tag{5.25}$$

where var_∞ indicates the limiting or asymptotic variance.

e.   Solving the ML equations

Solution of the ML equations, (5.19), for $\boldsymbol{\beta}$ is usually performed by an iterative weighted least squares method. This can be derived as an example of the use of Fisher scoring (Searle et al., 1992, p. 295). Fisher scoring is an iterative method for maximizing a likelihood and it takes the form

$$\boldsymbol{\theta}^{(m+1)} = \boldsymbol{\theta}^{(m)} + \mathbf{I}(\boldsymbol{\theta}^{(m)})^{-1}\frac{\partial l}{\partial\boldsymbol{\theta}}\bigg|_{\boldsymbol{\theta}=\boldsymbol{\theta}^{(m)}}, \tag{5.26}$$

where (m) indicates the mth iteration, $\mathbf{I}(\boldsymbol{\theta})$ is the information matrix and $\boldsymbol{\theta}$ is the entire parameter vector.

Using (5.24), (5.23), and (5.18) the portion of the equation for $\boldsymbol{\beta}$ (see E 5.8) is of the form

$$\boldsymbol{\beta}^{(m+1)} = \boldsymbol{\beta}^{(m)} + (\mathbf{X}'\mathbf{W}\mathbf{X})^{-1}\mathbf{X}'\mathbf{W}\boldsymbol{\Delta}(\mathbf{y}-\boldsymbol{\mu}), \tag{5.27}$$

where it is understood that $\mathbf{W}$, $\boldsymbol{\Delta}$, and $\boldsymbol{\mu}$ are evaluated at $\boldsymbol{\beta}^{(m)}$.

How does this relate to the working variate of (5.4)? We have

$$\mathbf{t} \;=\; \mathbf{X}\boldsymbol{\beta} + \boldsymbol{\Delta}(\mathbf{y}-\boldsymbol{\mu}) \tag{5.28}$$

so that

$$
\begin{aligned}
\operatorname{var}(\mathbf{t}) &= \operatorname{var}[\boldsymbol{\Delta}(\mathbf{y} - \boldsymbol{\mu})] \\
&= \left\{ {}_d \tau^2 v(\mu_i) g_\mu^2(\mu_i) \right\}, \qquad \text{using (5.14)} \\
&= \tau^2 \mathbf{W}^{-1}, \qquad\qquad\qquad\qquad\quad (5.29)
\end{aligned}
$$

so a weighted regression of $\mathbf{t}$ on $\mathbf{X}$ using weights equal to the inverse of the variance of $\mathbf{t}$ gives

$$
\begin{aligned}
\boldsymbol{\beta}^{(m+1)} &= (\mathbf{X}'\mathbf{W}\mathbf{X})^{-1}\mathbf{X}'\mathbf{W}[\mathbf{X}\boldsymbol{\beta}^{(m)} + \boldsymbol{\Delta}(\mathbf{y} - \boldsymbol{\mu})] \\
&= \boldsymbol{\beta}^{(m)} + (\mathbf{X}'\mathbf{W}\mathbf{X})^{-1}\mathbf{X}'\mathbf{W}\boldsymbol{\Delta}(\mathbf{y} - \boldsymbol{\mu}), \qquad (5.30)
\end{aligned}
$$

which is the same as (5.27).

f. Example: Potato flour dilutions

Finney (1971) gives an example of the growth of spores in a potato flour suspension. For each of 10 dilutions, five plates are tested for positive growth. The data are given in Table 5.1. As the flour suspensions get more concentrated, the probability of growth (i.e., proportion of positive plates) increases. Figure 5.1 shows that the probability of response, as a function of the natural logarithm of dilution, follows a roughly sigmoidal shape, so we might entertain a logistic regression model. Let y_i denote the number of plates out of five that show a positive response. A possible model is

$$
\mathrm{E}[y_i] = 5\pi(x_i) = 5\frac{1}{1 + e^{-(\alpha + \beta x_i)}} \qquad (5.31)
$$

$$
y_i \sim \text{indep. binomial } [5, \pi(x_i)].
$$

The log likelihood for this model is given by

$$
\begin{aligned}
l &= \sum \left[\log \binom{5}{y_i} + y_i(\alpha + \beta x_i) - 5\log(1 + e^{\alpha + \beta x_i}) \right] \\
&= c + \alpha \sum y_i + \beta \sum y_i x_i - 5\sum \log(1 + e^{\alpha + \beta x_i}), \qquad (5.32)
\end{aligned}
$$

Table 5.1: Potato Flour Data

Dilution	Spore Growth		Proportion
(g/100 ml	No. of Plates	No. Positive	of Residual Plates
1/128	5	0	0.0
1/64	5	0	0.0
1/32	5	2	0.4
1/16	5	2	0.4
1/8	5	3	0.6
1/4	5	4	0.8
1/2	5	5	1.0
1	5	5	1.0
2	5	5	1.0
4	5	5	1.0

where $c = \sum \binom{5}{y_i}$ is a function of the y_i but is constant in α and β. Figure 5.2 shows the log likelihood, (5.32), plotted as a function of α and β, from which the approximate maximizing values can be read. The ML equations are given by

$$\sum y_i = \sum \frac{5}{1 + e^{-(\hat{\alpha} + \hat{\beta} x_i)}}$$

$$\sum y_i x_i = \sum \frac{5 x_i}{1 + e^{-(\hat{\alpha} + \hat{\beta} x_i)}}. \qquad (5.33)$$

With $\sum y_i = 31$ and $\sum y_i x_i \doteq -17.329$ it is merely tedious arithmetic to verify that $\hat{\alpha} = 4.17$ and $\hat{\beta} = 1.62$ solve these equations to within rounding error. Figure 5.3 shows a plot of the data and the model, (5.31), using the ML estimates for α and β.

To illustrate the large-sample variance calculation note that

$$\tau^2 = 1$$

$$v(\mu_i) = \mu_i(1 - \mu_i)$$

$$g_\mu(\mu_i) = 1/v(\mu_i)$$

Figure 5.1: Proportion of positive spore growth plotted against log dilution for the potato flour data.

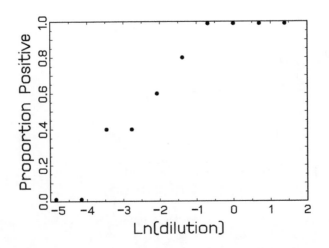

Figure 5.2: Log likelihood plotted against parameters for the potato flour data.

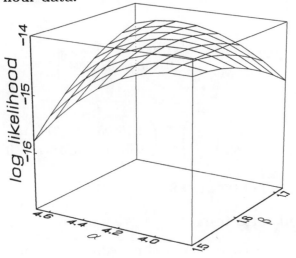

Figure 5.3: Proportion positive versus log dilution for the potato flour data.

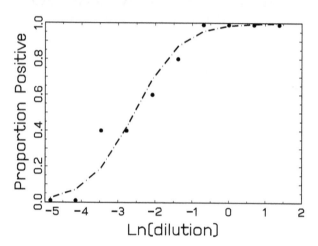

so that $\mathbf{W} = \left\{ _d \mu_i(1 - \mu_i) \right\}$. We thus have

$$\mathbf{X'WX} = \left[\begin{array}{cc} \sum \mu_i(1 - \mu_i) & \sum x_i\mu_i(1 - \mu_i) \\ \sum x_i\mu_i(1 - \mu_i) & \sum x_i^2\mu_i(1 - \mu_i) \end{array} \right]$$

$$= \left[\begin{array}{cc} 4.38306 & -11.09943 \\ -11.09943 & 32.89365 \end{array} \right]$$

with inverse

$$(\mathbf{X'WX})^{-1} = \left[\begin{array}{cc} 1.56805 & 0.52911 \\ 0.52911 & 0.20894 \end{array} \right].$$

This gives

$$\left(\begin{array}{c} \hat{\alpha} \\ \hat{\beta} \end{array} \right) \sim \mathcal{AN} \left[\left(\begin{array}{c} \alpha \\ \beta \end{array} \right), \left(\begin{array}{cc} 1.56805 & 0.52911 \\ 0.52911 & 0.20894 \end{array} \right) \right].$$

5.5   TESTS OF HYPOTHESES

a.   Likelihood ratio tests

Likelihood ratio tests follow the usual prescription of comparing the maximized values of the log likelihood both under H_0 and not restricted to H_0. If the difference is large (i.e., the unrestricted model fit is much better) then we reject H_0.

When there are multiple parameters we will often be interested in hypotheses concerning only a subset of the parameters. Accordingly, let the parameter vector $\boldsymbol{\theta}$ be partitioned into two components $\boldsymbol{\theta}' = (\boldsymbol{\theta}_1', \boldsymbol{\theta}_2')$ and suppose interest focuses on $\boldsymbol{\theta}_1$ while $\boldsymbol{\theta}_2$ is left unspecified. $\boldsymbol{\theta}_2$ is often called a *nuisance parameter*. Either or both of $\boldsymbol{\theta}_1$ and $\boldsymbol{\theta}_2$ could be vector-valued and, if the entire parameter vector is of interest, $\boldsymbol{\theta}_2$ could be null.

Suppose our hypothesis is of the form $H_0 : \boldsymbol{\theta}_1 = \boldsymbol{\theta}_{1,0}$, where $\boldsymbol{\theta}_{1,0}$ is a specified value of $\boldsymbol{\theta}_1$ and let $\hat{\boldsymbol{\theta}}_{2,0}$ be the MLE of $\boldsymbol{\theta}_2$ under the restriction that $\boldsymbol{\theta}_1 = \boldsymbol{\theta}_{1,0}$. The likelihood ratio test statistic is given by

$$-2\log\Lambda = -2\left[l(\boldsymbol{\theta}_{1,0}, \hat{\boldsymbol{\theta}}_{2,0}) - l(\hat{\boldsymbol{\theta}}_1, \hat{\boldsymbol{\theta}}_2)\right], \tag{5.34}$$

where $\hat{\boldsymbol{\theta}}' = (\hat{\boldsymbol{\theta}}_1', \hat{\boldsymbol{\theta}}_2')$ and the large-sample critical region of the test is to reject H_0 in favor of the alternative when

$$-2\log\Lambda > \chi^2_{\nu,1-\alpha}, \tag{5.35}$$

where ν is the dimension of $\boldsymbol{\theta}_1$.

b. Wald tests

An alternative method of testing is to use the large-sample normality of the ML estimator in order to form a test. From standard results (Appendix S)

$$\hat{\boldsymbol{\theta}}' \sim \mathcal{AN}[\boldsymbol{\theta}, \mathbf{I}^{-1}(\boldsymbol{\theta})], \tag{5.36}$$

where $\mathbf{I}(\boldsymbol{\theta})$ is the Fisher information for $\hat{\boldsymbol{\theta}}$. Again, if we write $\boldsymbol{\theta}' = (\boldsymbol{\theta}_1', \boldsymbol{\theta}_2')$, and write conformably

$$\mathbf{I}(\boldsymbol{\theta}) = \begin{bmatrix} \mathbf{I}_{11} & \mathbf{I}_{12} \\ \mathbf{I}_{21} & \mathbf{I}_{22} \end{bmatrix}, \tag{5.37}$$

then standard matrix algebra for partitioned matrices (Searle, 1982, p. 354) and multivariate normal calculations show that the large-sample variance of $\hat{\boldsymbol{\theta}}_1$ is given by

$$\text{var}_\infty(\hat{\boldsymbol{\theta}}_1) = \left(\mathbf{I}_{11} - \mathbf{I}_{12}\mathbf{I}_{22}^{-1}\mathbf{I}_{21}\right)^{-1}. \tag{5.38}$$

To test $H_0 : \boldsymbol{\theta}_1 = \boldsymbol{\theta}_{1,0}$ we form the Wald statistic

$$W = (\hat{\boldsymbol{\theta}}_1 - \boldsymbol{\theta}_{1,0})'[\text{var}_\infty(\hat{\boldsymbol{\theta}}_1)]^{-1}(\hat{\boldsymbol{\theta}}_1 - \boldsymbol{\theta}_{1,0}), \tag{5.39}$$

which, under H_0, has the same large-sample χ^2 distribution as the LRT with degrees of freedom equal to the dimension of $\boldsymbol{\theta}_1$. More explicitly we would reject the $H_0 : \boldsymbol{\theta}_1 = \boldsymbol{\theta}_{1,0}$ if

$$W > \chi^2_{\nu,1-\alpha}. \tag{5.40}$$

Both the LRT and the Wald tests are available to test the same hypotheses and have the same limiting distribution. What are the differences? For large samples, and if the deviation from the null hypothesis is not too extreme, the two test statistics will give similar, though not identical results (Bishop et al., 1975, Sect. 14.9). However, for small samples or for extreme deviations, they can differ. Generally, investigations have shown (Cox and Hinkley, 1974; McCullagh and Nelder, 1989) that use of the large sample-distribution for the LRT gives a more accurate approximation for small and moderate-sized samples than for the Wald test. LRTs also connect directly with statistical theory for optimal hypothesis testing and this approach often leads to test with desirable properties, such as maximal power. The LRT is thus to be preferred. The Wald test does, however, have a computational advantage since it does not require calculation of $\hat{\boldsymbol{\theta}}_{2,0}$.

c.   Illustration of tests

We use the potato flour data to illustrate these tests for the null hypothesis $H_0 : \beta = 0$, that is, no relationship between spore growth and log dilution. To perform the likelihood ratio test we must maximize the likelihood under the null hypothesis, that is, when the probability of growth is constant. Under $H_0, \hat{\alpha} \doteq 0.4896$ (see E 5.6). We thus have

$$l(\boldsymbol{\theta}_{1,0}, \hat{\boldsymbol{\theta}}_{2,0}) = l(0.50, 0) = -33.20,$$

while

$$l(\hat{\boldsymbol{\theta}}_1, \hat{\boldsymbol{\theta}}_2) = l(\hat{\alpha}, \hat{\beta}) = l(4.17, 1.62) = -14.214.$$

The LRT statistic is thus $-2 \log \Lambda = -2[-33.20 - (-14.21)] = 37.88$. The statistic has 1 degree of freedom, which is the dimension of β. So we easily reject H_0 at any usual level of significance and the p-value is $P\{\chi^2_1 \geq 37.88\} \doteq 0$.

The Wald test statistic uses $\hat{\beta} = 1.62$ from below (5.33) and the large sample variance, $\text{var}(\hat{\beta})_\infty = 0.2089$, from the end of Section 5.4. Substituting in (5.39) we then have $W = (1.62)(0.2089)^{-1}(1.62) = 1.62^2/0.2089 = 12.6$. This has a p-value of $P\{\chi_1^2 \geq 12.6\} \doteq 0.0004$, which again corresponds to rejection of the null hypothesis at the usual significance levels. This illustrates that the two test statistics need not be numerically similar for large deviations from the null hypothesis. However the same qualitative conclusion would ordinarily be reached.

d. Confidence intervals

Either the LRT or Wald test can be used to construct large-sample confidence intervals for $\boldsymbol{\theta}_1$. For the Wald test we include in the confidence set all values of $\boldsymbol{\theta}_1$ such that

$$(\hat{\boldsymbol{\theta}}_1 - \boldsymbol{\theta}_1)'[\text{var}_\infty(\hat{\boldsymbol{\theta}}_1)]^{-1}(\hat{\boldsymbol{\theta}}_1 - \boldsymbol{\theta}_1) \leq \chi_{\nu,1-\alpha}^2. \tag{5.41}$$

For the LRT we include in the confidence set all values $\boldsymbol{\theta}_1$ such that

$$-2\left[l(\boldsymbol{\theta}_1, \hat{\boldsymbol{\theta}}_{2,0}) - l(\hat{\boldsymbol{\theta}}_1, \hat{\boldsymbol{\theta}}_2)\right] \leq \chi_{\nu,1-\alpha}^2. \tag{5.42}$$

In (5.42) $\hat{\boldsymbol{\theta}}_2$ represents the MLE of $\boldsymbol{\theta}_2$ for each value of $\boldsymbol{\theta}_1$ checked for inclusion in the set. The computational burden of the likelihood-based confidence interval is thus larger than that for the Wald-based interval. However, as with LRTs, the small and moderate-sized sample performance of the LRT-based confidence region has generally been found to be better.

e. Illustration of confidence intervals

The Wald-based confidence interval for β is straightforward since it is based on

$$\hat{\beta} \sim \mathcal{AN}(\beta, 0.2089),$$

which gives a confidence interval of $1.62 \pm 1.96(0.2089)^{1/2} = (0.72, 2.52)$.

The likelihood-based confidence interval must solve for the values of β such that

$$-2[l(\beta, \hat{\alpha}_\beta) - l(\hat{\alpha}, \hat{\beta})] \leq 3.84,$$

where $\hat{\alpha}_\beta$ denotes the MLE of α when β is fixed at some value. Numerical calculations give the interval as $(0.90, 2.76)$. This interval is approximately the same length as the Wald interval but is not symmetrically placed about the MLE, an indication of the non-normality of the sampling distribution.

5.6   MAXIMUM QUASI-LIKELIHOOD

a.   Introduction

In some statistical investigations, such as the potato flour example of Section 5.5, we know the distribution of the data (binomial with $n = 5$ in that instance). In others we are less certain. For example, in analyzing data on costs of hospitalization we know the data are positive (though it would be nice to be paid for some hospital ordeals!) and they are invariably skewed right. With a little more experience with such data we would know that the variance increases with the mean and we might have a rough idea as to how quickly it increases. However, we are unlikely to know *a priori* exactly what distributional form is correct or even likely to fit well. But not knowing the distribution makes it impossible to construct a likelihood and thus to use such techniques as maximum likelihood and likelihood ratio tests.

It would therefore be useful to have inferential methods which work as well or almost as well as ML but without having to make specific distributional assumptions. This is the basic idea behind *quasi-likelihood*: to derive a likelihood-like quantity whose construction requires few assumptions.

What are the important characteristics of likelihood which are required to generate workable estimators? It turns out to be easier to mimic the properties of the derivative of the log likelihood (also called the *score function*) rather than the likelihood itself.

b.   Definition

We define an analog of likelihood using (5.9) and (5.10), except that we differentiate with respect to μ_i instead of γ_i. First, from (5.9) we want

$$E\left[\frac{\partial \log f_{Y_i}(y_i)}{\partial \mu_i}\right] = 0. \tag{5.43}$$

Then we observe that by the chain rule, what we will denote as v^* is

$$v^* = \text{var}\left(\frac{\partial \log f_{Y_i}(y_i)}{\partial \mu_i}\right) = \text{var}\left(\frac{\partial \log f_{Y_i}(y_i)}{\partial \gamma_i}\frac{\partial \gamma_i}{\partial \mu_i}\right)$$

$$= \left[\mathrm{var}\left(\frac{\partial \log f_{Y_i}(y_i)}{\partial \gamma_i}\right)\right]\left(\frac{\partial \gamma_i}{\partial \mu_i}\right)^2 \quad (5.44)$$

and using (5.10)

$$= \left(-E\left[\frac{\partial^2 \log f_{Y_i}(y_i)}{\partial \gamma_i^2}\right]\right)\left(\frac{\partial \gamma_i}{\partial \mu_i}\right)^2.$$

Now, by the nature of $f_{Y_i}(y_i)$ in (5.5), with $b(\gamma_i)$ containing no data this is

$$v^* = \frac{1}{\tau^2}\left[\frac{\partial^2 b(\gamma_i)}{\partial \gamma_i^2}\right]\left(\frac{\partial \gamma_i}{\partial \mu_i}\right)^2, \quad (5.45)$$

and from the definition of $v(\mu_i)$ below (5.14) this becomes

$$v^* = \frac{v(\mu_i)}{\tau^2}\left(\frac{\partial \gamma_i}{\partial \mu_i}\right)^2$$

$$= \frac{v(\mu_i)}{\tau^2}\frac{1}{v(\mu_i)^2} \quad \text{from (5.15).} \quad (5.46)$$

Thus

$$\mathrm{var}\left(\frac{\partial \log f_{Y_i}(y_i)}{\partial \mu_i}\right) = \frac{1}{\tau^2 v(\mu_i)}, \quad (5.47)$$

or, by (5.10) and using $\partial \mu_i$ in place of $\partial \gamma_i$,

$$\mathrm{var}\left(\frac{\partial \log f_{Y_i}(y_i)}{\partial \mu_i}\right) = -E\left[\frac{\partial^2 \log f_{Y_i}(y_i)}{\partial \mu_i^2}\right] = \frac{1}{\tau^2 v(\mu_i)}. \quad (5.48)$$

Observe that (5.43) and (5.48) are the analogs of (5.9) and (5.10).

We thus seek a quantity in place of $\partial \log f_{Y_i}(y_i)/\partial \mu_i$ which has properties (5.43) and (5.48). It is straightforward to verify (see E 5.9) that

$$q_i = \frac{y_i - \mu_i}{\tau^2 v(\mu_i)} \quad (5.49)$$

satisfies these same conditions, where we assume that $\mathrm{var}(y_i) \propto v(\mu_i)$. The τ that appears in (5.49) is merely the (unspecified) constant of proportionality relating $\mathrm{var}(y_i)$ to $v(\mu_i)$, which is not exactly the same as the τ that appears in the density (5.5). However, we will use the same notation since, as we see below, they play the same role.

Since the contribution to the log likelihood from y_i is the integral with respect to μ_i of $\partial \log f_{Y_i}(y_i)/\partial \mu_i$, we define the log quasi-likelihood via the contribution y_i makes to it:

$$Q_i = \int_{y_i}^{\mu_i} \frac{y_i - t}{\tau^2 v(t)} dt, \tag{5.50}$$

which, by definition, has derivative with respect to μ_i equal to q_i. Finally, to find the *maximum quasi-likelihood* (MQL) estimator of β we solve the *maximum quasi-likelihood equations*

$$\frac{\partial}{\partial \beta} \sum Q_i = \mathbf{0}. \tag{5.51}$$

Evaluating the derivative in (5.51) gives

$$\sum \frac{y_i - \mu_i}{\tau^2 v(\mu_i)} \frac{\partial \mu_i}{\partial \beta} = \mathbf{0},$$

which, using (5.16), is the same as

$$\sum \frac{y_i - \mu_i}{\tau^2 v(\mu_i) g_\mu(\mu_i)} \mathbf{x}_i' = \mathbf{0}, \tag{5.52}$$

or, in matrix notation,

$$\frac{1}{\tau^2} \mathbf{X}' \mathbf{W} \boldsymbol{\Delta} (\mathbf{y} - \boldsymbol{\mu}) = \mathbf{0}, \tag{5.53}$$

the same as (5.18). Note that by defining maximum quasi-likelihood estimators as solutions to the maximum quasi-likelihood equations, (5.51), we avoid a true maximization problem or even the definition of a quasi-likelihood or log quasi-likelihood itself.

In some ways this is a remarkable result. Q_i is constructed using only information about how the variance changes with the mean and nothing more. And, it is often the case that if we specify a mean-to-variance relationship, we obtain maximum quasi-likelihood equations which are exactly the same as those corresponding to a legitimate likelihood.

For example, suppose we are willing to assume the mean and variance are equal, so that what we build into quasi-likelihood is the fact that $v(\mu_i) = \mu_i$. Note that this allows the variance to be merely proportional to the mean rather than exactly equal to it, so that

$$Q_i = \int_{y_i}^{\mu_i} \frac{y_i - t}{\tau^2 t} dt,$$

$$= \left. \frac{1}{\tau^2}(y_i \log t - t) \right|_{y_i}^{\mu_i}$$

$$= \frac{1}{\tau^2}(y_i \log \mu_i - \mu_i - y_i \log y_i + y_i) \tag{5.54}$$

and the MQL equations for β are

$$\frac{\partial}{\partial \beta} \sum (y_i \log \mu_i - \mu_i) = \mathbf{0} \tag{5.55}$$

(the other terms dropping out). Instead of merely assuming that $v(\mu_i) = \mu_i$ suppose we make the assumption that $y_i \sim \text{Poisson}(\mu_i)$, which would actually force $\text{var}(y_i) = \mu_i$ as well. Then $\log f_{Y_i}(y_i) = y_i \log \mu_i - \mu_i - \log(y_i!)$ and the ML equations would be

$$\frac{\partial}{\partial \beta} \sum (y_i \log \mu_i - \mu_i) = \mathbf{0}, \tag{5.56}$$

which are the same as the MQL equations, (5.55)! In this case MQL and ML would give exactly the same estimates and hence MQL would be fully efficient. In other cases (see E 5.3) ML does not give equations of the form (5.19) and, in those cases, MQL may not be fully efficient. See E 5.11 for some simple calculations and Firth (1987) for more detail.

MQL has important advantages over ML. To explain, consider again the specific situation of regression with a Poisson-distributed response. ML would assume $\text{var}(y_i) = v(\mu_i)$. However, in practice it is often true that data appear selected from a distribution in which the variance is larger than the mean. If the variance is proportional to the mean, the specification of the model under quasi-likelihood is still correct because the assumption is only that $\text{var}(y_i) = \tau^2 v(\mu_i)$; that is, $\text{var}(y_i)$ is proportional to $v(\mu_i)$, not necessarily equal.

Thus MQL affords us two degrees of robustness. First, we need not make a distributional assumption and second, we have only to specify the mean-to-variance relationship up to a proportionality constant which can be estimated from the data (see below).

Inference using MQL proceeds much as ML for β. Under mild conditions (McCullagh, 1983) it can be shown that

$$\tilde{\beta} \sim \mathcal{AN}\left[\beta, \tau^2 (\mathbf{X'WX})^{-1}\right], \tag{5.57}$$

where $\tilde{\beta}$ is the MQL estimator of β and $\mathbf{W} = \left\{_d [v(\mu_i)g_\mu^2(\mu_i)]^{-1}\right\}$.

However, τ is usually handled differently and estimated via a moment estimator (McCullagh and Nelder, 1989, p. 328):

$$\hat{\tau}^2 = \frac{1}{n-p} \sum \frac{(y_i - \hat{\mu}_i)^2}{v(\mu_i)}, \tag{5.58}$$

where n is the number of observations and p is the dimension of β.

5.7   EXERCISES

E 5.1 Show that $\dfrac{\phi(p_i)^2}{\Phi(p_i)[1 - \Phi(p_i)]}$ is the inverse of an estimate of $\text{var}(t_i)$, where t_i is defined in (5.3).

E 5.2 Show that the binomial, Poisson and gamma distributions can be written in the form (5.5). *Hint for the gamma distribution*: Write the density in terms of the mean and coefficient of variation.

E 5.3 Suppose $y \sim \mathcal{N}(e^\theta, e^\theta)$, i.e., y is normal with equal mean and variance. Show that the distribution of y is *not* of the form (5.5).

E 5.4 Derive the log likelihood in (5.20).

E 5.5 Show that (5.21) can be written in the form (5.18).

E 5.6 Suppose $y_i \sim$ indep. Binomial(n, p) for $i = 1, 2, \ldots, m$, where $p = 1/(1 + e^{-\alpha})$. Show that the MLE of α is $\log[\sum y_i / (mn - \sum y_i)]$.

E 5.7 Using (5.24) verify that the large-sample variance of $\hat{\beta}$ is given by (5.25).

E 5.8 Derive (5.27) from (5.26).

E 5.9 Show that q_i of (5.49) satisfies (5.51), (5.52), and (5.53).

E 5.10 For binary (Bernoulli) and Poisson distributed data, in (5.19) show that $\mathbf{W\Delta} = \mathbf{I}$ and hence it simplifies to

$$\mathbf{X'y} = \mathbf{X'\mu}.$$

E 5.11 *Efficiency of MQL*: Suppose that $y_i \sim \mathcal{N}(\mu_i, \sigma_i^2)$ for $i = 1, 2, \ldots n$, where $\log \mu_i = x_i \beta$ and $v(\mu_i) = \mu_i$. Calculate the ratio of the large-sample variances of $\tilde{\beta}$, the MQL estimator of β and $\hat{\beta}$, the MLE of β. For concreteness, assume that $n/2$ of the observations have $x_i = 5$ and $n/2$ are 10. Do the calculations for β equal to 0.1, 1, and 10.

Chapter 6

LINEAR MIXED MODELS (LMMs)

6.1 A GENERAL MODEL

a. Introduction

Chapter 4 deals with linear models (LMs), $E(\mathbf{y}) = \mathbf{X}\boldsymbol{\beta}$, where elements of $\boldsymbol{\beta}$ are fixed effects, that is, unknown constants. An example is $E[y_{ij}] = \mu + \alpha_i$ where μ is a general mean and (in Section 1.3a) α_1 and α_2 represent effects on the response variable of a patient receiving the placebo or the drug progabide, respectively. Each of μ, α_1 and α_2 is a fixed effect, and in $E[\mathbf{y}] = \mathbf{X}\boldsymbol{\beta}$ the $\boldsymbol{\beta}$ is $[\mu \quad \alpha_1 \quad \alpha_2]'$.

In contrast, in Section 1.5a we discuss the model

$$E[y_{ij}] = \mu + a_i + \beta_j + c_{ij}, \tag{6.1}$$

where a_i is a random effect representing clinic i, β_j is a fixed effect for dose j of a drug, and c_{ij} is a random effect for interaction. This, with its mixture of fixed and random effects, is a *linear mixed model* (LMM). A special case of an LMM is when there are no fixed effects (except μ), whereupon it is called a *random model*.

In linear models, fixed effects are used for modeling the mean of $\mathbf{y}$ while random effects govern the variance-covariance structure of $\mathbf{y}$. In fact, a prime reason for having random effects is to simplify the otherwise difficult task of specifying the $N(N+1)/2$ distinct elements of $\mathrm{var}(\mathbf{y}_{N\times1})$. Without using random effects we would have to deal with elements of $\mathrm{var}(\mathbf{y})$ being a variety of forms; but with random factors we can conveniently deal with variances and covariances attributable to factors acknowledged to be affecting the data. Since the two kinds of effects (fixed and random) are different and so get treated differently when analyzing data, we need to know, for our data, how to decide for each factor whether it is to be deemed to be a fixed effects factor or a random effects factor. The making of this decision is discussed in Section 1.6. Having so decided, the procedures for an LMM are as follows.

b.   Basic properties

The starting point for an LM is $E[\mathbf{y}] = \mathbf{X}\beta$ with β being fixed effects; for an LMM we still use $\mathbf{X}\beta$ for fixed effects but add to it $\mathbf{Z}\mathbf{u}$ where $\mathbf{Z}$, like $\mathbf{X}$, is a known (model) matrix and $\mathbf{u}$ is the vector of random effects that occur in the data vector $\mathbf{y}$. Although the elements of $\mathbf{u}$ are random variables it is convenient to specify the model conditional on their unobservable but realized values. Thus we write not $E[\mathbf{y}]$ as $\mathbf{X}\beta + \mathbf{Z}\mathbf{u}$ but

$$E[\mathbf{y}|\mathbf{u}] = \mathbf{X}\beta + \mathbf{Z}\mathbf{u}, \qquad (6.2)$$

meaning that for the realized $\mathbf{u}$, (6.2) is the conditional mean. Were we to use $\mathbf{U}$ for random variables and $\mathbf{u}$ for their realized values, we would in place of $E[\mathbf{y}|\mathbf{u}]$ write $E[\mathbf{y}|\mathbf{U} = \mathbf{u}]$ — but the clumsiness of this is distracting, so we stay with $E[\mathbf{y}|\mathbf{u}]$.

In order to handle first and second moments of $\mathbf{y}$, those of $\mathbf{u}$ are needed. They get specified by

$$\mathbf{u} \sim (\mathbf{0}, \mathbf{D}), \text{ meaning that } E[\mathbf{u}] = \mathbf{0} \text{ and } \mathrm{var}(\mathbf{u}) = \mathbf{D}. \qquad (6.3)$$

There is no loss of generality in taking $E[\mathbf{u}]$ to be $\mathbf{0}$, because if it was otherwise, $E[\mathbf{u}] = \tau$, say, then $E[\mathbf{y}|\mathbf{u}] = \mathbf{X}\beta + \mathbf{Z}\mathbf{u}$ could be rewritten as $E[\mathbf{y}|\mathbf{u}] = \mathbf{X}\beta + \mathbf{Z}\tau + \mathbf{Z}(\mathbf{u}-\tau)$. Defining $\mathbf{X}^* = [\mathbf{X}\ \mathbf{Z}]$ and $\beta^* = [\beta'\ \tau']'$ gives $\mathbf{X}\beta + \mathbf{Z}\tau$ as $\mathbf{X}^*\beta^*$; and the further defining of $\mathbf{u} - \tau$ as $\mathbf{u}^*$ gives $E[\mathbf{y}|\mathbf{u}] = \mathbf{X}^*\beta^* + \mathbf{Z}\mathbf{u}^*$ which, with $E[\mathbf{u}^*] = \mathbf{0}$, has exactly the same form as $E[\mathbf{y}|\mathbf{u}] = \mathbf{X}\beta + \mathbf{Z}\mathbf{u}$.

For specifying $\mathrm{var}(\mathbf{y})$ we have $\mathrm{var}(\mathbf{u}) = \mathbf{D}$ from (6.3) and now define

$$\mathrm{var}(\mathbf{y}|\mathbf{u}) = \mathbf{R}. \qquad (6.4)$$

With $E[\mathbf{u}] = \mathbf{0}$ applied to (6.2), this gives (see E 6.1)

$$\mathbf{y} \sim (\mathbf{X}\beta,\ \mathbf{Z}\mathbf{D}\mathbf{Z}' + \mathbf{R}), \qquad (6.5)$$

showing that the fixed effects enter only the mean whereas the random effects model matrix and the variance of the random effects enter only the variance of **y**.

6.2 ATTRIBUTING STRUCTURE TO VAR(y)

The expression for var(**y**) in (6.5) is

$$\mathbf{V} = \text{var}(\mathbf{y}) = \mathbf{ZDZ'} + \mathbf{R}. \tag{6.6}$$

Some simplifications are now described, using the following example for illustration.

a. Example

Suppose data are scores on a mathematics exam given to 20 ninth-grade classes of fifteen high schools in New York City. Aside from differences between boys and girls (which would be modelled by fixed effects) there will undoubtedly be three sources of variability: (i) among schools, (ii) among classes within each school, and (iii) among pupils within each class. Let the exam score of pupil k (of gender t) in class j of school i be y_{tijk}. Then a model equation could be

$$\text{E}[y_{tijk}|s_i, c_{ij}] = \beta_t + s_i + c_{ij}. \tag{6.7}$$

β_t is the fixed effect for gender t. The school effects, s_i for $i = 1, 2, \ldots, 15$, and the class effects c_{ij} for $1, \ldots, 4$ for school i, would be treated as random effects. So would p_{tijk}, representing everything not accounted for by β_i, s_i and c_{ij} for the individual pupil. Thus $\boldsymbol{\beta}$ of $\mathbf{X}\boldsymbol{\beta}$ in (6.2) will have two elements, β_m and β_f for male and female, respectively; and **u** of **Zu** will have the 15 s_i effects and the 60 c_{ij} effects.

b. Taking covariances between factors as zero

For the example it is convenient to partition **u** into two sub-vectors $\mathbf{u}_1$ and $\mathbf{u}_2$, with $\mathbf{u}_1$ having all fifteen s_i effects as elements, and $\mathbf{u}_2$ having all 60 ($= 4 \times 15$) c_{ij} effects. Thus

$$\mathbf{u} = \begin{bmatrix} \mathbf{u}_1 \\ \mathbf{u}_2 \end{bmatrix} = \begin{bmatrix} \left\{ _c s_i \right\}_{i=1}^{15} \\ \left\{ \left\{ _c \left\{ _c c_{ij} \right\}_{j=1}^{4} \right\}_{i=1}^{15} \right\} \end{bmatrix}.$$

Partition $\mathbf{Z}$ correspondingly as

$$\mathbf{Z} = [\mathbf{Z}_1 \ \ \mathbf{Z}_2] \qquad \text{so that} \qquad \mathbf{Zu} = \mathbf{Z}_1\mathbf{u}_1 + \mathbf{Z}_2\mathbf{u}_2.$$

Also partition $\mathbf{D}$ as

$$\mathbf{D} = \mathrm{var}(\mathbf{u}) = \begin{bmatrix} \mathrm{var}(\mathbf{u}_1) & \mathrm{cov}(\mathbf{u}_1, \mathbf{u}_2') \\ \mathrm{cov}(\mathbf{u}_2, \mathbf{u}_1') & \mathrm{var}(\mathbf{u}_2) \end{bmatrix} = \begin{bmatrix} \mathbf{D}_1 & \mathbf{D}_{12} \\ \mathbf{D}_{21} & \mathbf{D}_2 \end{bmatrix},$$

with $\mathbf{D}_{21} = \mathbf{D}_{12}'$. Then (6.2) becomes

$$E[\mathbf{y}|\mathbf{u}] = \mathbf{X}\boldsymbol{\beta} + \mathbf{Z}_1\mathbf{u}_1 + \mathbf{Z}_2\mathbf{u}_2 \tag{6.8}$$

and $\mathbf{V} = \mathbf{ZDZ}' + \mathbf{R}$ gets to be

$$\mathbf{V} = \mathbf{Z}_1\mathbf{D}_1\mathbf{Z}_1' + \mathbf{Z}_2\mathbf{D}_2\mathbf{Z}_2' + \mathbf{Z}_1\mathbf{D}_{12}\mathbf{Z}_2' + \mathbf{Z}_2\mathbf{D}_{21}\mathbf{Z}_1' + \mathbf{R}. \tag{6.9}$$

$\mathbf{R}$ is defined in (6.4) as $\mathrm{var}(\mathbf{y}|\mathbf{u})$. For the example this is the variance-covariance matrix of the p_{tijk} terms in (6.7). These represent not only the variability among pupils but also any variability not attributable to s_i and c_{ij}.

The preceding notations extend very directly from the two random factors of the example to having r random factors, so that with

$$\mathbf{u} = \left\{ {}_c\, \mathbf{u}_i \right\}_{i=1}^r \qquad \text{and} \quad \mathbf{D} = \mathrm{var}(\mathbf{u}) = \left\{ {}_m\, \mathbf{D}_{ii'} \right\}_{i,i'=1}^r$$

and

$$\mathbf{Z} = [\mathbf{Z}_1 \ \ \mathbf{Z}_2 \ \cdots \ \mathbf{Z}_r] = \left\{ {}_r\, \mathbf{Z}_i \right\}_{i=1}^r,$$

(6.8) and (6.9) become

$$E[\mathbf{y}|\mathbf{u}] \ = \ \mathbf{X}\boldsymbol{\beta} + \sum_{i=1}^r \mathbf{Z}_i\mathbf{u}_i$$

and

$$\mathbf{V} \ = \ \sum_{i=1}^r \mathbf{Z}_i\mathbf{D}_{ii}\mathbf{Z}_i' + \sum_{\substack{i=1 \\ }}^r \sum_{\substack{i'=1 \\ i \neq i'}}^r \mathbf{Z}_i\mathbf{D}_{ii'}\mathbf{Z}_{i'}' + \mathbf{R}, \tag{6.10}$$

where

$$\mathbf{D}_{ii} \ = \ \mathrm{var}(\mathbf{u}_i) \qquad \text{and} \qquad \mathbf{D}_{ii'} = \mathrm{cov}(\mathbf{u}_i, \mathbf{u}_{i'}').$$

Thus, in the example, $\mathbf{D}_{11}$ is the variance-covariance matrix of schools and $\mathbf{D}_{12}$ is the matrix of covariances between schools and classes. In point of fact, it is reasonable to take those covariances as zero. Some of them are covariances between a school effect s_i and the class effects $c_{i'j}$ of classes in a different school; and there would seem to be no reason for thinking those covariances (correlations) are anything but zero. And other covariances in $\mathbf{D}_{12}$ are covariances between a school effect s_i and the class effects c_{ij} of classes within that school; and since we use random effects with the thought that they capture all the variability in the data, we assume those covariances are zero too; that is, $\mathbf{D}_{12} = \mathbf{0}$. This extends very directly to the general case of (6.10), so that for $i \neq i'$ we take $\mathbf{D}_{ii'} = \mathbf{0}$. Hence, on writing $\mathbf{D}_i$ for $\mathbf{D}_{ii}$,

$$\mathbf{D} = \left\{ {}_{d}\, \mathbf{D}_i \right\}_{i=1}^{r} \qquad \text{and} \qquad \mathbf{V} = \sum_{i=1}^{r} \mathbf{Z}_i \mathbf{D}_i \mathbf{Z}_i' + \mathbf{R}.$$

c. The traditional variance components model

– i. *Customary notation*

If in the example we assume that there is no covariance between schools and that schools exhibit homogeneity of variance, then for the 15 schools (6.9) has

$$\mathbf{D}_1 = \sigma_s^2 \mathbf{I}_{15}. \tag{6.11}$$

Similar assumptions for the four classes within school i would give

$$\text{var}[c_{i1} \quad c_{i2} \quad c_{i3} \quad c_{i4}]' = \sigma_i^2 \mathbf{I}_4.$$

And making the very reasonable assumption that the classes in one school are independent of those in every other school gives

$$\mathbf{D}_2 = \left\{ {}_{d}\, \text{var}(\mathbf{u}_i) \right\}_{i=1}^{15} = \left\{ {}_{d}\, \sigma_i^2 \mathbf{I}_4 \right\}_{i=1}^{15}.$$

An even simpler assumption is that the four classes within a school have the same variance for all 15 schools, that is, that $\sigma_i^2 = \sigma_c^2 \ \forall\, i$, so giving

$$\mathbf{D}_2 = \sigma_c^2 \mathbf{I}_{60} \tag{6.12}$$

a form that is similar to (6.11). Thus (6.11) and (6.12), and a similar form for $\mathbf{R}$, namely

$$\mathbf{R} = \sigma^2 \mathbf{I}_{1200},$$

makes up the standard structure for the traditional variance components model.

The general case of r random effects factors then has

$$\mathbf{D} = \left\{ {}_d\, \sigma_i^2 \mathbf{I}_{q_i} \right\}_{i=1}^r$$

for random factor i having q_i effects in the data (i.e., $\mathbf{u}_i$ of order $q_i \times 1$) and

$$\mathbf{V} = \sum_{i=1}^r \mathbf{Z}_i \mathbf{Z}_i' \sigma_i^2 + \sigma^2 \mathbf{I}_N \qquad (6.13)$$

for $\mathbf{y}$ of order $N \times 1$.

– ii.   *Amended notation*

An amendment to the preceding notation suggested by Hartley and Rao (1967) amounts to redefining $\mathbf{D}$ so as to include $\sigma^2 \mathbf{I}_N$. This is achieved by defining

$$\sigma_0^2 \equiv \sigma^2, \quad \mathbf{Z}_0 = \mathbf{I}_N \quad \text{and} \quad q_0 = N.$$

Then we define

$$\mathbf{D}_* = \begin{bmatrix} \sigma_0^2 \mathbf{I}_{q_0} & \mathbf{0} \\ \mathbf{0} & \mathbf{D} \end{bmatrix} = \left\{ {}_d\, \sigma_i^2 \mathbf{I}_{q_i} \right\}_{i=0}^r \qquad (6.14)$$

and from (6.13)

$$\mathbf{V} = \sum_{i=0}^r \mathbf{Z}_i \mathbf{Z}_i' \sigma_i^2, \qquad (6.15)$$

which can be written as

$$\mathbf{V} = \mathbf{Z}_* \mathbf{D}_* \mathbf{Z}_*' \quad \text{for} \quad \mathbf{Z}_* = [\mathbf{Z}_0 \ \mathbf{Z}_1 \ \mathbf{Z}_2 \cdots \mathbf{Z}_r] = [\mathbf{Z}_0 \ \mathbf{Z}]. \qquad (6.16)$$

Corresponding to $\mathbf{Z}_0$ will be $\mathbf{u}_0$ of order $N \times 1$, familiarly thought of in the context of analysis of variance models as the residual error term.

The variances σ_i^2 for $i = 0, 1, \ldots, r$ are called variance components because they are the components of the variance of an individual observation; that is, for the example

$$\text{var}(y_{tijk}) = \sigma_s^2 + \sigma_c^2 + \sigma^2.$$

d. An LMM for longitudinal data

Longitudinal data are successive observations on each of a collection of observational units (often people). An example is blood pressure measurements taken weekly on a group of patients. If $\mathbf{y}_i$ is the vector of n_i measurements on patient i, a model equation suggested by Laird and Ware (1982) is

$$E[\mathbf{y}_i|\mathbf{u}_i] = \mathbf{X}_i\boldsymbol{\beta} + \mathbf{Z}_i\mathbf{u}_i$$

with vectors $\boldsymbol{\beta}$ and $\mathbf{u}_i$ consisting of fixed and random effects, respectively. $\boldsymbol{\beta}$ is the same for all patients and $\mathbf{u}_i$ is specific to patient i.

Suppose that there are m such patients. Then for

$$\mathbf{y} = \left\{_c \mathbf{y}_i\right\}, \quad \mathbf{X} = \left\{_c \mathbf{X}_i\right\}, \quad \mathbf{Z} = \left\{_d \mathbf{Z}_i\right\} \quad \text{and} \quad \mathbf{u} = \left\{_c \mathbf{u}_i\right\}$$

we have

$$\left\{_c E[\mathbf{y}_i|\mathbf{u}_i]\right\}_{i=1}^{m} = E[\mathbf{y}|\mathbf{u}] = \mathbf{X}\boldsymbol{\beta} + \mathbf{Z}\mathbf{u}.$$

And the variance structure suggested by Laird and Ware (1982) is $\mathbf{V} = \mathbf{Z}\mathbf{D}\mathbf{Z}' + \mathbf{R}$ with

$$\mathbf{D}_{ii} = \mathbf{D} \,\forall\, i, \qquad \mathbf{D}_{ii'} = \mathbf{0} \,\forall\, i \neq i' \qquad \text{and} \qquad \mathbf{R} = \left\{_d \mathbf{R}_i\right\}$$

so that

$$\mathbf{V} = \text{var}(\mathbf{y}) = \left\{_d \mathbf{Z}_i\mathbf{D}\mathbf{Z}_i' + \mathbf{R}_i\right\}.$$

More details for this model are described in Chapter 8.

6.3 ESTIMATING FIXED EFFECTS FOR V KNOWN

We take $\mathbf{y}$ to be normally distributed,

$$\mathbf{y} \sim \mathcal{N}(\boldsymbol{\mu} = \mathbf{X}\boldsymbol{\beta}, \mathbf{V}),$$

so that the log likelihood is

$$l = -\frac{1}{2}(\mathbf{y} - \boldsymbol{\mu})'\mathbf{V}^{-1}(\mathbf{y} - \boldsymbol{\mu}) - \frac{1}{2}\log|\mathbf{V}| - \frac{N}{2}\log 2\pi. \qquad (6.17)$$

Then from (6.69) (Section 6.12) we use

$$\frac{\partial l}{\partial \theta} = \frac{\partial \boldsymbol{\mu}'}{\partial \theta}\mathbf{V}^{-1}(\mathbf{y} - \boldsymbol{\mu}) \qquad (6.18)$$

with $\mu = \mathbf{X}\boldsymbol{\beta}$ and $\boldsymbol{\theta} = \boldsymbol{\beta}$. Making those substitutions in (6.18) and equating the result to $\mathbf{0}$ with $\boldsymbol{\beta}$ written as $\boldsymbol{\beta}^0$ gives

$$\mathbf{X}'\mathbf{V}^{-1}\mathbf{X}\boldsymbol{\beta}^0 = \mathbf{X}'\mathbf{V}^{-1}\mathbf{y} \qquad \text{so that} \qquad \boldsymbol{\beta}^0 = (\mathbf{X}'\mathbf{V}^{-1}\mathbf{X})^-\mathbf{X}'\mathbf{V}^{-1}\mathbf{y}.$$
(6.19)

Because $\boldsymbol{\beta}^0$ varies with the choice of $(\mathbf{X}'\mathbf{V}^{-1}\mathbf{X})^-$, we confine attention to $\mathbf{X}\boldsymbol{\beta}^0$ which is invariant because $\mathbf{X}(\mathbf{X}'\mathbf{V}^{-1}\mathbf{X})^-\mathbf{X}'\mathbf{V}^{-1}$ (see Section M.4c of Appendix M) is. Thus

$$\mathrm{ML}(\mathbf{X}\boldsymbol{\beta}) = \mathbf{X}\boldsymbol{\beta}^0 = \mathbf{X}(\mathbf{X}'\mathbf{V}^{-1}\mathbf{X})^-\mathbf{X}'\mathbf{V}^{-1}\mathbf{y} \qquad (6.20)$$

is the ML estimator of $\mathbf{X}\boldsymbol{\beta}$; and so $\boldsymbol{\lambda}'\mathbf{X}\boldsymbol{\beta}^0$ is the ML estimator of $\boldsymbol{\lambda}'\mathbf{X}\boldsymbol{\beta}$ for any $\boldsymbol{\lambda}$.

With $\mathrm{var}(\mathbf{y}) = \mathbf{V}$ it is easily seen that

$$\mathrm{var}(\mathbf{X}\boldsymbol{\beta}^0) = \mathbf{X}(\mathbf{X}'\mathbf{V}^{-1}\mathbf{X})^-\mathbf{X}'\mathbf{V}^{-1}\mathbf{X}(\mathbf{X}'\mathbf{V}^{-1}\mathbf{X})^{-'}\mathbf{X}'.$$

Then, because $(\mathbf{X}'\mathbf{V}^{-1}\mathbf{X})^{-'}$ is a generalized inverse of $(\mathbf{X}'\mathbf{V}^{-1}\mathbf{X})$, and also because of the invariance property referred to prior to (6.20), $\mathrm{var}(\mathbf{X}\boldsymbol{\beta}^0)$ reduces to

$$\mathrm{var}(\mathbf{X}\boldsymbol{\beta}^0) = \mathbf{X}(\mathbf{X}'\mathbf{V}^{-1}\mathbf{X})^-\mathbf{X}'. \qquad (6.21)$$

To test the null hypothesis $H_0 : \mathbf{K}'\mathbf{X}\boldsymbol{\beta} = \mathbf{m}$, where $\mathbf{K}'$ is of full row rank $(r_{\mathbf{K}} \leq r_{\mathbf{X}})$, we can derive a chi-square statistic using

$$\chi^2 = (\mathbf{K}'\mathbf{X}\boldsymbol{\beta}^0 - \mathbf{m})'[\mathbf{K}'\mathbf{X}(\mathbf{X}'\mathbf{V}^{-1}\mathbf{X})^-\mathbf{X}'\mathbf{K}]^{-1}(\mathbf{K}'\mathbf{X}\boldsymbol{\beta}^0 - \mathbf{m}). \qquad (6.22)$$

Under H_0, X^2 has a central χ^2 distribution with $r_{\mathbf{K}} = \mathrm{rank}(\mathbf{K})$ degrees of freedom.

More typically $\mathbf{V}$ is known only up to a scalar multiple. To emphasize the connections with Chapter 5 and for simplicity of notation we therefore write $\mathbf{V}$ in terms of a weight matrix $\mathbf{W}$, which is the inverse of $\mathbf{V}$ up to a scalar multiple, that is, $\mathbf{V} = \sigma^2\mathbf{W}^{-1}$, where $\mathbf{W}$ is assumed known. In such a case the following statistic can be derived as the likelihood ratio test and is also the uniformly most powerful invariant test (Lehmann, 1986):

$$F = \frac{(\mathbf{K}'\mathbf{X}\boldsymbol{\beta}^0 - \mathbf{m})'[\mathbf{K}'\mathbf{X}(\mathbf{X}'\mathbf{W}\mathbf{X})^-\mathbf{X}'\mathbf{K}]^{-1}(\mathbf{K}'\mathbf{X}\boldsymbol{\beta}^0 - \mathbf{m})}{r_{\mathbf{K}}\hat{\sigma}^2}, \qquad (6.23)$$

where

$$\hat{\sigma}^2 = \frac{\mathbf{y}'[\mathbf{W} - \mathbf{W}\mathbf{X}(\mathbf{X}'\mathbf{W}\mathbf{X})^-\mathbf{X}'\mathbf{W}]\mathbf{y}}{N - r_{\mathbf{X}}}.$$

Under the null hypothesis, F has an $\mathcal{F}$-distribution on $r_{\mathbf{K}}$ and $N - r_{\mathbf{X}}$ degrees of freedom. The null hypothesis is rejected at significance level α when F exceeds $\mathcal{F}_{N-r_{\mathbf{X}},1-\alpha}^{r_{\mathbf{K}}}$.

6.4   ESTIMATING FIXED EFFECTS FOR V UNKNOWN

a.   Estimation

With **V** unknown but not being a function of β, the log likelihood function l of (6.17) has to be maximized with respect to elements of both $\boldsymbol{\mu}$ and **V**. For $\boldsymbol{\mu} = \mathbf{X}\beta$ and $\boldsymbol{\theta} = \beta$, setting $\partial l/\partial\boldsymbol{\theta}$ to **0** will lead to the same result for β^0 as in (6.19), only with **V** therein being replaced by the solution $\hat{\mathbf{V}}$ coming from maximizing l with respect to the parameters in **V**. No matter what $\hat{\mathbf{V}}$ is, the ML estimator of $\mathbf{X}\beta$ will be

$$\mathrm{ML}(\mathbf{X}\beta) = \mathbf{X}\hat{\beta} = \mathbf{X}(\mathbf{X}'\hat{\mathbf{V}}^{-1}\mathbf{X})^{-}\mathbf{X}'\hat{\mathbf{V}}^{-1}\mathbf{y}, \qquad (6.24)$$

where we here introduce the symbol $\hat{\beta}$ to represent β^0 of (6.19) but with **V** replaced by $\hat{\mathbf{V}}$, which is **V** with its parameters replaced by their ML estimators. The ML equations for **V** are obtained from equating to **0** the expression

$$\frac{\partial l}{\partial\varphi_k} = -\tfrac{1}{2}\left[\mathrm{tr}\left(\mathbf{V}^{-1}\frac{\partial\mathbf{V}}{\partial\varphi_k}\right) - (\mathbf{y}-\boldsymbol{\mu})'\mathbf{V}^{-1}\frac{\partial\mathbf{V}}{\partial\varphi_k}\mathbf{V}^{-1}(\mathbf{y}-\boldsymbol{\mu})\right] \quad (6.25)$$

of (6.70) using φ for each parameter in **V**; and in doing this $\boldsymbol{\mu}$ is replaced by $\mathbf{X}\hat{\beta}$ of (6.24). On writing

$$\left.\frac{\partial\mathbf{V}}{\partial\varphi_k}\right|_{\mathbf{V}=\hat{\mathbf{V}}} \qquad \text{as} \qquad \hat{\mathbf{V}}_{\varphi_k},$$

this gives

$$\mathrm{tr}(\hat{\mathbf{V}}^{-1}\hat{\mathbf{V}}_{\varphi_k}) = \mathbf{y}'\hat{\mathbf{P}}\hat{\mathbf{V}}_{\varphi_k}\hat{\mathbf{P}}\mathbf{y}, \qquad (6.26)$$

where

$$\mathbf{P} = \mathbf{V}^{-1} - \mathbf{V}^{-1}\mathbf{X}(\mathbf{X}'\mathbf{V}^{-1}\mathbf{X})^{-}\mathbf{X}'\mathbf{V}^{-1} \qquad (6.27)$$

and $\hat{\mathbf{P}}$ is **P** with **V** replaced by $\hat{\mathbf{V}}$. For the case of the parameters in **V** being variance components, as in (6.15), we describe ML estimation in Section 6.8.

b.   Sampling variance

Instead of dealing with the variance (matrix) of a vector $\mathbf{X}\hat{\beta}$ we consider the simpler case of the scalar $\boldsymbol{\ell}'\hat{\beta}$ for estimable $\boldsymbol{\ell}'\beta$ (i.e., $\boldsymbol{\ell}' = \mathbf{t}'\mathbf{X}$ for some $\mathbf{t}'$).

For known **V** we have from (6.21) that $\mathrm{var}(\boldsymbol{\ell}'\beta^0) = \boldsymbol{\ell}'(\mathbf{X}'\mathbf{V}^{-1}\mathbf{X})^{-}\boldsymbol{\ell}$. A replacement for this when **V** is not known is to use $\boldsymbol{\ell}(\mathbf{X}'\hat{\mathbf{V}}^{-1}\mathbf{X})^{-}\boldsymbol{\ell}$, which is an estimate of $\operatorname{var}(\boldsymbol{\ell}'\boldsymbol{\beta}^0) = \operatorname{var}[\boldsymbol{\ell}'(\mathbf{X}'\mathbf{V}^{-1}\mathbf{X})^-\mathbf{X}'\mathbf{V}^{-1}\mathbf{y}]$. But this is *not* an estimate of $\operatorname{var}(\boldsymbol{\ell}'\hat{\boldsymbol{\beta}}) = \operatorname{var}[\boldsymbol{\ell}'(\mathbf{X}'\hat{\mathbf{V}}^{-1}\mathbf{X})^-\mathbf{X}'\hat{\mathbf{V}}^{-1}\mathbf{y}]$. The latter requires taking account of the variability in $\hat{\mathbf{V}}$ as well as that in $\mathbf{y}$. To deal with this, Kackar and Harville (1984, p. 854) observe that (in our notation) $\boldsymbol{\ell}'\hat{\boldsymbol{\beta}} - \boldsymbol{\ell}'\boldsymbol{\beta}$ can be expressed as the sum of two independent parts, $\boldsymbol{\ell}'\hat{\boldsymbol{\beta}} - \boldsymbol{\ell}'\boldsymbol{\beta}^0$ and $\boldsymbol{\ell}'\boldsymbol{\beta}^0 - \boldsymbol{\ell}'\boldsymbol{\beta}$. This leads to $\operatorname{var}(\boldsymbol{\ell}'\hat{\boldsymbol{\beta}})$ being expressed as a sum of two variances which we write as

$$\operatorname{var}(\boldsymbol{\ell}'\hat{\boldsymbol{\beta}}) = \operatorname{var}(\boldsymbol{\ell}'\boldsymbol{\beta}^0 - \boldsymbol{\ell}'\boldsymbol{\beta}) + \operatorname{var}(\boldsymbol{\ell}'\hat{\boldsymbol{\beta}} - \boldsymbol{\ell}'\boldsymbol{\beta}^0)$$

$$\approx \boldsymbol{\ell}'(\mathbf{X}'\mathbf{V}^{-1}\mathbf{X})^-\boldsymbol{\ell} + \boldsymbol{\ell}'\mathbf{T}\boldsymbol{\ell}, \tag{6.28}$$

where, in the words of Kenward and Roger (1997, p. 985), "the component $\mathbf{T}$ [in our notation] represents the amount by which the asymptotic variance-covariance matrix underestimates (in a matrix sense) $\operatorname{var}(\hat{\boldsymbol{\beta}})$." The matrix $\mathbf{T}$ in (6.28) is defined in adapting Kenward and Roger's equation (1) for the variance components model of Section 6.2c to write

$$\boldsymbol{\ell}'\mathbf{T}\boldsymbol{\ell} = \boldsymbol{\ell}'(\mathbf{X}'\mathbf{V}^{-1}\mathbf{X})^- \tag{6.29}$$

$$\times \left\{ \sum_{i=0}^{r}\sum_{j=0}^{r} c_{ij}\left[\mathbf{G}_{ij} - \mathbf{F}_i(\mathbf{X}'\mathbf{V}^{-1}\mathbf{X})^-\mathbf{F}_j\right] \right\}(\mathbf{X}'\mathbf{V}^{-1}\mathbf{X})^-\boldsymbol{\ell},$$

where c_{ij} is an element of the asymptotic variance-covariance matrix of the vector of estimated variance components; that is,

$$\mathbf{C} = \left\{_m c_{ij}\right\}_{i=0,\,j=0}^{r\quad r} = \left\{_m \operatorname{cov}_\infty(\hat{\sigma}_i^2, \hat{\sigma}_j^2)\right\}_{i=0,\,j=0}^{r\quad r},$$

which is (6.64) of Section 6.8c. Also in (6.29)

$$\mathbf{G}_{ij} = \mathbf{X}'\mathbf{V}^{-1}\mathbf{Z}_i\mathbf{Z}_i'\mathbf{V}^{-1}\mathbf{Z}_j\mathbf{Z}_j'\mathbf{V}^{-1}\mathbf{X} \text{ and } \mathbf{F}_i = -\mathbf{X}'\mathbf{V}^{-1}\mathbf{Z}_i\mathbf{Z}_i'\mathbf{V}^{-1}\mathbf{X}. \tag{6.30}$$

We now use (6.30) in (6.29) together with both

$$\mathbf{h}' = \boldsymbol{\ell}'(\mathbf{X}'\mathbf{V}^{-1}\mathbf{X})^-\mathbf{X}'\mathbf{V}^{-1} \tag{6.31}$$

and the general result for matrices $\mathbf{S}_{ij}$ and vectors $\mathbf{t}$ that

$$\mathbf{t}'\sum_i\sum_j c_{ij}\mathbf{S}_{ij}\mathbf{t} = \sum_i\sum_j c_{ij}\mathbf{t}'\mathbf{S}_{ij}\mathbf{t} = \operatorname{tr}\left[\mathbf{C}\left\{_m \mathbf{t}'\mathbf{S}_{ij}\mathbf{t}\right\}_{i,j=0}^{r}\right]. \tag{6.32}$$

This gives

$$\ell'\mathbf{T}\ell = \sum_{i=0}^{r}\sum_{j=0}^{r} c_{ij}\mathbf{h}'\mathbf{Z}_i\mathbf{Z}_i'\mathbf{P}\mathbf{Z}_j\mathbf{Z}_j'\mathbf{h} = \text{tr}\left[\mathbf{C}\left\{{}_{m}\mathbf{h}'\mathbf{Z}_i\mathbf{Z}_i'\mathbf{P}\mathbf{Z}_j\mathbf{Z}_j'\mathbf{k}\right\}_{i,j=0}^{r}\right].$$
(6.33)

Thus (6.28) becomes

$$\text{var}(\ell'\hat{\boldsymbol{\beta}}) \approx \ell'(\mathbf{X}'\mathbf{V}^{-1}\mathbf{X})\ell + \text{tr}\left[\mathbf{C}\left\{{}_{m}\mathbf{h}'\mathbf{Z}_i\mathbf{Z}_i'\mathbf{P}\mathbf{Z}_j\mathbf{Z}_j'\mathbf{k}\right\}_{i,j=0}^{r}\right].$$
(6.34)

To use (6.34) it does, of course, have to be calculated with $\hat{\mathbf{V}}$ in place of $\mathbf{V}$, meaning also $\hat{\mathbf{P}}$ in place of $\mathbf{P}$ — and these replacements also have to be made in (6.31).

c. Bias in the variance

Kenward and Roger (1997) additionally point out that $\ell'(\mathbf{X}'\hat{\mathbf{V}}^{-1}\mathbf{X})^{-}\ell$ is a biased estimate of $\ell'(\mathbf{X}'\mathbf{V}^{-1}\mathbf{X})^{-}\ell$, and they investigate that bias for nonsingular $\mathbf{X}'\mathbf{V}^{-1}\mathbf{X}$ with unstructured $\mathbf{V}$. We adapt their methods for the variance components model having $\mathbf{V} = \sum_{i=0}^{r}\mathbf{Z}_i\mathbf{Z}_i'\sigma_i^2$. For investigating the bias a starting point is a two-term Taylor series expansion:

$$\ell'(\mathbf{X}'\hat{\mathbf{V}}^{-1}\mathbf{X})^{-}\ell \doteq \ell'\left[(\mathbf{X}'\mathbf{V}^{-1}\mathbf{X})^{-} + (\hat{\sigma}^2 - \sigma^2)'\frac{\partial(\mathbf{X}'\mathbf{V}^{-1}\mathbf{X})^{-}}{\partial\sigma^2}\right.$$

$$\left. + \frac{1}{2}\sum_{i=0}^{r}\sum_{j=0}^{r}(\hat{\sigma}_i^2 - \sigma_i^2)(\hat{\sigma}_j^2 - \sigma_j^2)\frac{\partial^2(\mathbf{X}'\mathbf{V}^{-1}\mathbf{X})^{-}}{\partial\sigma_i^2\partial\sigma_j^2}\right]$$
(6.35)

This has expected value

$$\text{E}[\ell'(\mathbf{X}'\hat{\mathbf{V}}^{-1}\mathbf{X})^{-}\ell] \approx \ell(\mathbf{X}'\mathbf{V}^{-1}\mathbf{X})^{-}\ell$$

$$ + \frac{1}{2}\sum_{i=0}^{r}\sum_{j=0}^{r}\text{cov}(\hat{\sigma}_i^2, \hat{\sigma}_j^2)\ell'\frac{\partial^2(\mathbf{X}'\mathbf{V}^{-1}\mathbf{X})^{-}}{\partial\sigma_i^2\partial\sigma_j^2}\ell.$$

On using for the derivative (6.76) from Section 6.12c,

$$\text{E}[\ell'(\mathbf{X}'\hat{\mathbf{V}}^{-1}\mathbf{X})^{-}\ell] - \ell'(\mathbf{X}'\mathbf{V}^{-1}\mathbf{X})^{-}\ell$$

$$\approx -\frac{1}{2}\sum_{i=0}^{r}\sum_{j=0}^{r} c_{ij}\ell'(\mathbf{X}'\mathbf{V}^{-1}\mathbf{X})^{-}\mathbf{X}'\mathbf{V}^{-1}$$

$$\times \ (\mathbf{Z}_i\mathbf{Z}_i'\mathbf{P}\mathbf{Z}_j\mathbf{Z}_j' + \mathbf{Z}_j\mathbf{Z}_j'\mathbf{P}\mathbf{Z}_i\mathbf{Z}_i')\mathbf{V}^{-1}\mathbf{X}(\mathbf{X}'\mathbf{V}^{-1}\mathbf{X})^-\boldsymbol{\ell}$$

$$= \ -\sum_{i=0}^{r}\sum_{j=0}^{r} c_{ij}\mathbf{h}'\mathbf{Z}_i\mathbf{Z}_i'\mathbf{P}\mathbf{Z}_j\mathbf{Z}_j'\mathbf{h}, \quad \text{on using (6.31)}$$

$$= \ -\boldsymbol{\ell}'\mathbf{T}\boldsymbol{\ell}, \quad \text{from (6.33).} \tag{6.36}$$

But in (6.28) we want to estimate

$$\text{var}(\boldsymbol{\ell}'\hat{\boldsymbol{\beta}}) \approx \boldsymbol{\ell}'(\mathbf{X}'\mathbf{V}^{-1}\mathbf{X})^-\boldsymbol{\ell} + \boldsymbol{\ell}'\mathbf{T}\boldsymbol{\ell} \tag{6.37}$$

and from (6.36) we have

$$\text{E}[\boldsymbol{\ell}'(\mathbf{X}'\hat{\mathbf{V}}^{-1}\mathbf{X})^-\boldsymbol{\ell}] \approx \boldsymbol{\ell}'(\mathbf{X}'\mathbf{V}^{-1}\mathbf{X})^-\boldsymbol{\ell} - \boldsymbol{\ell}'\mathbf{T}\boldsymbol{\ell}. \tag{6.38}$$

Therefore an approximately unbiased estimator for (6.37) is

$$\boldsymbol{\ell}'(\mathbf{X}'\hat{\mathbf{V}}^{-1}\mathbf{X})^-\boldsymbol{\ell} + 2\boldsymbol{\ell}'\mathbf{T}\boldsymbol{\ell} \tag{6.39}$$

with everything calculated using $\hat{\mathbf{V}}$ and $\hat{\mathbf{P}}$ in place of $\mathbf{V}$ and $\mathbf{P}$.

d. Approximate F-statistics

The F-statistic in (6.23) has an $\mathcal{F}$-distribution; it is for the case of $\mathbf{V} = \sigma^2\mathbf{W}^{-1}$ with $\mathbf{W}$ known. But when $\mathbf{V}$ is not of this simple form, it has to be replaced in $\mathbf{K}'\mathbf{X}\hat{\boldsymbol{\beta}}$ and in F by an estimate, $\hat{\mathbf{V}}$, and the resulting value of F, call it $\hat{F}$, has an unknown distribution. If we assume that $\hat{F}$ is distributed approximately as $\mathcal{F}$, one way of making this approximation is to assume that

$$\lambda\hat{F} \sim \mathcal{F}_d^r, \tag{6.40}$$

where

$$r = r_{\mathbf{K}} \quad \text{for} \quad \mathbf{K} \text{ of } H_0 : \mathbf{K}'\mathbf{X}\boldsymbol{\beta} = \mathbf{m}.$$

Then, similar to Satterthwaite (1946), λ and d are derived by equating first and second moments of both sides of (6.40). This gives

$$\lambda\text{E}[\hat{F}] \ = \ d/(d-2)$$

and

$$\lambda^2\text{var}(\hat{F}) \ = \ 2d^2(r+d-2)/r(d-2)^2(d-4).$$

These lead to

$$\lambda = \frac{d}{(d-2)\mathrm{E}[\hat{F}]}$$

and

$$d = \frac{1+2/r}{\mathrm{var}(\hat{F}/(2\{\mathrm{E}[\hat{F}]\}^2) - 1/r}.$$

Kenward and Roger (1997) derive these results in their equations (7) and (8) and give some extensions. The calculation of $\mathrm{E}[\hat{F}]$ and $\mathrm{var}(\hat{F})$ is tedious.

6.5 PREDICTING RANDOM EFFECTS FOR V KNOWN

The assumptions about random effects differ from those for fixed effects and so treatment of the two kinds of effects is not the same. A fixed effect is considered to be a constant, which we wish to estimate. But a random effect is considered as just an effect coming from a population of effects. It is this population that is an extra assumption, compared to fixed effects, and we would hope it would lead to an estimation method for random effects being an improvement over that for fixed effects. To emphasize this distinction we use the term *prediction* of random effects rather than estimation.

For instance, in Example 4 of Section 1.4a, we treat clinic i as being from a population of clinics, with $\mathrm{E}[y_{ij}|a_i] = \mu + a_i$ being the ith clinic's true response. We may wish to predict the value of a_i to gain information about the performance of that particular clinic. Alternatively, we may want to use the predicted values of the a_i from several clinics in order to rank the clinics, or to select the best ones. Since they are all assumed to be selected from the same distribution it makes sense that their predicted values will have some degree of similarity and be less variable than might be anticipated without such an assumption.

Using the assumption that the realized random effects which determine the data are just a random selection from a conceptual population of such effects, it is not difficult to show that the "best" prediction of a_i (best in the sense of minimized mean squared error of prediction — see Chapter 13) is the conditional mean $\mathrm{E}[a_i|\mathbf{y}]$. In using this as the predictor of a_i we are using the expected value of the random effect in light of the data. An example of this is in the dairy farming industry where bulls are selected for use in artificial breeding on the basis of their daughters' average milk yield. Suppose the kth bull has a daughter with average milk yield $\bar{y}_k$. It is perfectly reasonable to think that in the population of bulls there will be bulls other than the kth that nevertheless have (or could have) the same daughter average, namely $\bar{y}_k$. Despite this, these bulls will not necessarily all have the same genetic values, let alone all the same as that of bull k. Therefore, since $\bar{y}_k$ is our data, and if a is the random effect representing bull genetic values, the best we can do for estimating bull k's genetic value is the conditional mean $E[a|\bar{y}_k]$. Not surprisingly, since predictors calculated as $E[a_i|\mathbf{y}]$ are "best", they have smaller mean squared error than would estimates based on assuming the random effects were fixed effects. They also have less variability and are sometimes called *shrinkage estimators*. This is because, just as in Section 1.4b–iv,

$$\text{var}(a) \;=\; \text{var}(E[a|\mathbf{y}]) + E[\text{var}(a|\mathbf{y})]$$

$$=\; \text{var}(\tilde{a}) + \text{ a positive value,}$$

where $\tilde{a} = E[a|\mathbf{y}]$ is the predictor. Thus

$$\text{var}(\tilde{a}) \leq \text{var}(a)$$

and so $\tilde{a}$ is said to be a *shrinkage estimator*.

From the preceding discussion we now turn to the general case of

$$\mathbf{y} \sim (\mathbf{X}\beta, \mathbf{V}) \text{ for } \mathbf{V} = \mathbf{ZDZ'} + \mathbf{R},$$

for which the conditional expected value $E[\mathbf{u}|\mathbf{y}]$ is, assuming $\mathbf{y}$ and $\mathbf{u}$ follow a jointly normal distribution,

$$E[\mathbf{u}|\mathbf{y}] = \mathbf{DZ'V}^{-1}(\mathbf{y} - \mathbf{X}\beta). \tag{6.41}$$

Replacing β by $\beta^0 = (\mathbf{X'V}^{-1}\mathbf{X})^-\mathbf{X'V}^{-1}\mathbf{y}$ of (6.19) gives what is called the best linear unbiased predictor (BLUP). Derivation of (6.41) and other results concerning prediction are detailed in Chapter 13.

We write

$$\tilde{\mathbf{u}} \;=\; \mathbf{DZ'V}^{-1}(\mathbf{y} - \mathbf{X}\beta) \tag{6.42}$$

and

$$\tilde{\mathbf{u}}^0 \;=\; \mathbf{DZ'V}^{-1}(\mathbf{y} - \mathbf{X}\beta^0) = \mathbf{DZ'Py}, \tag{6.43}$$

using **P** of (6.27). Then

$$\text{var}(\tilde{\mathbf{u}}) = \mathbf{DZ'V}^{-1}\mathbf{ZD} \quad \text{and} \quad \text{var}(\tilde{\mathbf{u}}^0) = \mathbf{DZ'PZD}, \quad (6.44)$$

the latter result using **PVP** = **P**. Thus with **D** and **V** known, (6.44) provides opportunities for testing hypotheses or deriving confidence intervals for elements of **u**. And use can also take advantage of the results:

$$\text{cov}(\boldsymbol{\beta}^0, \tilde{\mathbf{u}}^{0\prime}) = 0, \qquad \text{cov}(\tilde{\mathbf{u}}^0, \tilde{\mathbf{u}}') = \text{var}(\tilde{\mathbf{u}}^0)$$

and

$$\text{var}(\tilde{\mathbf{u}}^0 - \mathbf{u}) = \mathbf{D} - \mathbf{DZ'PZD}. \quad (6.45)$$

6.6 PREDICTING RANDOM EFFECTS FOR V UNKNOWN

a. Estimation

When **D** and **V** are unknown, they are typically replaced by $\hat{\mathbf{D}}$ and $\hat{\mathbf{V}}$ in $\tilde{\mathbf{u}}$ of (6.42), giving what could be called the estimated best predictor, to be denoted $\hat{\mathbf{u}}$:

$$\hat{\mathbf{u}} = \hat{\mathbf{D}}\mathbf{Z}'\hat{\mathbf{V}}^{-1}(\mathbf{y} - \mathbf{X}\hat{\boldsymbol{\beta}}) = \hat{\mathbf{D}}\mathbf{Z}'\hat{\mathbf{P}}\mathbf{y}.$$

b. Sampling variance

Kackar and Harville (1984) give extensive discussion of deriving the mean squared error of $\boldsymbol{\ell}'\hat{\boldsymbol{\beta}}^0 + \mathbf{m}'\hat{\mathbf{u}}$; Prasad and Rao (1990) suggest an alternative approximation and apply it to their three special cases of small-area estimators. And although the Kenward and Roger (1997) form of var($\boldsymbol{\ell}'\hat{\boldsymbol{\beta}}$) given by (6.28) and (6.29) is very different looking from Kackar and Harville (1984), they both ultimately reduce to the same thing. Using similar methods of reduction for the variance components model, var($\mathbf{m}'\hat{\mathbf{u}} - \mathbf{m}'\mathbf{u}$) comes from Kackar and Harville (1984) as

$$\text{var}(\mathbf{m}'\hat{\mathbf{u}} - \mathbf{m}'\mathbf{u}) = \mathbf{m}'\mathbf{Dm} - \mathbf{m}'\mathbf{DZ'PZDm} \quad (6.46)$$

$$+ \sum_{i=0}^{r}\sum_{j=0}^{r} c_{ij}[(\mathbf{m}_i' - \mathbf{m}'\mathbf{DZ'PZ}_i)\mathbf{Z}_i'\mathbf{PZ}_j(\mathbf{m}_j - \mathbf{Z}_j'\mathbf{PZDm})],$$

where $\mathbf{m}' = [\mathbf{m}_0' \ \mathbf{m}_1' \cdots \mathbf{m}_i' \cdots \mathbf{m}_r']$. We write

$$\Delta_{ij} = (\mathbf{m}_i' - \mathbf{m}'\mathbf{DZ'PZ}_i)\mathbf{Z}_i'\mathbf{PZ}_j(\mathbf{m}_j - \mathbf{Z}_j'\mathbf{PZDm})$$

so that

$$\text{var}(\mathbf{m}'\hat{\mathbf{u}} - \mathbf{m}'\mathbf{u}) \approx \mathbf{m}'\mathbf{D}\mathbf{m} - \mathbf{m}'\mathbf{D}\mathbf{Z}'\mathbf{P}\mathbf{Z}\mathbf{D}\mathbf{m} + \sum_{i=0}^{r}\sum_{j=0}^{r} c_{ij}\Delta_{ij}. \quad (6.47)$$

c. Bias in the variance

The procedure used for obtaining the bias of $\widehat{\text{var}}(\boldsymbol{\ell}'\hat{\boldsymbol{\beta}})$ in Section 6.4c can be applied similarly to $\widehat{\text{var}}(\mathbf{m}'\hat{\mathbf{u}}^0 - \mathbf{m}'\mathbf{u})$ starting from an expression like that of (6.35), namely

$$E[\mathbf{m}'(\hat{\mathbf{D}} - \hat{\mathbf{D}}\mathbf{Z}'\hat{\mathbf{P}}\mathbf{Z}\hat{\mathbf{D}})\mathbf{m}]$$

$$\approx \quad \mathbf{m}'(\mathbf{D} - \mathbf{D}\mathbf{Z}'\mathbf{P}\mathbf{Z}\mathbf{D})\mathbf{m} - \sum_{i=0}^{r}\sum_{j=0}^{r} c_{ij}\mathbf{m}'\frac{\partial^2(\mathbf{D} - \mathbf{D}\mathbf{Z}'\mathbf{P}\mathbf{Z}\mathbf{D})}{\partial\sigma_i^2\partial\sigma_j^2}\mathbf{m}. \quad (6.48)$$

The result, after using (6.78) in Section 6.12, is

$$E[\mathbf{m}'(\hat{\mathbf{D}} - \hat{\mathbf{D}}\mathbf{Z}'\hat{\mathbf{P}}\mathbf{Z}\hat{\mathbf{D}})\mathbf{m}] \approx \mathbf{m}'(\mathbf{D} - \mathbf{D}\mathbf{Z}'\mathbf{P}\mathbf{Z}\mathbf{D})\mathbf{m} - \sum_{i=0}^{r}\sum_{j=0}^{r} c_{ij}\Delta_{ij}. \quad (6.49)$$

Thus, by comparison with (6.47),

$$\mathbf{m}'(\hat{\mathbf{D}} \quad - \quad \hat{\mathbf{D}}\mathbf{Z}'\hat{\mathbf{P}}\mathbf{Z}\hat{\mathbf{D}})\mathbf{m} + 2\sum_{i=0}^{r}\sum_{j=0}^{r} c_{ij}\Delta_{ij} \quad\quad (6.50)$$

is an approximately unbiased estimator of $\text{var}(\mathbf{m}'\hat{\mathbf{u}} - \mathbf{m}'\mathbf{u})$. As with $\text{var}(\boldsymbol{\ell}'\hat{\boldsymbol{\beta}})$, of course, (6.50) must be calculated with $\hat{\mathbf{D}}$, $\hat{\mathbf{V}}$ and $\hat{\mathbf{P}}$ replacing $\mathbf{D}$, $\mathbf{V}$ and $\mathbf{P}$.

6.7 ANOVA ESTIMATION OF VARIANCE COMPONENTS

For random effects we want not only to predict them, as just discussed, but also to estimate their variances. Methods for doing this are detailed in many books and papers, the two main methods being ANOVA and ML. We here briefly outline the main ideas of the ANOVA methodology which, although seldom applicable to GLMMs of later chapters, is important for LMMs, both historically and for its practicality in certain circumstances.

a. Balanced data

Suppose in the one-way classification random model, with model equation $E[y_{ij}|a_i] = \mu + a_i$, as in Section 2.1, that we have $i = 1, 2, \ldots, m$ and $j = 1, 2, \ldots, n$; that is, n observations in every one of the m classes, the simplest example of balanced data. Then $\mathbf{V}$ of $\mathbf{V} = \mathbf{ZDZ'} + \mathbf{R}$ turns out to be

$$\mathbf{V} = \left\{ {}_d \; \sigma^2 \mathbf{I}_n + \sigma_a^2 \mathbf{J}_n \right\}_{i=1}^m. \tag{6.51}$$

For this, the traditional analysis of variance table contains the following two sums of squares:

$$\text{SSA} = n \sum_{i=1}^m (\bar{y}_{i\cdot} - \bar{y}_{\cdot\cdot})^2 \quad \text{and} \quad \text{SSE} = \sum_{i=1}^m \sum_{j=1}^n (y_{ij} - \bar{y}_{i\cdot})^2. \tag{6.52}$$

The expected values of these, based on $\mathbf{V}$ of (6.51), are

$$E[\text{SSA}] = (m-1)(n\sigma_a^2 + \sigma^2) \quad \text{and} \quad E[\text{SSE}] = m(n-1)\sigma^2. \tag{6.53}$$

The ANOVA method of estimating variance components is to equate expected values, such as (6.53), to the corresponding calculated values, (6.52). The solutions for the variance components are taken as the ANOVA estimates thereof. This gives

$$\tilde{\sigma}^2 = \frac{\text{SSE}}{m(n-1)} \tag{6.54}$$

and

$$\tilde{\sigma}_a^2 = \frac{1}{n}\left(\frac{\text{SSA}}{m-1} - \tilde{\sigma}^2\right). \tag{6.55}$$

This method of estimation extends very directly to analysis of variance of balanced (equal subclass numbers) data where there are more sums of squares than just the two of the preceding example. The resulting estimators are always unbiased, although they can yield negative estimates, as is possible, for example, in (6.55). The estimators are also minimum variance quadratic unbiased. On assuming normality for $\mathbf{y}$ they are minimum variance unbiased and their sampling variances and unbiased estimators thereof are readily available. Searle et al. (1992, Chapter 4) has extensive details for the case of balanced data.

b.   Unbalanced data

For unbalanced data (unequal subclass numbers) the ANOVA method can still be applied, but its utility is severely limited. This is because with many cases of unbalanced data there is more than one set of sums of squares that might be laid out as an analysis of variance (see Chapter 5). In such cases there is therefore no unique set of sums of squares as there is with balanced data. Consequently there is no unique set of equations such as (6.54) and (6.55) and no unique estimators. Related methods include LaMotte's various quadratic estimators, and Rao's minimum norm quadratic unbiased estimators, some of which utilize *a priori* values of the variance components. Searle et al., (1992, Section 11.3) discuss these methods in some detail and give extensive references.

From a theoretical statistics perspective, ANOVA estimators are not always based on sufficient statistics; and minimal, complete, sufficient statistics do not exist (see E 6.7). As a consequence, there are no uniformly optimal ANOVA estimators. One way out of this dilemma may be to invoke further criteria for choosing quadratic forms for estimating variance components. In Section 14.2b we show that restricted ML (REML) estimation suggests using quadratic forms coming from best linear unbiased predicted (BLUP) values.

A consequence of all this is that ANOVA estimation of variance components is losing some (much) of its popularity. This includes the three well-known methods of Henderson (1953), a landmark paper in its time, which definitively motivated much interest in estimating variance components from unbalanced data and which provided methodology that was pivotal and widely used for some thirty years or more. Extensive details of these three methods, of their application to the two-way classification random model, and of ANOVA estimation from unbalanced data are given in the sixty pages of Chapter 6 of Searle et al. (1992).

6.8   MAXIMUM LIKELIHOOD (ML) ESTIMATION

a.   Estimators

In place of the numerous forms of ANOVA estimation (with its deficiencies) the method now widely preferred is maximum likelihood (ML) estimation or variations thereof. In applying it to

$$\mathbf{y} \sim \mathcal{N}\left(\mathbf{X}\beta, \mathbf{V} = \sum_{i=0}^{r} \mathbf{Z}_i \mathbf{Z}_i' \sigma_i^2\right) \tag{6.56}$$

we simultaneously seek ML estimators of β and $\mathbf{V}$. Section 1.7a–iii describes the need for distinguishing between solutions of ML equations and the ML estimators derived therefrom. In keeping with that, the solution $\dot{\beta}$ for β is as in (6.24), only with the solution $\dot{\mathbf{V}}$ replacing $\mathbf{V}$, so that

$$\mathbf{X}\dot{\beta} = \mathbf{X}(\mathbf{X}\dot{\mathbf{V}}^{-1}\mathbf{X})^{-}\mathbf{X}\dot{\mathbf{V}}^{-1}\mathbf{y}. \tag{6.57}$$

And for σ^2 we use $\partial \mathbf{V}/\partial\sigma_i^2 = \partial(\sum_j \mathbf{Z}_j\mathbf{Z}_j'\sigma_i^2)/\partial\sigma_i^2 = \mathbf{Z}_i\mathbf{Z}_i'$ in $\partial l/\partial\varphi$ of (6.25). This gives

$$\frac{\partial l}{\partial\sigma_i^2} = -\frac{1}{2}\mathrm{tr}(\mathbf{V}^{-1}\mathbf{Z}_i\mathbf{Z}_i') - \frac{1}{2}(\mathbf{y} - \mathbf{X}\beta)'\mathbf{V}^{-1}\mathbf{Z}_i\mathbf{Z}_i'\mathbf{V}^{-1}(\mathbf{y} - \mathbf{X}\beta).$$

In equating this to 0 for each $i = 0, 1, \ldots, r$ and continuing with the $\dot{\beta}$ and $\dot{\mathbf{V}}$ notation we have

$$\mathrm{tr}(\dot{\mathbf{V}}^{-1}\mathbf{Z}_i\mathbf{Z}_i') = (\mathbf{y} - \mathbf{X}\dot{\beta})'\dot{\mathbf{V}}^{-1}\mathbf{Z}_i\mathbf{Z}_i'\dot{\mathbf{V}}^{-1}(\mathbf{y} - \mathbf{X}\dot{\beta}). \tag{6.58}$$

Now (6.57) and (6.58) have to be solved for $\dot{\beta}$ and $\dot{\sigma}^2$, so leading to $\dot{\mathbf{V}}$. But the right-hand side of (6.58) involves $\mathbf{X}\dot{\beta}$ of (6.57); and that equation involves $\dot{\mathbf{V}}$. So somehow these equations must simultaneously be solved numerically (often by iteration).

In point of fact we can reduce the two equations (6.57) and (6.58) by observing from (6.57) that $\dot{\mathbf{V}}^{-1}(\mathbf{y} - \mathbf{X}\dot{\beta})$ needed for (6.58) is

$$\dot{\mathbf{V}}^{-1}(\mathbf{y} - \mathbf{X}\dot{\beta}) = \left[\dot{\mathbf{V}}^{-1} - \dot{\mathbf{V}}^{-1}\mathbf{X}(\mathbf{X}\dot{\mathbf{V}}^{-1}\mathbf{X})^{-}\mathbf{X}'\dot{\mathbf{V}}^{-1}\right]\mathbf{y} = \dot{\mathbf{P}}\mathbf{y} \tag{6.59}$$

for $\mathbf{P}$ of (6.27) with, of course, $\dot{\mathbf{P}}$ being $\mathbf{P}$ with $\dot{\mathbf{V}}$ in place of $\mathbf{V}$. Therefore (6.58) can be written as

$$\left\{ _c\ \mathrm{tr}(\dot{\mathbf{V}}^{-1}\mathbf{Z}_i\mathbf{Z}_i') \right\}_{i=0}^{r} = \left\{ _c\ \mathbf{y}'\dot{\mathbf{P}}\mathbf{Z}_i\mathbf{Z}_i'\dot{\mathbf{P}}\mathbf{y} \right\}_{i=0}^{r}. \tag{6.60}$$

So now, for obtaining a solution $\dot{\sigma}^2$, we need concentrate only on (6.60) using (6.59). And when a solution is obtained, (6.57) will yield $\mathbf{X}\dot{\beta}$. Of course, solving (6.60) is not simple; for a few experiment designs yielding balanced data it does have straightforward algebraic solutions, as shown in Searle et al. (1992, Sect. 4.7).

At this point we can evaluate the profile likelihood for $\mathbf{V}$, denoted l_P, which is the likelihood for a given value of $\mathbf{V}$ with the maximizing value of β for that $\mathbf{V}$ inserted. Using (6.59) in the log likelihood, (6.17), gives the profile log likelihood of

$$\log l_P(\mathbf{V}) = -\tfrac{1}{2}\mathbf{y}'\mathbf{P}\mathbf{y} - \tfrac{1}{2}\log|\mathbf{V}| - \tfrac{N}{2}\log(2\pi). \tag{6.61}$$

Although (6.60) and (6.61) do not appear to involve the σ^2s, they do, of course, because the σ^2s are embedded in $\mathbf{V}$ and $\mathbf{P}$. For unbalanced data and even for some balanced data (e.g., two-way crossed classification, random model, see Searle et al., 1992, Section 4.7d), solving (6.60) or maximizing (6.61) has to be achieved by numerical methods. A prime difficulty is to obtain solutions within the range of the parameters (see Section 2.2b-iii). For the variance components model this means $\dot{\sigma}_i^2 \geq 0$ for $i > 0$ and $\dot{\sigma}_0^2 > 0$. On achieving this, the corresponding $\check{\mathbf{V}}$ will be the ML estimator $\hat{\mathbf{V}}$, and $\mathrm{ML}(\mathbf{X}\beta)$ will be $\mathbf{X}\hat{\beta} = \mathbf{X}(\mathbf{X}'\hat{\mathbf{V}}^{-1}\mathbf{X})^{-}\mathbf{X}\hat{\mathbf{V}}^{-1}\mathbf{y}$ as in (6.24). Further discussion of computing techniques for finding the ML estimates will be found in Chapter 14.

b.   Information matrix

Asymptotic sampling variances of ML estimators are obtained from the inverse of the information matrix, which is minus the expected value of the matrix of second derivatives (with respect to the parameters) of the likelihood. Expressions for this, for the general model where $\boldsymbol{\mu}$ depends on a parameter $\boldsymbol{\theta}$ and $\mathbf{V}$ depends on a parameter $\boldsymbol{\varphi}$ and where we assume $\mathbf{y} \sim \mathcal{N}[\boldsymbol{\mu}(\boldsymbol{\theta}), \mathbf{V}(\boldsymbol{\varphi})]$, are given in Section 6.12a–iii. For $\mathbf{y} \sim \mathcal{N}(\mathbf{X}\beta, \sum_{i=0}^{r} \mathbf{Z}_i \mathbf{Z}_i' \sigma_i^2)$ where for (6.74) $\boldsymbol{\mu} = \mathbf{X}\beta, \boldsymbol{\theta} = \beta, \mathbf{V} = \sum_{i=0}^{r} \mathbf{Z}_i \mathbf{Z}_i' \sigma_i^2$ and each element of $\boldsymbol{\varphi}$ is a σ_i^2, (6.74) gives the information matrix as

$$\mathbf{I}\begin{bmatrix} \beta \\ \sigma^2 \end{bmatrix} = \begin{bmatrix} \mathbf{X}'\mathbf{V}^{-1}\mathbf{X} & \mathbf{0} \\ \mathbf{0} & \frac{1}{2}\left\{_m \mathrm{tr}[\mathbf{Z}_i'\mathbf{V}^{-1}\mathbf{Z}_j(\mathbf{Z}_i'\mathbf{V}^{-1}\mathbf{Z}_j)']\right\}_{i,j=0}^{r} \end{bmatrix}.$$
$$(6.62)$$

c.   Asymptotic sampling variances

Inverting (6.62) but using a generalized inverse for $\mathbf{X}'\mathbf{V}^{-1}\mathbf{X}$ gives

$$\mathrm{var}_\infty \begin{bmatrix} \mathbf{X}\hat{\beta} \\ \hat{\sigma}^2 \end{bmatrix}$$

$$= \begin{bmatrix} \mathbf{X}(\mathbf{X}'\mathbf{V}^{-1}\mathbf{X})^{-}\mathbf{X}' & \mathbf{0} \\ \mathbf{0} & 2\left[\left\{_m \mathrm{tr}[\mathbf{Z}_i'\mathbf{V}^{-1}\mathbf{Z}_j(\mathbf{Z}_i'\mathbf{V}^{-1}\mathbf{Z}_j)']\right\}_{i,j=0}^{r}\right]^{-1} \end{bmatrix}$$

so that

$$\mathrm{var}_\infty(\mathbf{X}\hat{\beta}) = \mathbf{X}(\mathbf{X}'\mathbf{V}^{-1}\mathbf{X})^{-}\mathbf{X}' \qquad (6.63)$$

$$\text{var}_\infty(\hat{\boldsymbol{\sigma}}^2) = 2\left[\left\{\,_m \text{tr}[\mathbf{Z}_i'\mathbf{V}^{-1}\mathbf{Z}_j(\mathbf{Z}_i'\mathbf{V}^{-1}\mathbf{Z}_j)']\right\}_{i,j=0}^{r}\right]^{-1} \tag{6.64}$$

and

$$\text{cov}_\infty(\mathbf{X}\hat{\boldsymbol{\beta}},\ \hat{\boldsymbol{\sigma}}^{2'}) = \mathbf{0}. \tag{6.65}$$

6.9   RESTRICTED MAXIMUM LIKELIHOOD (REML)

For estimating variance components an alternative maximum likelihood procedure, known as *restricted* (or *residual*) *maximum likelihood* (REML), maximizes the likelihood of linear combinations of elements of $\mathbf{y}$. They are chosen as $\mathbf{k}'\mathbf{y}$ (for vector $\mathbf{k}$) so that $\mathbf{k}'\mathbf{y}$ contains none of the fixed effects in $\boldsymbol{\beta}$. This means having $\mathbf{k}'$ such that $\mathbf{k}'\mathbf{X} = \mathbf{0}$. For optimality we use the maximum number, $N - r_\mathbf{X}$, of linearly independent vectors $\mathbf{k}'$ and write $\mathbf{K} = [\mathbf{k}_1 \quad \mathbf{k}_2 \quad \cdots \quad \mathbf{k}_{N-r_\mathbf{X}}]$. This results in doing maximum likelihood on $\mathbf{K}'\mathbf{y}$ instead of $\mathbf{y}$, where $\mathbf{K}'\mathbf{X} = \mathbf{0}$ and $\mathbf{K}'$ has full row rank $N - r_\mathbf{X}$. (These results are described in Sections M.4f and g)

a.   Estimation

For $\mathbf{y} \sim \mathcal{N}(\mathbf{X}\boldsymbol{\beta}, \mathbf{V})$ with $\mathbf{K}'\mathbf{X} = \mathbf{0}$, we have

$$\mathbf{K}'\mathbf{y} \sim \mathcal{N}(\mathbf{0}, \mathbf{K}'\mathbf{V}\mathbf{K}).$$

ML equations for $\mathbf{K}'\mathbf{y}$ can therefore be derived from those for $\mathbf{y} \sim \mathcal{N}(\mathbf{X}\boldsymbol{\beta}, \mathbf{V})$, namely (6.58), by replacing

$$
\begin{array}{llll}
\mathbf{y} & \text{with} & \mathbf{K}'\mathbf{y}; & \\
\mathbf{Z} & \text{with} & \mathbf{K}'\mathbf{Z} &
\end{array}
\quad \text{and} \quad
\begin{array}{lll}
\mathbf{X} & \text{with} & \mathbf{K}'\mathbf{X} = \mathbf{0}; \\
\mathbf{V} & \text{with} & \mathbf{K}'\mathbf{V}\mathbf{K}.
\end{array}
\tag{6.66}
$$

On using (6.77) of Section 6.12, namely

$$\mathbf{P} = \mathbf{K}(\mathbf{K}'\mathbf{V}\mathbf{K})^{-1}\mathbf{K}',$$

the ML equations for $\mathbf{K}'\mathbf{y}$ reduce to

$$\left\{\,_c \text{tr}(\dot{\mathbf{P}}\mathbf{Z}_i\mathbf{Z}_i')\right\}_{i=0}^{r} = \left\{\,_c \mathbf{y}'\dot{\mathbf{P}}\mathbf{Z}_i\mathbf{Z}_i'\dot{\mathbf{P}}\mathbf{y}\right\}_{i=0}^{r}. \tag{6.67}$$

These are the REML equations, to be solved for $\dot{\boldsymbol{\sigma}}^2$ which occurs in $\dot{\mathbf{P}}$. It is easily seen that they are the same as the ML equations (6.60) except for $\dot{\mathbf{V}}$ on the left-hand side being replaced by $\dot{\mathbf{P}}$ in (6.67).

b. Sampling variances

If the replacements noted in (6.66) are made to $\text{var}_\infty(\hat{\sigma}^2)$ of (6.64) the result is the variance-covariance matrix of the REML estimators:

$$\text{var}_\infty(\hat{\sigma}^2_{\text{REML}}) = 2\left[\left\{_m \ \text{tr}[\mathbf{Z}'_i\mathbf{PZ}_j(\mathbf{Z}'_i\mathbf{PZ}_j)']\right\}^{r}_{i,j=0}\right]^{-1}.$$

6.10   NOTES AND EXTENSIONS

a.   ML or REML?

An oft-asked question is: Should one use ML or REML? Searle et al. (1992, Sect. 6.8) definitely prefer each of ML and REML over ANOVA estimation. We firmly endorse that preference, particularly because, as has already been mentioned, ANOVA methods do not apply satisfactorily to generalized linear mixed models.

For addressing "ML versus REML" there are a number of features of the two methods that can easily be stated. Both have the merit of being based on the well-respected maximum likelihood principle. This does have the problem that if any of the maximizing solutions are negative, one has to adjust those solutions to yield estimators in the parameter space (see, e.g., Sections 2.2b–ii and –iii). On the other hand, the maximum likelihood principle yields asymptotic sampling variances of the variance components estimators; but it also has the demerit of being difficult to compute. ML provides estimation of fixed effects, but REML itself does not. Nevertheless, overriding these features there seems to be a growing preference for REML, influenced by its following merits. First, it is sensible for balanced data for which REML *solutions* (not estimators) are the ANOVA estimators — and these, despite their ability to be negative, have the substantive merit of being minimal variance unbiased, under normality — and even minimum variance quadratic unbiased otherwise. But there is no guarantee that properties of this nature apply to REML solutions from unbalanced data. Second, REML estimators are based on taking into account the degrees of freedom for the fixed effects in the model. Sections 1.7a–ii, 2.1b, 2.2b–vi and 3.2c show examples of this for balanced data. This is particularly important when the rank of $\mathbf{X}$ is large in relation to the sample size. And, although REML for unbalanced data yields no clean algebraic results, presumably this degree-of-freedom feature occurs with unbalanced data too. Third, because $\boldsymbol{\beta}$ is not involved in

REML, the resulting estimators (of variance components) are invariant to the value of $\boldsymbol{\beta}$. Change $\boldsymbol{\beta}$ (but with $\mathbf{X}$ unchanged) and one does not alter REML estimators. Finally, REML estimators do not seem to be as sensitive to outliers in the data (see Verbyla, 1993) as are ML estimators.

b.   Other methods for estimating variances

Several other methods for estimating variance components are mentioned briefly in Searle et al., (1992, Chap. 11), and even more briefly here. The methods are referred to mostly by their acronyms, which indicate their primary properties. The best known is MINQUE: minimum norm quadratic unbiased estimation, a method which is based on a pre-assigned value of $\boldsymbol{\sigma}^2$. As such, the resulting estimates depend on that value; and for this reason we feel it is of little appeal. Variants of MINQUE are MINQU(0), which takes the pre-assigned value of $\boldsymbol{\sigma}^2$ as having every σ_i^2 (except σ^2) as zero; and I-MINQUE, iterated MINQUE, the estimates from which are identical to REML solutions (which, of course, can be negative); and conversely, solutions from the first iteration of REML are a set of MINQUE estimates. There is also MIVQUE, minimum variance quadratic unbiased estimation — and several variants of it.

6.11   APPENDIX FOR CHAPTER 6

a.   Differentiating a log likelihood

– i.   *A general likelihood under normality*

For the general model under normality,

$$\mathbf{y} \sim \mathcal{N}(\boldsymbol{\mu}, \mathbf{V}) \quad \text{with} \quad \mathrm{E}[\mathbf{y}] = \boldsymbol{\mu} \quad \text{and} \quad \mathrm{var}(\mathbf{y}) = \mathbf{V},$$

the density function is

$$f(\mathbf{y}|\boldsymbol{\mu}, \mathbf{V}) = \frac{\exp[-\frac{1}{2}(\mathbf{y} - \boldsymbol{\mu})'\mathbf{V}^{-1}(\mathbf{y} - \boldsymbol{\mu})]}{(2\pi)^{\frac{1}{2}N}|\mathbf{V}|^{\frac{1}{2}}}.$$

Thus the log likelihood is

$$l = -\tfrac{1}{2}(\mathbf{y} - \boldsymbol{\mu})'\mathbf{V}^{-1}(\mathbf{y} - \boldsymbol{\mu}) - \tfrac{1}{2}\log|\mathbf{V}| - \tfrac{1}{2}N\log 2\pi. \qquad (6.68)$$

We consider a general parameterization of $\boldsymbol{\mu}$ and $\mathbf{V}$ such that each element of $\boldsymbol{\mu}$ is a function of elements of a parameter vector $\boldsymbol{\theta}$; and, similarly, each element of $\mathbf{V}$ is a function of elements of a parameter vector $\boldsymbol{\varphi}$ which is unrelated to $\boldsymbol{\theta}$. Thus we write

$$\boldsymbol{\mu} = \boldsymbol{\mu}(\boldsymbol{\theta}) \qquad \text{and} \qquad \mathbf{V} = \mathbf{V}(\boldsymbol{\varphi})$$

and so have, after ignoring $N/2 \log 2\pi$,

$$l = -\tfrac{1}{2}[\mathbf{y} - \boldsymbol{\mu}(\boldsymbol{\theta})]'[\mathbf{V}(\boldsymbol{\varphi})]^{-1}[\mathbf{y} - \boldsymbol{\mu}(\boldsymbol{\theta})] - \tfrac{1}{2}\log|\mathbf{V}(\boldsymbol{\varphi})|.$$

– ii.   *First derivatives*

Direct differentiation of l (which, in application to vectors demands careful consideration of conformability, as seen in Section M.5) gives

$$\frac{\partial l}{\partial \boldsymbol{\theta}} = \frac{\partial \boldsymbol{\mu}'}{\partial \boldsymbol{\theta}} \mathbf{V}^{-1} (\mathbf{y} - \boldsymbol{\mu}). \tag{6.69}$$

And, for φ_k being an element of $\boldsymbol{\varphi}$ in $\mathbf{V}(\boldsymbol{\varphi})$

$$\frac{\partial l}{\partial \varphi_k} = -\tfrac{1}{2}\left[\text{tr}\left(\mathbf{V}^{-1}\frac{\partial \mathbf{V}}{\partial \varphi_k}\right) - (\mathbf{y} - \boldsymbol{\mu})'\mathbf{V}^{-1}\frac{\partial \mathbf{V}}{\partial \varphi_k}\mathbf{V}^{-1}(\mathbf{y} - \boldsymbol{\mu})\right]. \tag{6.70}$$

Using $\boldsymbol{\mu} = \boldsymbol{\mu}(\boldsymbol{\theta})$ and $\mathbf{V} = \mathbf{V}(\boldsymbol{\varphi})$ in each of these, and equating them to zero gives the ML equations. In the case of (6.70) there will be one such equation for each φ_k of $\boldsymbol{\varphi}$.

– iii.   *Information matrix*

$$\frac{\partial^2 l}{\partial \boldsymbol{\theta} \, \partial \boldsymbol{\theta}'} \;\; = \;\; \frac{\partial}{\partial \boldsymbol{\theta}}\left(\frac{\partial l}{\partial \boldsymbol{\theta}'}\right) = \frac{\partial}{\partial \boldsymbol{\theta}}\left(\left[\frac{\partial \boldsymbol{\mu}'}{\partial \boldsymbol{\theta}}\mathbf{V}^{-1}(\mathbf{y} - \boldsymbol{\mu})\right]'\right), \qquad \text{using (6.69)}$$

$$= \;\; \frac{\partial}{\partial \boldsymbol{\theta}}\left[(\mathbf{y} - \boldsymbol{\mu})'\,\mathbf{V}^{-1}\frac{\partial \boldsymbol{\mu}}{\partial \boldsymbol{\theta}'}\right]$$

$$= \;\; -\frac{\partial \boldsymbol{\mu}'}{\partial \boldsymbol{\theta}}\mathbf{V}^{-1}\frac{\partial \boldsymbol{\mu}}{\partial \boldsymbol{\theta}'} + \left\{{}_c\,(\mathbf{y} - \boldsymbol{\mu})'\,\mathbf{V}^{-1}\frac{\partial^2 \boldsymbol{\mu}}{\partial \theta_i \, \partial \boldsymbol{\theta}'}\right\}$$

and so

$$-\text{E}\left(\frac{\partial^2 l}{\partial \boldsymbol{\theta} \, \partial \boldsymbol{\theta}}\right) = \frac{\partial \boldsymbol{\mu}'}{\partial \boldsymbol{\theta}}\mathbf{V}^{-1}\frac{\partial \boldsymbol{\mu}}{\partial \boldsymbol{\theta}}. \tag{6.71}$$

Also

$$
\frac{\partial^2 l}{\partial \varphi_k \, \partial \boldsymbol{\theta}'} = \frac{\partial}{\partial \varphi_k} \left[(\mathbf{y} - \boldsymbol{\mu})' \mathbf{V}^{-1} \frac{\partial \boldsymbol{\mu}}{\partial \boldsymbol{\theta}'} \right] = (\mathbf{y} - \boldsymbol{\mu})' \frac{\partial \mathbf{V}^{-1}}{\partial \varphi_k} \frac{\partial \boldsymbol{\mu}}{\partial \boldsymbol{\theta}'}
$$

$$
= - (\mathbf{y} - \boldsymbol{\mu})' \mathbf{V}^{-1} \frac{\partial \mathbf{V}}{\partial \varphi_k} \mathbf{V}^{-1} \frac{\partial \boldsymbol{\mu}}{\partial \boldsymbol{\theta}'} .
$$

Since $E[\mathbf{y} - \boldsymbol{\mu}] = \mathbf{0}$

$$
-E \left[\frac{\partial^2 l}{\partial \varphi_k \, \partial \boldsymbol{\theta}'} \right] = \mathbf{0}. \tag{6.72}
$$

Next, on differentiating (6.70) with respect to φ_s,

$$
\frac{\partial^2 l}{\partial \varphi_s \, \partial \varphi_k} =
$$

$$
-\tfrac{1}{2} \left\{ \text{tr} \left(-\mathbf{V}^{-1} \frac{\partial \mathbf{V}}{\partial \varphi_s} \mathbf{V}^{-1} \frac{\partial \mathbf{V}}{\partial \varphi_k} + \mathbf{V}^{-1} \frac{\partial^2 \mathbf{V}}{\partial \varphi_s \, \partial \varphi_k} \right) \right.
$$

$$
+ (\mathbf{y} - \boldsymbol{\mu})' \left[(-1) \mathbf{V}^{-1} \frac{\partial \mathbf{V}}{\partial \varphi_s} \mathbf{V}^{-1} \frac{\partial \mathbf{V}}{\partial \varphi_k} \mathbf{V}^{-1} + \mathbf{V}^{-1} \frac{\partial^2 \mathbf{V}}{\partial \varphi_s \, \partial \varphi_k} \mathbf{V}^{-1} \right.
$$

$$
\left. - \mathbf{V}^{-1} \frac{\partial \mathbf{V}}{\partial \varphi_k} \mathbf{V}^{-1} \frac{\partial \mathbf{V}}{\partial \varphi_s} \mathbf{V}^{-1} \right] (\mathbf{y} - \boldsymbol{\mu}) \right\} .
$$

Now for any $\mathbf{A}$

$$
E[(\mathbf{y} - \boldsymbol{\mu})' \mathbf{A} (\mathbf{y} - \boldsymbol{\mu})] = \text{tr} \{ \mathbf{A} E[(\mathbf{y} - \boldsymbol{\mu})(\mathbf{y} - \boldsymbol{\mu})'] \} = \text{tr}(\mathbf{A}\mathbf{V}).
$$

Therefore

$$
-E \left(\frac{\partial^2 l}{\partial \varphi_s \, \partial \varphi_k} \right) = \tfrac{1}{2} \left\{ \text{tr} \left(-\mathbf{V}^{-1} \frac{\partial \mathbf{V}}{\partial \varphi_s} \mathbf{V}^{-1} \frac{\partial \mathbf{V}}{\partial \varphi_k} + \mathbf{V}^{-1} \frac{\partial^2 \mathbf{V}}{\partial \varphi_s \, \partial \varphi_k} \right) \right.
$$

$$
+ \text{tr} \left[+ \mathbf{V}^{-1} \frac{\partial \mathbf{V}}{\partial \varphi_s} \mathbf{V}^{-1} \frac{\partial \mathbf{V}}{\partial \varphi_k} \right.
$$

$$
\left. \left. - \mathbf{V}^{-1} \frac{\partial^2 \mathbf{V}}{\partial \varphi_s \, \partial \varphi_k} + \mathbf{V}^{-1} \frac{\partial \mathbf{V}}{\partial \varphi_k} \mathbf{V}^{-1} \frac{\partial \mathbf{V}}{\partial \varphi_s} \right] \right\}
$$

$$
= \tfrac{1}{2} \text{tr} \left(\mathbf{V}^{-1} \frac{\partial \mathbf{V}}{\partial \varphi_k} \mathbf{V}^{-1} \frac{\partial \mathbf{V}}{\partial \varphi_s} \right) . \tag{6.73}
$$

Therefore, on assembling (6.71), (6.72) and (6.73), the information matrix is

$$
-\mathrm{E}
\begin{bmatrix}
\dfrac{\partial^2 l}{\partial\theta\,\partial\theta'} & \dfrac{\partial^2 l}{\partial\theta\,\partial\varphi'} \\[2ex]
\left(\dfrac{\partial^2 l}{\partial\theta\,\partial\varphi'}\right)' & \dfrac{\partial^2 l}{\partial\varphi\,\partial\varphi'}
\end{bmatrix}
$$

$$
=
\begin{bmatrix}
\dfrac{\partial\mu'}{\partial\theta}\mathbf{V}^{-1}\dfrac{\partial\mu}{\partial\theta} & \mathbf{0} \\[2ex]
\mathbf{0} & \frac{1}{2}\left\{\operatorname*{tr}_m\left(\mathbf{V}^{-1}\dfrac{\partial\mathbf{V}}{\partial\varphi}\mathbf{V}^{-1}\dfrac{\partial\mathbf{V}}{\partial\varphi_k}\right)\right\}
\end{bmatrix}.
\qquad (6.74)
$$

b.   Differentiating a generalized inverse

Suppose $\mathbf{A}$ is a function of the scalar x, and that we write $d\mathbf{A}/dx$ as $d\mathbf{A}$. Then for $\mathbf{A}^-$ being a generalized inverse of $\mathbf{A}$ defined by

$$
\mathbf{A}\mathbf{A}^-\mathbf{A} = \mathbf{A}
$$

we have

$$
(d\mathbf{A})\mathbf{A}^-\mathbf{A} + \mathbf{A}(d\mathbf{A}^-)\mathbf{A} + \mathbf{A}\mathbf{A}^-(d\mathbf{A}) = d\mathbf{A}.
$$

Post-multiplying by $\mathbf{A}^-\mathbf{A}$ gives

$$
(d\mathbf{A})\mathbf{A}^-\mathbf{A} + \mathbf{A}(d\mathbf{A}^-)\mathbf{A} + \mathbf{A}\mathbf{A}^-(d\mathbf{A})\mathbf{A}^-\mathbf{A} = (d\mathbf{A})\mathbf{A}^-\mathbf{A}
$$

which is

$$
\mathbf{A}(d\mathbf{A}^-)\mathbf{A} = -\mathbf{A}\mathbf{A}^-(d\mathbf{A})\mathbf{A}^-\mathbf{A}.
\qquad (6.75)
$$

Recalling that equations $\mathbf{A}\mathbf{x} = \mathbf{y}$ can be solved as $\mathbf{x} = \mathbf{A}^-\mathbf{y}$ we can "solve" (6.75) as

$$
d\mathbf{A}^- = -\mathbf{A}^-\mathbf{A}\mathbf{A}^-(d\mathbf{A})\mathbf{A}^-\mathbf{A}\mathbf{A}^-.
$$

When $\mathbf{A}$ is nonsingular this gives $d\mathbf{A}^{-1}$. When $\mathbf{A}$ is singular we can either assume that $\mathbf{A}^-$ is reflexive (i.e., that $\mathbf{A}^-\mathbf{A}\mathbf{A}^- = \mathbf{A}^-$), or if $\mathbf{A}^-$ is not reflexive, use $\mathbf{A}^* = \mathbf{A}^-\mathbf{A}\mathbf{A}^-$ in its place, and $\mathbf{A}^*$ *is* reflexive. In either case we then get

$$
d\mathbf{A}^- = -\mathbf{A}^-(d\mathbf{A})\mathbf{A}^-.
$$

c. Differentiation for the variance components model

For (6.35) and the expected value that follows it, we want the partial derivative $\partial^2(\mathbf{X'V}^{-1}\mathbf{X})/\partial\sigma_i^2\,\partial\sigma_j^2$. Its derivation proceeds as follows. Recall from (6.14) and (6.16) that

$$\mathbf{D}_* = \operatorname{var}(\mathbf{u}) = \operatorname{var}\left\{{}_c\,\mathbf{u}_i\right\}_{i=0}^{r} = \left\{{}_d\,\sigma_i^2\mathbf{I}_{q_i}\right\}_{i=0}^{r}$$

and

$$\mathbf{V} = \mathbf{Z}_*\mathbf{D}_*\mathbf{Z}_*' = \sum_{i=0}^{r}\sigma_i^2\mathbf{Z}_i\mathbf{Z}_i'.$$

As here, and in all that follows, we have $i = 0, 1, \ldots, r$. From $\mathbf{D}_*$ and $\mathbf{V}$ we then have the following results:

$$\frac{\partial\mathbf{D}_*}{\partial\sigma_i^2} = \left\{{}_d\,\delta_{ij}\mathbf{I}_{q_j}\right\}_{j=0}^{r}\ \text{with}\ \delta_{ii} = 1\ \text{and}\ \delta_{ij} = 0\ \text{for}\ j \neq i.$$

$$\frac{\partial\mathbf{V}}{\partial\sigma_i^2} = \mathbf{Z}_i\mathbf{Z}_i'$$

$$\frac{\partial\mathbf{V}^{-1}}{\partial\sigma_i^2} = -\mathbf{V}^{-1}\mathbf{Z}_i\mathbf{Z}_i'\mathbf{V}^{-1}$$

$$\frac{\partial(\mathbf{X'V}^{-1}\mathbf{X})^-}{\partial\sigma_i^2} = (\mathbf{X'V}^{-1}\mathbf{X})^-\mathbf{X'V}^{-1}\mathbf{Z}_i\mathbf{Z}_i'\mathbf{V}^{-1}\mathbf{X}(\mathbf{X'V}^{-1}\mathbf{X})^-$$

$$\frac{\partial^2(\mathbf{X'V}^{-1}\mathbf{X})^-}{\partial\sigma_i^2\,\partial\sigma_j^2} = (\mathbf{X'V}^{-1}\mathbf{X})^-\mathbf{X'V}^{-1}\mathbf{Z}_j\mathbf{Z}_j'\mathbf{V}^{-1}\mathbf{X}(\mathbf{X'V}^{-1}\mathbf{X})^-$$

$$\times\ \mathbf{X'V}^{-1}\mathbf{Z}_i\mathbf{Z}_i'\mathbf{V}^{-1}\mathbf{X}(\mathbf{X'V}^{-1}\mathbf{X})^-$$

$$-\ (\mathbf{X'V}^{-1}\mathbf{X})^-\mathbf{X'V}^{-1}\mathbf{Z}_j\mathbf{Z}_j'\mathbf{V}^{-1}\mathbf{Z}_i\mathbf{Z}_i'\mathbf{V}^{-1}\mathbf{X}(\mathbf{X'V}^{-1}\mathbf{X})^-$$

$$-\ (\mathbf{X'V}^{-1}\mathbf{X})^-\mathbf{X'V}^{-1}\mathbf{Z}_i\mathbf{Z}_i'\mathbf{V}^{-1}\mathbf{Z}_j\mathbf{Z}_j'\mathbf{V}^{-1}\mathbf{X}(\mathbf{X'V}^{-1}\mathbf{X})^-$$

$$+\ (\mathbf{X'V}^{-1}\mathbf{X})^-\mathbf{X'V}^{-1}\mathbf{Z}_i\mathbf{Z}_i'\mathbf{V}^{-1}\mathbf{X}(\mathbf{X'V}^{-1}\mathbf{X})^-$$

$$\times\ \mathbf{X'V}^{-1}\mathbf{Z}_j\mathbf{Z}_j'\mathbf{V}^{-1}\mathbf{X}(\mathbf{X'V}^{-1}\mathbf{X})^-.$$

Then for

$$\mathbf{P} = \mathbf{V}^{-1} - \mathbf{V}^{-1}\mathbf{X}(\mathbf{X}'\mathbf{V}^{-1}\mathbf{X})^{-}\mathbf{X}'\mathbf{V}^{-1}$$

$$\frac{\partial^2(\mathbf{X}'\mathbf{V}^{-1}\mathbf{X})^{-}}{\partial\sigma_i^2\,\partial\sigma_j^2} = -(\mathbf{X}'\mathbf{V}^{-1}\mathbf{X})^{-}\mathbf{X}'\mathbf{V}^{-1}(\mathbf{Z}_i\mathbf{Z}_i'\mathbf{P}\mathbf{Z}_j\mathbf{Z}_j' + \mathbf{Z}_j\mathbf{Z}_j'\mathbf{P}\mathbf{Z}_i\mathbf{Z}_i')$$

$$\times \mathbf{V}^{-1}\mathbf{X}(\mathbf{X}'\mathbf{V}^{-1}\mathbf{X})^{-}. \qquad (6.76)$$

To finish, Δ_{ij} of (6.47) comes from $\partial^2(\mathbf{m}'\mathbf{DZ}'\mathbf{PZ}'\mathbf{Dm})/\partial\sigma_i^2\,\partial\sigma_j^2$ just as (6.76) was needed for (6.35) to yield (6.36). First recall (e.g., Searle et al., 1992, Sect. M.4f) for $\mathbf{X}$ having N rows and rank $r_{\mathbf{X}}$ that for $\mathbf{K}'$ satisfying $\mathbf{K}'\mathbf{X} = \mathbf{0}$ with $\mathbf{K}'$ having full row rank $N - r_{\mathbf{X}}$ we have

$$\mathbf{P} = \mathbf{V}^{-1} - \mathbf{V}^{-1}\mathbf{X}(\mathbf{X}'\mathbf{V}^{-1}\mathbf{X})^{-}\mathbf{X}'\mathbf{V}^{-1} = \mathbf{K}(\mathbf{K}'\mathbf{V}\mathbf{K})^{-1}\mathbf{K}'.$$

$$(6.77)$$

Therefore

$$\frac{\partial\mathbf{P}}{\partial\sigma_i^2} = -\mathbf{K}(\mathbf{K}'\mathbf{V}\mathbf{K})^{-1}\mathbf{K}'\mathbf{Z}_i\mathbf{Z}_i'\mathbf{K}(\mathbf{K}'\mathbf{V}\mathbf{K})^{-1}\mathbf{K}' = -\mathbf{P}\mathbf{Z}_i\mathbf{Z}_i'\mathbf{P}.$$

Finally, for $\mathbf{E}_i = \partial\mathbf{D}_*/\partial\sigma_i^2$, which is a block diagonal matrix with the ith block being $\mathbf{I}_{q_i}$ and the rest $\mathbf{0}$,

$$\frac{\partial}{\partial\sigma_i^2}(\mathbf{m}'\mathbf{DZ}'\mathbf{PZDm})$$

$$= \mathbf{m}'\mathbf{E}_i\mathbf{Z}'\mathbf{PZDm} - \mathbf{m}'\mathbf{DZ}'\mathbf{PZ}_i\mathbf{Z}_i'\mathbf{PZDm} + \mathbf{m}'\mathbf{DZ}'\mathbf{PZE}_i\mathbf{m}$$

$$= \mathbf{m}_i'\mathbf{Z}_i'\mathbf{PZDm} - \mathbf{m}'\mathbf{DZ}'\mathbf{PZ}_i\mathbf{Z}_i'\mathbf{PZDm} + \mathbf{m}'\mathbf{DZ}'\mathbf{PZ}_i\mathbf{m}_i$$

$$= 2\mathbf{m}_i'\mathbf{Z}_i'\mathbf{PZDm} - \mathbf{m}'\mathbf{DZ}'\mathbf{PZ}_i\mathbf{Z}_i'\mathbf{PZDm}$$

because each term is a scalar and so equals its transpose. Therefore twice the value of Δ_{ij} of (6.47) is

$$2\Delta_{ij} = \frac{\partial^2(\mathbf{m}'\mathbf{DZ}'\mathbf{PZDm})}{\partial\sigma_i^2\,\partial\sigma_j^2}$$

$$= -2\mathbf{m}'_i\mathbf{Z}'_i\mathbf{PZ}_j\mathbf{Z}'_j\mathbf{PZDm} + 2\mathbf{m}'_i\mathbf{Z}'_i\mathbf{PZ}_j\mathbf{m}_j$$

$$- \mathbf{m}'_j\mathbf{Z}'_j\mathbf{PZ}_i\mathbf{Z}'_i\mathbf{PZDm} + \mathbf{m}'\mathbf{DZ}'\mathbf{PZ}_j\mathbf{Z}'_j\mathbf{PZ}_i\mathbf{Z}'_i\mathbf{PZDm}$$

$$+ \mathbf{m}'\mathbf{DZ}'\mathbf{PZ}_i\mathbf{Z}'_i\mathbf{PZ}_j\mathbf{Z}'_j\mathbf{PZDm} - \mathbf{m}'\mathbf{DZ}'\mathbf{PZ}_i\mathbf{Z}'_i\mathbf{PZ}_j\mathbf{m}_j$$

$$= 2(-\mathbf{m}'_i\mathbf{Z}'_i\mathbf{PZ}_j\mathbf{Z}'_j\mathbf{PZDm} + \mathbf{m}'_i\mathbf{Z}'_i\mathbf{PZ}_j\mathbf{m}_j - \mathbf{m}'\mathbf{DZ}'\mathbf{PZ}_i\mathbf{Z}'_i\mathbf{PZ}_j\mathbf{m}_j$$

$$+ \mathbf{m}'\mathbf{DZ}'\mathbf{PZ}_i\mathbf{Z}'_i\mathbf{PZ}_j\mathbf{Z}'_j\mathbf{PZDm})$$

$$= 2(\mathbf{m}'_i - \mathbf{m}'\mathbf{DZ}'\mathbf{PZ}_i)\mathbf{Z}'_i\mathbf{PZ}_j(\mathbf{m}_j - \mathbf{Z}'_j\mathbf{PZDm}). \qquad (6.78)$$

as required.

6.12 EXERCISES

E 6.1 Use (6.2), (6.3), and (1.14) to prove (6.6).

E 6.2 (a) Why does $\mathbf{V}^{-1} = \mathbf{L}'\mathbf{L}$ for some non-singular $\mathbf{L}$?

   (b) Through using $\mathbf{L}$ of (a), explain why $\mathbf{X}\boldsymbol{\beta}^0$ of (6.19) is invariant to $(\mathbf{X}'\mathbf{V}^{-1}\mathbf{X})^-$.

E 6.3 For the linear model $E[\mathbf{y}] = \mathbf{X}\boldsymbol{\beta}$, where $\mathbf{X}$ is of full column rank, show that all linear combinations of $\boldsymbol{\beta}$ are estimable.

E 6.4 In the linear model $E[\mathbf{y}] = \mathbf{X}\boldsymbol{\beta}$, with $\mathrm{var}(\mathbf{y}) = \mathbf{V}$, where $\mathbf{V}$ is known and nonsingular invertible, show that the MLE of an estimable function $\mathbf{c}'\boldsymbol{\beta}$ is $\mathbf{c}'(\mathbf{X}'\mathbf{V}^{-1}\mathbf{X})^-\mathbf{X}'\mathbf{V}^{-1}\mathbf{y}$. Why is it reasonable to assume that $\mathbf{V}$ is nonsingular whereas it is not reasonable to assume $\mathbf{X}'\mathbf{V}^{-1}\mathbf{X}$ is?

E 6.5 Consider the balanced one-way random model:

$$y_{ij}|a_i \ \sim \ \text{indep.} \ \mathcal{N}(\mu + a_i, \sigma^2) \ i = 1, 2, \dots, m; \ j = 1, 2, \dots, n$$

$$a_i \ \sim \ \text{i.i.d.} \ \mathcal{N}(0, \sigma_a^2).$$

Find a $100(1 - \alpha)\%$ confidence interval for the intraclass correlation coefficient.

E 6.6 Write the following models in matrix notation and in each case
      determine the marginal mean and variance of y. If a factor is not
      specified, assume it is fixed.

   (a) $y_{ij}|a_i \sim$ indep. $\mathcal{N}(\mu + a_i + \beta_j, \sigma^2); a_i \sim$ i.i.d. $\mathcal{N}(0, \sigma_a^2);$
       $i = 1, 2, \ldots, m; \; j = 1, 2, \ldots, n.$ (Two-way mixed).

   (b) $y_{ij}|a_i, b_j \sim$ indep. $\mathcal{N}(\mu + a_i + b_j, \sigma_a^2); a_i \sim$ i.i.d. $\mathcal{N}(0, \sigma_a^2);$
       $b_j \sim$ i.i.d. $\mathcal{N}(0, \sigma_b^2); a_i$ and b_j indep. ; $i = 1, 2, \ldots, m;$
       $j = 1, 2, \ldots, n.$ (Two-way random).

   (c) $y_{ijk}|a_i, g_{ij} \sim$ indep. $\mathcal{N}(\mu + a_i + \beta_j + g_{ij}, \sigma^2); a_i \sim$ i.i.d. $\mathcal{N}(0, \sigma_a^2);$
       $g_{ij} \sim$ i.i.d. $\mathcal{N}(0, \sigma_g^2); a_i$ and g_{ij} indep.; $i = 1, 2, \ldots, m;$
       $j = 1, 2, \ldots, n; \; k = 1, 2, \ldots, r.$ (Two-way mixed with inter-
       action).

E 6.7 Consider the unbalanced one-way random model:

$$y_{ij}|a_i \quad \sim \quad \text{indep.} \; \mathcal{N}(\mu + a_i, \sigma^2); \; i = 1, 2, \ldots, m; \; j = 1, 2, \ldots, n_i$$

$$a_i \quad \sim \quad \text{i.i.d.} \; \mathcal{N}(0, \sigma_a^2).$$

Show that the sufficient statistics are not complete. One way to
do this is to develop an unbiased estimator of zero, or equivalently
two different unbiased estimators of the same parameter based
on the sufficient statistics. Do this by constructing two different
estimators of σ_a^2. *Hint:* Consider the "usual" sums of squares
for treatments in a one-way ANOVA, $\Sigma_i n_i(\bar{y}_{i\cdot} - \bar{y}_{\cdot\cdot})^2$, and the
unweighted version, $\Sigma_i(\bar{y}_{i\cdot} - \bar{y}_u)^2$, where $\bar{y}_u = (1/m)\Sigma_i \bar{y}_{i\cdot}$. An
important implication of this result is that there is no UMVUE
for σ_a^2.

E 6.8 (a) For $\mathbf{K}'\mathbf{y}$ of Section 6.9 (REML) write the log likelihood; de-
          note it as l_1.

     (b) Kenward and Roger write the log likelihood as

         $$l_2 = \text{constant} \; - \frac{1}{2}\log|\mathbf{V}| - \frac{1}{2}\log|\mathbf{X}'\mathbf{V}^{-1}\mathbf{X}| - \frac{1}{2}\mathbf{y}'\mathbf{P}\mathbf{y}$$

         for $\mathbf{P} = \mathbf{V}^{-1} - \mathbf{V}^{-1}\mathbf{X}(\mathbf{X}'\mathbf{V}^{-1}\mathbf{X})^{-}\mathbf{X}'\mathbf{V}^{-1}$. Show that the
         quadratic forms in $\mathbf{y}$ are the same in l_1 and l_2.

     (c) If $\mathbf{V}$ is a function of t, write $\partial\mathbf{V}/\partial t$ as $\mathbf{V}_t$. Show that $\partial l_1/\partial t = \partial l_2/\partial t$.

E 6.9 For $\mathbf{K}$ of Section 6.9, determine the effect on $\mathbf{Z'Py}$ of the replacements listed in (6.66).

E 6.10 The ML and REML equations for estimating σ^2 are (6.60) and (6.67), respectively. Use those equations to derive ML and REML solutions for the following models. In each case $i = 1, \ldots, m$, and $j = 1, \ldots, n$

(a) $\mathrm{E}[y_{ij}] = \mu, \mathbf{V} = \sigma^2 \mathbf{I}_N$.

(b) $\mathrm{E}[y_{ij}] = \mu + \alpha_i, \mathbf{V} = \sigma^2 \mathbf{I}_N$.

(c) $\mathrm{E}[y_{ij}] = \mu, \mathbf{V} = \sigma^2 \mathbf{I}_N + \left\{ _d \sigma_a^2 \mathbf{J}_n \right\}_{i=1}^m$.

(d) $\mathrm{E}[y_{ij}] = \mu + bx_i, \mathbf{V} = \sigma^2 \mathbf{I}_N$.

E 6.11 In line with definitions (6.21), define $\mathbf{u}_0$ and $\mathbf{u}_* = \begin{bmatrix} \mathbf{u}_0 \\ \mathbf{u} \end{bmatrix}$ so that

$$\mathbf{Z}_* \mathbf{u}_* = \begin{bmatrix} \mathbf{Z}_0 & \mathbf{Z} \end{bmatrix} \begin{bmatrix} \mathbf{u}_0 \\ \mathbf{u} \end{bmatrix} = \mathbf{Z}_0 \mathbf{u}_0 + \mathbf{Z} \mathbf{u}.$$

For $\boldsymbol{\beta}$ of (6.29) and $\tilde{\mathbf{u}}^0$ of (6.57) show that $\mathbf{y} - \mathbf{X}\boldsymbol{\beta}^0 - \mathbf{Z}\tilde{\mathbf{u}}^0$ is identical to the predictor of $\mathbf{u}_0^0$ derived from $\mathbf{D}_* \mathbf{Z}'_* \mathbf{Py}$.

Chapter 7

GENERALIZED LINEAR MIXED MODELS (GLMMs)

7.1 INTRODUCTION

The use of random factors is not restricted to linear mixed models, the topic of Chapter 6. For many of the same reasons as seen there, we may want to incorporate random factors into nonlinear models. That is, we may wish to build a model that accommodates correlated data, or to consider the levels of a factor as selected from a population of levels in order to make inference to that population.

For example, suppose we wish to study factors affecting cost of hospitalization by taking a random sample of patient records from each of 15 research hospitals. The costs within a hospital almost certainly must be regarded as correlated. They will be similar because of the general costs of running the hospital, billing practices, costs of nearby, competing hospitals, and so on. Also, a goal may be to make inferences to a larger population of research hospitals. Both of these could be accommodated by incorporating random hospital effects into the model. And a potential benefit could be gained by using best prediction technology to improve the predictions for individual hospitals.

How should the random effects be incorporated? For many problems the decision between treating a factor as fixed or random is a subtle one. As illustration, if we change the example of hospitalization costs slightly and study only three hospitals, all of which are unique and whose effects cannot easily be regarded as a random sample, we must treat them as fixed. But this would not fundamentally change the way in which they would be incorporated into the mean of the response. This line of argument suggests that random factors should be incorporated in the same manner and in the same portion of the model as the fixed factors. This is exactly the approach of Chapter 6. Our basic linear model there had mean $E[\mathbf{y}] = \mathbf{X}\boldsymbol{\beta}$. We incorporated random effects by enlarging the model to be $E[\mathbf{y}|\mathbf{u}] = \mathbf{X}\boldsymbol{\beta} + \mathbf{Z}\mathbf{u}$. If we write a combined model matrix $\mathbf{X}^* = [\mathbf{X} \quad \mathbf{Z}]$ and an enlarged "parameter" vector $\boldsymbol{\beta}^* = [\boldsymbol{\beta}' \quad \mathbf{u}']'$ it is easy to see that $E[\mathbf{y}|\mathbf{u}] = \mathbf{X}^*\boldsymbol{\beta}^*$.

This suggests a straightforward extension of the generalized linear models of Chapter 5: Append the random effects in the form $\mathbf{Z}\mathbf{u}$ to the linear predictor $\mathbf{X}\boldsymbol{\beta}$. This will achieve the two main goals of incorporating correlation and allowing broader inference. However, the nonlinear nature of the model creates complications not encountered in Chapter 6.

In the remainder of the chapter we define the generalized linear mixed model (GLMM), explore the consequences of adding random factors and discuss a variety of inferential methods. The issue of prediction of random effects we leave to Chapter 13. Models in which the random effects cannot be incorporated in a linear predictor are dealt with briefly in Chapter 11.

7.2 STRUCTURE OF THE MODEL

a. Conditional distribution of y

To specify the model we start with the conditional distribution of $\mathbf{y}$ given $\mathbf{u}$. As in (5.5) and (5.6), the response vector $\mathbf{y}$ is typically, but not necessarily, assumed to consist of conditionally independent elements, each with a distribution with density from the exponential family or similar to the exponential family:

$$y_i|\mathbf{u} \quad \sim \quad \text{indep.} \ f_{Y_i|\mathbf{u}}(y_i|\mathbf{u})$$

$$f_{Y_i|\mathbf{u}}(y_i|\mathbf{u}) \quad = \quad \exp\{[y_i\gamma_i - b(\gamma_i)]/\tau^2 - c(y_i, \tau)\}. \tag{7.1}$$

Next we model a transformation of the mean, which would be some function of γ_i, as a linear model in both the fixed and random factors:

$$E[y_i|\mathbf{u}] \quad = \quad \mu_i$$

$$g(\mu_i) \;=\; \mathbf{x}_i'\boldsymbol{\beta} + \mathbf{z}_i'\mathbf{u}. \tag{7.2}$$

As in Chapter 5, $g(\cdot)$ is a known function, called the *link function* (since it links together the conditional mean of y_i and the linear form of predictors), $\mathbf{x}_i'$ is the ith row of the model matrix for the fixed effects, and $\boldsymbol{\beta}$ is the fixed effects parameter vector. To that specification we have added $\mathbf{z}_i'$, which is the ith row of the model matrix for the random effects, and $\mathbf{u}$, the random effects vector. Note that we are using μ_i here to denote the conditional mean of y_i given $\mathbf{u}$, not the unconditional mean. To complete the specification we assign a distribution to the random effects:

$$\mathbf{u} \sim f_{\mathbf{U}}(\mathbf{u}). \tag{7.3}$$

In light of the fact that the conditional distribution of $\mathbf{y}$ given $\mathbf{u}$ is just a notational extension of the generalized linear model of Chapter 5 (i.e., μ_i represents the conditional mean rather than the marginal or unconditional mean; otherwise, all is the same), many of the relationships derived there will hold. Correspondingly, as below (5.14), we denote the conditional variance of y_i given $\mathbf{u}$ as $\tau^2 v(\mu_i)$ in order to display its dependence on the conditional mean μ_i.

7.3   CONSEQUENCES OF HAVING RANDOM EFFECTS

a.   Marginal versus conditional distribution

Since the model specification in (7.1) and (7.2) is made conditional on the value of $\mathbf{u}$, we now derive aspects of the marginal distribution of $\mathbf{y}$ in order to understand what has been assumed for the observed data.

b.   Mean of y

The mean of $\mathbf{y}$ can be calculated by the usual device of iterated expectation:

$$
\begin{aligned}
\mathrm{E}[y_i] \;&=\; \mathrm{E}\left[\mathrm{E}[y_i|\mathbf{u}]\right] \\[2mm]
&=\; \mathrm{E}[\mu_i] \\[2mm]
&=\; \mathrm{E}[g^{-1}(\mathbf{x}_i'\boldsymbol{\beta} + \mathbf{z}_i'\mathbf{u})].
\end{aligned}
\tag{7.4}
$$

This cannot, in general, be simplified, due to the nonlinear function $g^{-1}(\cdot)$.

To illustrate for a particular $g(\cdot)$, suppose we have a log link so that $g(\mu) = \log \mu$ and $g^{-1}(x) = \exp\{x\}$. Then we have

$$
\begin{aligned}
\mathrm{E}[y_i] &= \mathrm{E}[\exp\{\mathbf{x}_i'\boldsymbol{\beta} + \mathbf{z}_i'\mathbf{u}\}] \\[2mm]
&= \exp\{\mathbf{x}_i'\boldsymbol{\beta}\}\mathrm{E}[\exp\{\mathbf{z}_i'\mathbf{u}\}] \\[2mm]
&= \exp\{\mathbf{x}_i'\boldsymbol{\beta}\}M_{\mathbf{u}}(\mathbf{z}_i), \quad (7.5)
\end{aligned}
$$

where $M_{\mathbf{u}}(\mathbf{z}_i)$ is the moment generating function of $\mathbf{u}$ evaluated at $\mathbf{z}_i$ (see Section S.1c).

Suppose further that $u_i \sim \mathcal{N}(0, \sigma_u^2)$ and that each row of $\mathbf{Z}$ has only a single non-zero entry, equal to 1. Then $M_u(\mathbf{z}_i) = \exp\{\sigma_u^2/2\}$ and

$$
\mathrm{E}[y_i] = \exp\{\mathbf{x}_i'\boldsymbol{\beta}\}\exp\{\sigma_u^2/2\}
$$

or

$$
\log \mathrm{E}[y_i] = \mathbf{x}_i'\boldsymbol{\beta} + \sigma_u^2/2. \quad (7.6)
$$

c.   Variances

To derive the marginal variance of $\mathbf{y}$ we use formula (1.14):

$$
\begin{aligned}
\mathrm{var}(y_i) &= \mathrm{var}(\mathrm{E}[y_i|\mathbf{u}]) + \mathrm{E}[\mathrm{var}(y_i|\mathbf{u})] \\[2mm]
&= \mathrm{var}(\mu_i) + \mathrm{E}[\tau^2 v(\mu_i)] \\[2mm]
&= \mathrm{var}(g^{-1}[\mathbf{x}_i'\boldsymbol{\beta} + \mathbf{z}_i'\mathbf{u}]) + \mathrm{E}\left[\tau^2 v(g^{-1}[\mathbf{x}_i'\boldsymbol{\beta} + \mathbf{z}_i'\mathbf{u}])\right], (7.7)
\end{aligned}
$$

which again cannot be simplified appreciably without making specific assumptions about the form of $g(\cdot)$ and/or the conditional distribution of $\mathbf{y}$.

To illustrate the derivation assume, as before, that we have a log link and now further assume that the elements of $\mathbf{y}$ are conditionally independent given $\mathbf{u}$ with a Poisson distribution. Hence the conditional variance of y_i given $\mathbf{u}$ is $\tau^2 v(\mu_i) = \mu_i$. Using these facts in (7.7) gives

$$
\begin{aligned}
\mathrm{var}(y_i) &= \mathrm{var}(\mu_i) + \mathrm{E}[\mu_i] \\[2mm]
&= \mathrm{var}(\exp\{\mathbf{x}_i'\boldsymbol{\beta} + \mathbf{z}_i'\mathbf{u}\}) + \mathrm{E}\left[\exp\{\mathbf{x}_i'\boldsymbol{\beta} + \mathbf{z}_i'\mathbf{u}\}\right]
\end{aligned}
$$

$$= \quad \mathrm{E}[(\exp\{2(\mathbf{x}_i'\boldsymbol{\beta} + \mathbf{z}_i'\mathbf{u})\})] - [\mathrm{E}(\exp\{\mathbf{x}_i'\boldsymbol{\beta} + \mathbf{z}_i'\mathbf{u}\})]^2$$

$$+ \mathrm{E}\left[\exp\{\mathbf{x}_i'\boldsymbol{\beta} + \mathbf{z}_i'\mathbf{u}\}\right] \tag{7.8}$$

$$= \quad \exp\{2\mathbf{x}_i'\boldsymbol{\beta}\}\left(M_u(2\mathbf{z}_i) - [M_u(\mathbf{z}_i)]^2 + \exp\{-\mathbf{x}_i'\boldsymbol{\beta}\}M_u(\mathbf{z}_i)\right).$$

If we make the further assumption that $u_i \sim \mathcal{N}(0, \sigma_u^2)$ and that each row of $\mathbf{Z}$ has only a single non-zero entry, equal to 1, then

$$\mathrm{var}(y_i) \quad = \quad \exp\{2\mathbf{x}_i'\boldsymbol{\beta}\}\left(\exp\{2\sigma_u^2\} - \exp\{\sigma_u^2\}\right) + \exp\{\mathbf{x}_i'\boldsymbol{\beta}\}\exp\{\sigma_u^2/2\}$$

$$= \quad \mathrm{E}[y_i]\left(\exp\{\mathbf{x}_i'\boldsymbol{\beta}\}\left[\exp\{3\sigma_u^2/2\} - \exp\{\sigma_u^2/2\}\right] + 1\right). \tag{7.9}$$

Since the term in parentheses in (7.9) is greater than 1, we see that the variance is larger than the mean. Therefore, although the conditional distribution of y_i given $\mathbf{u}$ is Poisson, the marginal distribution cannot be. In fact, under these assumptions, it will always be overdispersed compared to the Poisson distribution. In this sense we can think of random effects as a way to model or attribute overdispersion to a particular source.

d.   Covariances and correlations

As noted before, the use of random effects introduces a correlation among observations which have any random effect in common. The same is true for generalized linear mixed models. Assuming conditional independence of the elements of $\mathbf{y}$ and using (1.16), we have

$$\mathrm{cov}(y_i, y_j) \quad = \quad \mathrm{cov}(\mathrm{E}[y_i|\mathbf{u}], \mathrm{E}[y_j|\mathbf{u}]) + \mathrm{E}[\mathrm{cov}(y_i, y_j|\mathbf{u})]$$

$$= \quad \mathrm{cov}(\mu_i, \mu_j) + \mathrm{E}[0]$$

$$= \quad \mathrm{cov}(g^{-1}[\mathbf{x}_i'\boldsymbol{\beta} + \mathbf{z}_i'\mathbf{u}], g^{-1}[\mathbf{x}_j'\boldsymbol{\beta} + \mathbf{z}_j'\mathbf{u}]). \tag{7.10}$$

If we have a log link, this can be evaluated as

$$\mathrm{cov}(y_i, y_j) \quad = \quad \mathrm{cov}(\exp\{\mathbf{x}_i'\boldsymbol{\beta} + \mathbf{z}_i'\mathbf{u}\}, \exp\{\mathbf{x}_j'\boldsymbol{\beta} + \mathbf{z}_j'\mathbf{u}\})$$

$$= \quad \exp\{\mathbf{x}_i'\boldsymbol{\beta} + \mathbf{x}_j'\boldsymbol{\beta}\}\mathrm{cov}(\exp\{\mathbf{z}_i'\mathbf{u}\}, \exp\{\mathbf{z}_j'\mathbf{u}\}) \tag{7.11}$$

$$= \quad \exp\{\mathbf{x}_i'\boldsymbol{\beta} + \mathbf{x}_j'\boldsymbol{\beta}\}\left[M_u(\mathbf{z}_i + \mathbf{z}_j) - M_u(\mathbf{z}_i)M_u(\mathbf{z}_j)\right].$$

Again we make further assumptions, namely that $\mathbf{u} \sim \mathcal{N}(\mathbf{0}, \mathbf{I}\sigma_u^2)$ and that each row of $\mathbf{Z}$ has only a single non-zero entry, equal to 1. Then

$$\text{cov}(y_i, y_j) = \exp\{\mathbf{x}_i'\boldsymbol{\beta} + \mathbf{x}_j'\boldsymbol{\beta}\} \left[\exp\{\sigma_u^2\}(\exp\{\mathbf{z}_i'\mathbf{z}_j\sigma_u^2\} - 1)\right], \quad (7.12)$$

which is equal to zero if $\mathbf{z}_i'\mathbf{z}_j = 0$ (i.e., if the two observations do not share a random effect) and is positive otherwise (in which case $\mathbf{z}_i'\mathbf{z}_j = 1$).

From (7.12) and (7.9), when $\mathbf{z}_i'\mathbf{z}_j = 1$, we can calculate the correlation (after canceling $\exp\{\mathbf{x}_i'\boldsymbol{\beta} + \mathbf{x}_j'\boldsymbol{\beta}\}$ in the numerator and denominator) as:

$$\text{corr}(y_i, y_j)$$

$$= \frac{e^{2\sigma_u^2} - e^{\sigma_u^2}}{\sqrt{\left(e^{2\sigma_u^2} - e^{\sigma_u^2} + e^{-\mathbf{x}_i'\boldsymbol{\beta} + \sigma_u^2/2}\right)\left(e^{2\sigma_u^2} - e^{\sigma_u^2} + e^{-\mathbf{x}_j'\boldsymbol{\beta} + \sigma_u^2/2}\right)}}$$

$$= \frac{1}{\sqrt{\left(1 + \eta e^{-\mathbf{x}_i'\boldsymbol{\beta}}\right)\left(1 + \eta e^{-\mathbf{x}_j'\boldsymbol{\beta}}\right)}}, \quad (7.13)$$

where η is given by $1/(e^{3\sigma_u^2/2} - e^{\sigma_u^2/2})$.

7.4 ESTIMATION BY MAXIMUM LIKELIHOOD

a. Likelihood

From (7.1), (7.2), and (7.3) it is straightforward to write down a formula for the likelihood:

$$L = \int \prod_i f_{Y_i|\mathbf{u}}(y_i|\mathbf{u}) f_{\mathbf{U}}(\mathbf{u}) \, d\mathbf{u}, \quad (7.14)$$

where, as before, the integration is over the q-dimensional distribution of $\mathbf{u}$.

As an example, consider modeling data in correlated clusters thought to come from a Poisson distribution. An example of such a situation is described in Diggle et al. (1994) in which they consider the analysis of the number of epileptic seizures in patients on a drug or placebo. In this context, the clusters would be repeated measurements taken on the same patients. Let y_{ij} denote the jth count taken on the ith cluster. We might therefore create a model as:

$$y_{ij}|\mathbf{u} \quad \sim \quad \text{indep. Poisson}(\mu_{ij}); \quad i = 1, 2, \ldots, m; j = 1, 2, \ldots n_i;$$

$$\log \mu_{ij} \quad = \quad \mathbf{x}'_{ij}\boldsymbol{\beta} + u_i \tag{7.15}$$

$$u_i \quad \sim \quad \text{i.i.d. } \mathcal{N}(0, \sigma_u^2).$$

This uses a log link and a normal distribution for the random cluster (patient) effects. The normal distribution assumption for the random effects is viable since the log link carries the range of the parameter space for μ_{ij} into the entire real line. The random effects u_i are shared among observations within the same cluster and hence those observations are being modeled as correlated.

The log likelihood can be simplified as follows (see E 7.3)

$$l \quad = \quad \log \left(\prod_{i=1}^{m} \int_{-\infty}^{\infty} \prod_{j=1}^{n_i} \frac{\mu_{ij}^{y_{ij}} e^{-\mu_{ij}}}{y_{ij}!} \frac{1}{\sqrt{2\pi\sigma^2}} e^{-\frac{1}{2\sigma^2}u_i^2} du_i \right)$$

$$= \quad \mathbf{y}'\mathbf{X}\boldsymbol{\beta} - \sum_{i,j} \log y_{ij}! \tag{7.16}$$

$$+ \sum_{i} \log \int_{-\infty}^{\infty} \exp \left\{ y_{i\cdot} u_i - \Sigma_j e^{\mathbf{x}'_{ij}\boldsymbol{\beta}+u_i} \right\} \frac{1}{\sqrt{2\pi\sigma^2}} e^{-\frac{1}{2\sigma^2}u_i^2} du_i.$$

Unfortunately, (7.16) cannot be simplified further or evaluated in closed form and hence maximizing values cannot be expressed in closed form either.

In the simplest cases, numerical integration for calculating the likelihood is straightforward and hence numerical maximization of the likelihood is not too difficult. For example, for (7.15), as seen in (7.16), the log likelihood is the sum of independent contributions from each cluster, each of which involves just a single-dimensional integral. This integral can be evaluated accurately using standard quadrature techniques, for example, Gauss-Hermite quadrature (see Chapter 14).

This "brute force" approach to maximum likelihood works relatively well in simple situations: a single random effect, two or perhaps three nested random effects, and random effects which come in clusters (e.g., longitudinal data with subjects having random intercepts and slopes).

However, for more complicated structures (e.g., crossed random factors) it fails.

b. Likelihood equations

– i. *For the fixed effects parameters*

Even though the likelihood equations are numerically difficult, we can write them in a simpler form. From (7.14)

$$l = \log \int f_{\mathbf{Y}|\mathbf{u}}(\mathbf{y}|\mathbf{u}) f_{\mathbf{U}}(\mathbf{u}) \, d\mathbf{u} = \log f_{\mathbf{Y}}(\mathbf{y}), \qquad (7.17)$$

so that

$$\frac{\partial l}{\partial \boldsymbol{\beta}} = \frac{\partial}{\partial \boldsymbol{\beta}} \int f_{\mathbf{Y}|\mathbf{u}}(\mathbf{y}|\mathbf{u}) f_{\mathbf{U}}(\mathbf{u}) \, d\mathbf{u} \Big/ f_{\mathbf{Y}}(\mathbf{y})$$

$$= \int \left[\frac{\partial}{\partial \boldsymbol{\beta}} f_{\mathbf{Y}|\mathbf{u}}(\mathbf{y}|\mathbf{u}) \right] f_{\mathbf{U}}(\mathbf{u}) \, d\mathbf{u} \Big/ f_{\mathbf{Y}}(\mathbf{y}), \qquad (7.18)$$

since $f_{\mathbf{U}}(\mathbf{u})$ does not involve $\boldsymbol{\beta}$. Noting that

$$\frac{\partial}{\partial \boldsymbol{\beta}} f_{\mathbf{Y}|\mathbf{u}}(\mathbf{y}|\mathbf{u}) = \left(\frac{1}{f_{\mathbf{Y}|\mathbf{u}}(\mathbf{y}|\mathbf{u})} \frac{\partial f_{\mathbf{Y}|\mathbf{u}}(\mathbf{y}|\mathbf{u})}{\partial \boldsymbol{\beta}} \right) f_{\mathbf{Y}|\mathbf{u}}(\mathbf{y}|\mathbf{u})$$

$$= \frac{\partial \log f_{\mathbf{Y}|\mathbf{u}}(\mathbf{y}|\mathbf{u})}{\partial \boldsymbol{\beta}} f_{\mathbf{Y}|\mathbf{u}}(\mathbf{y}|\mathbf{u}), \qquad (7.19)$$

we can rewrite (7.18) as

$$\frac{\partial l}{\partial \boldsymbol{\beta}} = \int \frac{\partial \log f_{\mathbf{Y}|\mathbf{u}}(\mathbf{y}|\mathbf{u})}{\partial \boldsymbol{\beta}} f_{\mathbf{Y}|\mathbf{u}}(\mathbf{y}|\mathbf{u}) f_{\mathbf{U}}(\mathbf{u}) \, d\mathbf{u} \Big/ f_{\mathbf{Y}}(\mathbf{y})$$

$$= \int \frac{\partial \log f_{\mathbf{Y}|\mathbf{u}}(\mathbf{y}|\mathbf{u})}{\partial \boldsymbol{\beta}} f_{\mathbf{U}|\mathbf{y}}(\mathbf{u}|\mathbf{y}) \, d\mathbf{u}. \qquad (7.20)$$

Using (5.18), which gives the derivative of the log likelihood for a GLM, in (7.20) gives

$$\frac{\partial l}{\partial \boldsymbol{\beta}} = \int \mathbf{X}' \mathbf{W}^* (\mathbf{y} - \boldsymbol{\mu}) f_{\mathbf{U}|\mathbf{y}}(\mathbf{u}|\mathbf{y}) \, d\mathbf{u}$$

$$= \mathbf{X}' \mathrm{E}[\mathbf{W}^*|\mathbf{y}]\mathbf{y} - \mathbf{X}' \mathrm{E}[\mathbf{W}^* \boldsymbol{\mu}|\mathbf{y}], \qquad (7.21)$$

where, using the notation below (5.18), $\mathbf{W}^* = \left\{ _d [\tau v(\mu_i) g_\mu(\mu_i)]^{-1} \right\}$.

The likelihood equation for β is therefore

$$\mathbf{X}'\mathrm{E}[\mathbf{W}^*|\mathbf{y}]\mathbf{y} = \mathbf{X}'\mathrm{E}[\mathbf{W}^*\boldsymbol{\mu}|\mathbf{y}], \qquad (7.22)$$

which is similar to (5.19), the difference being that $\mathbf{W}^*$ and $\mathbf{W}^*\boldsymbol{\mu}$ are replaced by their conditional expected values given $\mathbf{y}$.

In cases like the Poisson example of (7.15), $\mathbf{W}^* = \mathbf{I}$ and the equations simplify to

$$\mathbf{X}'\mathbf{y} = \mathbf{X}'\mathrm{E}[\boldsymbol{\mu}|\mathbf{y}]. \qquad (7.23)$$

Computing issues related to solving these equations are described in Chapter 14.

– ii.   *For the random effects parameters*

A result similar to (7.20) can be derived for the ML equations for the parameters in the distribution of $f_\mathbf{U}(\mathbf{u})$. Let φ denote those parameters so that

$$\frac{\partial l}{\partial \varphi} = \int \frac{\partial \log f_\mathbf{U}(\mathbf{u})}{\partial \varphi} f_{\mathbf{U}|\mathbf{y}}(\mathbf{u}|\mathbf{y}) \, d\mathbf{u}$$

$$= \mathrm{E}\left[\frac{\partial \log f_\mathbf{U}(\mathbf{u})}{\partial \varphi} \middle| \mathbf{y} \right]. \qquad (7.24)$$

Further simplifications are not possible without specifying a form for the random effects distribution.

7.5   OTHER METHODS OF ESTIMATION

The difficulty in evaluating the likelihood for models such as (7.15) and the fact that data analysts must resort to numerical maximization has led to both alternative approaches and to a body of research for effective ways to compute and maximize the likelihoods. Chapter 14 treats the latter topic; here we introduce some of the alternative methods of estimation and, in Chapter 9, discuss marginal models which are often computationally less intensive.

a.   Penalized quasi-likelihood

For the generalized linear models (GLMs) of Chapter 5, the use of working variates, as in (5.4) and (5.28), and the principle of quasi-likelihood

(Section 5.6) are highly useful concepts. Quasi-likelihood is attractive because of its ability to generate highly efficient estimators without making precise distributional assumptions. Working variates form the basis of efficient computing algorithms for both maximum likelihood and maximum quasi-likelihood. A natural question is whether they can be adapted for use in GLMMs.

Working variates for GLMs begin with a Taylor expansion of the link function around the mean of y_i :

$$t_i \equiv \mathbf{x}'_i\boldsymbol{\beta} + g_\mu(\mu_i)(y_i - \mu_i).$$

The working variate thus follows a linear model and can be used to form a provisional estimate of $\boldsymbol{\beta}$. This local approximation is repeated at each update of an iterative algorithm.

The direct analog of working variates for the GLMM specification in (7.1) and (7.2) would be an expansion around the conditional mean of y_i :

$$t_i \equiv \mathbf{x}'_i\boldsymbol{\beta} + \mathbf{z}'_i\mathbf{u} + g_\mu(\mu_i)(y_i - \mu_i)$$

or

$$\mathbf{t} = \mathbf{X}\boldsymbol{\beta} + \mathbf{Z}\mathbf{u} + \boldsymbol{\Delta}(\mathbf{y} - \boldsymbol{\mu}), \tag{7.25}$$

where $\boldsymbol{\Delta} = \left\{ _d\, g_\mu(\mu_i) \right\}$. To derive a local approximation, the next step would be to calculate the variance of $\mathbf{t}$. But this approach quickly becomes complicated since $\boldsymbol{\Delta}$ (through its dependence on $\boldsymbol{\mu}$) and $\boldsymbol{\mu} = E[\mathbf{y}|\mathbf{u}]$ itself are random functions of $\mathbf{u}$ and their variances are not easily calculated.

A possible simplification is to set $\mathbf{u}$ in $\boldsymbol{\Delta}$ equal to its mean, $\mathbf{0}$, simplifying (7.25) to be

$$\mathbf{t} = \mathbf{X}\boldsymbol{\beta} + \mathbf{Z}\mathbf{u} + \boldsymbol{\Delta}^*(\mathbf{y} - \boldsymbol{\mu}), \tag{7.26}$$

where $\boldsymbol{\Delta}^* = \left\{ _c\, g_\mu[g^{-1}(\mathbf{x}'_i\boldsymbol{\beta})] \right\}$.

Under this simplification

$$\mathrm{var}(\mathbf{t}) = \mathbf{Z}\mathbf{D}\mathbf{Z}' + \boldsymbol{\Delta}^*\mathrm{var}(\mathbf{y} - \boldsymbol{\mu})\boldsymbol{\Delta}^*$$

$$\equiv \mathbf{Z}\mathbf{D}\mathbf{Z}' + \mathbf{R}. \tag{7.27}$$

That is, the working variate $\mathbf{t}$ approximately follows a linear mixed model (LMM) as in Chapter 6. This suggests an iterative algorithm (Schall, 1991) in which an LMM is fitted to get estimates of $\boldsymbol{\beta}$ and $\mathbf{u}$. These are then used to recalculate the working variate, and so on.

A completely different justification of this approach is via what is called *penalized quasi-likelihood* (PQL). Recall that quasi-likelihood does not specify a distribution, only the mean-to-variance relationship. This is not a sufficient basis on which to estimate the variance-covariance structure. One suggestion (Green and Silverman, 1994) to remedy this defect is to add a penalty function to the quasi-likelihood of the form $\frac{1}{2}\mathbf{u}'\mathbf{D}^{-1}\mathbf{u}$, that is

$$\text{PQL} = \sum Q_i - \frac{1}{2}\mathbf{u}'\mathbf{D}^{-1}\mathbf{u}, \tag{7.28}$$

where Q_i is defined in (5.50).

The maximum quasi-likelihood equations would come from differentiating (7.28) with respect to $\boldsymbol{\beta}$ and $\mathbf{u}$ and would be [compare (5.53)]

$$\frac{1}{\tau^2}\mathbf{X}'\mathbf{W}\boldsymbol{\Delta}(\mathbf{y} - \boldsymbol{\mu}) = \mathbf{0}$$

and

$$\frac{1}{\tau^2}\mathbf{Z}'\mathbf{W}\boldsymbol{\Delta}(\mathbf{y} - \boldsymbol{\mu}) - \mathbf{D}^{-1}\mathbf{u} = \mathbf{0}. \tag{7.29}$$

These lead (Breslow and Clayton, 1993) to a computational algorithm similar to that of Schall (1991). Yet another justification for this approach is via Laplace approximations (see Chapter 14 and Wolfinger, 1994).

Despite the number of ways in which basically the same approach has been justified, it has not been found to work well in practice, especially for binary data in small clusters (Breslow and Clayton, 1993; Breslow and Lin, 1995; Lin and Breslow, 1996). We therefore recommend that unmodified penalized quasi-likelihood not be used in practice. More detail is given in Chapter 14.

b.   Conditional likelihood

An approach very different in nature to integrating random effects out of the distribution and working with the marginal distribution is to consider a conditional likelihood and construct conditional estimators and tests as in Section 3.8e. In the conditional approach we start with the conditional distribution of the data given the random effects, but instead of hypothesizing a distribution for them and integrating them out, they are treated as fixed parameters. The sufficient statistics for them are derived and the conditional distribution given the sufficient statistics (which, by definition, is free of the random effects) is used for inferential purposes.

A classic application of the conditional approach is to the case of matched pairs binary data. For example, suppose we wish to ask whether cancer patients get more effective treatment in major cancer centers than in community hospitals. We cannot compare remission rates directly since patient populations might be drastically different. For example, major cancer centers might appear to provide poorer treatment merely because they treat the most difficult cases.

A possible solution is to employ a matched pairs design: A patient from a cancer center is matched with a patient from a community hospital on the basis of treatment date, type of treatment received, and patient's age. Suppose that the response variable is whether or not there is a sizable shrinkage in tumor size within 90 days and let $y_{ij} = 1$ for shrinkage and 0 otherwise. Here i indexes pairs $i = 1, 2, \ldots, n$ and j indexes type of hospital (with $j = 1$ representing a cancer center and $j = 2$ representing a community hospital). Also, let x_{ij} be 0 when $j = 1$ and 1 when $j = 2$.

A possible model for y_{ij} is:

$$y_{ij}|u_i \sim \text{indep. Bernoulli}[\pi(x_{ij})]$$

$$\text{logit}[\pi(x_{ij})] = \alpha + u_i + \beta x_{ij}.$$

Primary interest focuses on β, which represents the log odds of tumor shrinkage for community hospitals as compared to cancer centers, which is assumed constant within each pair. The u_i represent the pair-to-pair differences in the probability of tumor shrinkage.

What happens if we treat the u_i as fixed effects and estimate them, along with β? Maximum likelihood gives (see E 7.6)

$$\hat{\beta} = 2\log\frac{N_{01}}{N_{10}}, \tag{7.30}$$

where N_{10} is the number of pairs with $y_{i1} = 1$ and $y_{i2} = 0$ and where

Table 7.1: Fate of Matched Pairs

Cancer	Community Hospital	
Center	Success	Failure
Success	N_{11}	N_{10}
Failure	N_{01}	N_{00}

N_{01} is the number of pairs with $y_{i1} = 0$ and $y_{i2} = 1$. This is perhaps easiest to visualize in a 2×2 format as in Table 7.1.

It is not hard to show that this ML estimator is twice what it "should" be in the sense that it converges to 2β (see E 7.7).

What is the remedy? A commonly used approach is that of conditional likelihood, in which we derive the sufficient statistics for the u_i and work with the conditional distribution given those sufficient statistics.

We follow the development in (3.134) through (3.138). If interest focuses on β, then we will want to base inferences on the conditional distribution of $T = \sum_{i,j} y_{ij} x_{ij}$ given $S_1 = y_1., S_2 = y_2., \ldots, S_m = y_m.$. In our example,

$$T = \sum_{i,j} y_{ij} x_{ij}$$

$$= \sum_i y_{i2}$$

$$= y_{\cdot 2}$$

$$= \text{number of successes in community hospitals,} \quad (7.31)$$

so we want the conditional distribution of the total number of successes in the community hospitals conditional on the number of successes in each pair. From Table 7.1, $T = N_{11} + N_{01}$. Now N_{11} is just the number of pairs that have two successes, so conditional on S_i, it is known and fixed. We therefore focus on the conditional distribution of N_{01} given the S_i. We build it up in two steps. First consider a pair for which $S_i = 1$.

If $S_i = 1$, there are two possibilities: $\{y_{i1} = 0, y_{i2} = 1\}$ or $\{y_{i1} = $

$1, y_{i2} = 0\}$. The conditional probability of the first event is

$$P\{y_{i1} = 1, y_{i2} = 0 | S_i = 1\} = \frac{P\{y_{i1} = 1, y_{i2} = 0\}}{P\{y_{i1} = 0, y_{i2} = 1\} + P\{y_{i1} = 1, y_{i2} = 0\}}$$

$$= \frac{\left(1 - \frac{1}{1+e^{-(\alpha+u_i)}}\right)\frac{1}{1+e^{-(\alpha+u_i+\beta)}}}{\left(1 - \frac{1}{1+e^{-(\alpha+u_i)}}\right)\frac{1}{1+e^{-(\alpha+u_i+\beta)}} + \frac{1}{1+e^{-(\alpha+u_i)}}\left(1 - \frac{1}{1+e^{-(\alpha+u_i+\beta)}}\right)}$$

$$= \frac{e^{-(\alpha+u_i+\beta)}}{e^{-(\alpha+u_i)} + e^{-(\alpha+u_i+\beta)}}$$

$$= \frac{1}{1+e^{-\beta}}. \tag{7.32}$$

So the conditional distribution of y_{i2} given $S_i = 1$ is Bernoulli with a probability of success having a logit of β.

Writing T as

$$T = N_{11} + N_{01}$$

$$= \sum_{S_i=2} 1 + \sum_{S_i=1} I_{\{y_{i2}=1\}}, \tag{7.33}$$

shows that the conditional distribution of $N_{01} = T - N_{11}$ given the S_i is the sum of independent Bernoullis, each with conditional probability of success of $(1 + e^{-\beta})^{-1}$, that is,

$$N_{01} | S \sim \text{binomial}\left(N_{01} + N_{10}, \frac{1}{1+e^{-\beta}}\right). \tag{7.34}$$

It is therefore straightforward to show that the maximizing value of the conditional distribution (the conditional MLE) is

$$\hat{\beta} = \log\frac{N_{01}}{N_{10}}, \tag{7.35}$$

which is a consistent estimator of β as $N_{11} + N_{01}$ increases.

Also, under $H_0 : \beta = 0$, the distribution is binomial$(N_{11} + N_{01}, \frac{1}{2})$, from which exact tests or p-values can easily be derived. To do so, we use as our test statistic N_{01}, the number of successes in the community hospitals out of the pairs for which $S_i = 1$. If we were testing against

Table 7.2: Matched Pairs Data

Cancer	Community Hospital		
Center	Success	Failure	Total
Success	501	157	658
Failure	146	132	278
Total	647	289	936

the alternative that $H_A : \beta > 0$, we would reject for large values of N_{01}. Therefore the p-value for the one-tailed test would be:

$$p\text{-value} = P\{X \geq N_{01}\}, \qquad (7.36)$$

where $X \sim \text{binomial}(N_{01} + N_{10}, \frac{1}{2})$.

As a numerical illustration, consider the data of Table 7.2.   The conditional analysis discards the $501 + 132 = 633$ responses for which the response in the cancer center and community hospital are the same and bases the analysis on the 303 remaining. The p-value for the one-tailed test is

$$p\text{-value} \ = \ P\{X \geq 157\}$$

$$\text{where } X \ \sim \ \text{binomial}(303, 1/2), \qquad (7.37)$$

which is approximately 0.283. A usual convention is to multiply the one-sided p-value by 2 to get the two-tailed p-value, so the answer is $2(0.283) = 0.566$, suggesting no difference between cancer centers and community hospitals.

By its nature the conditional approach has three potentially serious drawbacks. First, because it treats the random effects as unknown parameters to be conditioned away, it is incapable of making inference to quantities involving the random effects: for example, their variances or predicted values. Second, it discards information that might be available by making only weak assumptions about the form of the distribution from which the random effects are chosen. Third, it removes any information that would be gained by comparing across levels of the random effects. In some situations, virtually *all* the information of interest is garnered from comparisons across levels of a random effect. Hence use of a conditional approach would be disastrous in such a context: It would eliminate all the information of interest due to the extreme manner in which the effects are handled.

Basically, the conditional approach is effective and attractive when interest centers almost exclusively on effects that can be measured *within* levels of a random factor. One important such setting is longitudinal studies (see Chapter 8) where interest focuses on change over time, a within-subject (within-cluster) covariate. When extensive information exists across levels of a random factor or when interest focuses on the random factor itself, the conditional approach cannot be used.

c.   Simpler models

To avoid the computational difficulties of GLMMs, other models have been considered, mostly on the basis of computational convenience. For example, the beta-binomial model of (2.72) has long been used for modeling correlated binary data. The most basic use of the beta-binomial is for a setting in which there are m groups, and within the ith group we have n_i observations, $y_{ij} \sim$ binomial(k_{ij}, p_{ij}), with j running from 1 to n_i. This specification is conditional on the values of the p_{ij}. To complete the specification we assume that $p_{ij} \sim$ beta(α_i, β_i). Given these distributional assumptions it is straightforward to show (see E 7.8) that the likelihood is given by

$$ L = \prod_{i=1}^{m} \prod_{j=1}^{n_i} \frac{B(\alpha_i + y_{ij}, \beta_i + k_{ij} - y_{ij})}{B(\alpha_i, \beta_i)}, \qquad (7.38) $$

where $B(\alpha, \beta)$ represents the beta function.

Working with the log likelihood, some simplifications occur as in (2.79). However, the likelihood still cannot be maximized in closed form, and numerical maximization is required. Likelihood ratio tests can be performed to test for dispersion or to compare the means of the various groups.

For data having Poisson distributions, a natural distribution to incorporate correlation is the gamma distribution. As with the beta-binomial model, we consider a situation with m groups, and within group i we have n_i observations, $y_{ij} \sim$ Poisson(λ_{ij}) with j running from 1 to n_i. This is conditional on the values of the λ_{ij}. To complete the specification we assume that $\lambda_{ij} \sim$ gamma(r_i, β_i). This allows for easy integration over the distribution of the parameters across groups, so that the likelihood takes a simple closed form.

However, like the beta-binomial model, it is limited in its application. It cannot handle models for the fixed effects (e.g., a regression situation), it cannot separate sources of variation in a crossed design, and it generally does not have the flexibility to tackle a wide variety of practical problems. Generalizations to handle more complicated covariate patterns and more complicated random effects structures have been considered by Lee and Nelder (1996), at the cost of additional computational complexity.

7.6   TESTS OF HYPOTHESES

The usual large-sample tools (see Chapter 2 and Sections S.4-5) are about the only techniques currently available for statistical inference.

a.   Likelihood ratio tests

The likelihood ratio test for nested models can be performed in the usual way by comparing $-2 \log \Lambda$ to a chi-square distribution. Testing whether a variance component is zero leads to the same boundary-of-the-parameter-space problem noted before in Chapter 2 [see (2.89)]. In the simple case where we are testing the null hypothesis that a single variance component is equal to zero, the large-sample distribution is a 50/50 mixture of the constant 0 and a χ_1^2 distribution. The critical values are thus given by (see E 7.4) $\chi_{1,1-2\alpha}^2$ for an α-level test.

Since the likelihood cannot, in general, be evaluated analytically the same is true of the likelihood ratio test statistic. It can be calculated only numerically for a given data set. In many cases it is a challenge even to perform the numerical maximization and calculation.

b.   Asymptotic variances

Again, with the difficulty of calculating the likelihood, even large-sample variances and standard errors can be a computational burden. Numerical methods are needed to calculate even the observed Fisher information (i.e., the negative of the second derivative matrix of the log likelihood).

c.   Wald tests

For large samples, when construction of the observed or expected information is possible, Wald tests can be formed by utilizing the large sample normality of estimators. This can be for an individual parameter:

$$\frac{\hat{\beta}_i - \beta_{i,0}}{\sqrt{\widehat{\mathrm{var}}_\infty(\hat{\beta}_i)}} \sim \mathcal{AN}(0,1) \qquad (7.39)$$

or for a set of linear combinations of the parameters,

$$\mathbf{K}'\hat{\boldsymbol{\beta}} - \mathbf{K}'\boldsymbol{\beta}_0 \sim \mathcal{AN}(0, \mathbf{K}'\mathbf{I}^{-1}\mathbf{K}), \qquad (7.40)$$

where $\mathbf{I}$ represents the observed or expected information matrix.

d. Score tests

For testing the presence of a single random effect or multiple random effects, score tests have also been proposed (Commenges et al., 1994; Jacqmin-Gadda and Commenges, 1995; Lin, 1997; Commenges and Jaqmin-Gadda, 1997). These have the advantage of not requiring the maximum likelihood estimators under the GLMM. However, they often have less power than the tests based on the random effects models.

7.7 ILLUSTRATION: CHESTNUT LEAF BLIGHT

The American chestnut tree used to be a predominant hardwood in the forests of the eastern United States, reaching 80 to 100 feet in height at maturity and providing timber and low-fat, high-protein nutrition for animals and humans in the form of chestnuts. In the early 1900s an imported fungal pathogen, which causes chestnut leaf blight, was introduced into the United States. The disease spread from infected trees in the New York City area and by 1950 had killed more than 3 billion trees and virtually eliminated the chestnut tree in the United States. Economic losses in both timber and nut production have been estimated in the hundreds of billions of dollars. As well, there are ecological impacts in eliminating a dominant species.

To try to bring this tree back to the U.S. forests, several methods have been explored, including the development of blight-resistant varieties. We focus instead on attempts to weaken the fungus by infecting it with a virus that reduces the fungus' virulence. The basic idea is to release these hypovirulent isolates of chestnut blight fungus and let the viruses infect the natural populations of the fungus, thereby allowing chestnuts trees to survive.

Michael Milgroom from the Department of Plant Pathology at Cornell University, and his colleague, Paolo Cortesi from the University of Milan, have studied this system (Cortesi et al., 1996; Cortesi and Milgroom, 1998). Viruses can spread between fungal individuals only when they come in contact and fuse together. A major obstacle in spreading this virus and thus controlling the disease is that different isolates of the fungus cannot necessarily transfer the virus to one another. Cortesi and Milgroom have worked with six incompatibility genes, which may block the transmission of this virus between isolates of the fungus. By developing lab isolates that are compatible with a wide variety of naturally occurring isolates (and thus able to transmit the hypovirulence) an avenue may be opened to biocontrol of this fungal disease.

To estimate the effects of these genes, Milgroom and Cortesi made extensive attempts to pair isolates which differ on the first gene only, the second gene only, the first and the second gene, and so on. For each combination of isolates they attempt transmission an average of 30 times and record a binary response of whether or not the attempt succeeded in transmitting the virus.

Questions of interest included whether pre-identified genes actually do have an influence on transmission of the virus (and if so, to what degree), whether there are other, as yet unidentified, genes that might affect transmission, and whether transmission is symmetric. By symmetry of transmission we mean the following: Suppose the infected fungus is type b at the locus for the first gene and the non-infected isolate (that we are trying to infect) is type B. The two isolates are the same at the other five loci. Is the probability of transmission the same as when using type B to try to infect type b?

a.   A random effects probit model

A common model in genetics for describing the presence or absence of a trait is the threshold model. This arises from assuming that a large number of genes each have a small and additive effect and that when the cumulative effect exceeds a threshold of zero, the trait is present in an individual. Letting $y = 1$ denote the presence of the trait, ϵ represent the additive genetic effect and $\mathbf{x}'\boldsymbol{\beta}$ represent either genetic or non-genetic fixed effects, we can appeal to the central limit theorem to give the probit model:

$$P\{y = 1\} \;=\; P\{\mathbf{x}'\boldsymbol{\beta} + \epsilon > 0\}$$

$$\epsilon \;\sim\; \mathcal{N}(0,1), \tag{7.41}$$

so that we have

$$\mathrm{P}\{y=1\} \;=\; \mathrm{P}\{-\epsilon < \mathbf{x}'\boldsymbol{\beta}\} = \Phi(\mathbf{x}'\boldsymbol{\beta})$$

or

$$\Phi^{-1}(\pi) \;=\; \mathbf{x}'\boldsymbol{\beta},$$

where $\pi = \mathrm{E}[y] = \mathrm{P}\{y=1\}$.

We use this model by letting y_i be equal to 1 if the attempt succeeds in transmitting the virus and 0 otherwise.

– i.   The fixed effects

We concentrate first on building the fixed effects portion of the model. With $\mathbf{x}'_i$ the ith row of the model matrix for the fixed effects, our model is

$$\mathbf{x}'_i\boldsymbol{\beta} \;=\; \mu + \sum_{j=1}^{6} \alpha_j MCH_j + \sum_{j=1}^{6} \gamma_j ASY_j, \tag{7.42}$$

where MCH_j is 1 if there is a mismatch at locus j and zero otherwise and ASY_j is $1/2$ if there is a mismatch at locus j with a b donor, $-1/2$ if it is a B donor and 0 if there is no mismatch. The effect of a mismatch on gene j (averaged over donor types b and B) is thus measured by α_j, and γ_j measures the difference between a mismatch with a donor type b and a donor type B.

– ii.   The random effects

The fact that different isolates of the fungus are used which may differ with regard to genes other than the six pre-identified suggests that we might model their effects as being selected from a normal distribution. Let $\mathbf{Z}_T$ be the model matrix for the transmission effects, that is, an incidence matrix identifying which donor isolates are used for which attempted transmissions. Similarly define $\mathbf{Z}_R$ for the recipient isolates, with the ith row of the matrices denoted respectively as $\mathbf{z}'_{iT}$ and $\mathbf{z}'_{iR}$. With $\mathbf{x}'_i$ defined by (7.42), $\mathbf{Z} = [\mathbf{Z}_T \; \mathbf{Z}_R]$, and $\mathbf{u}' = [\mathbf{u}'_T \; \mathbf{u}'_R]$, a reasonable model might then be:

$$P\{y_i = 1|\mathbf{u}\} = \Phi(\mathbf{x_i'}\boldsymbol{\beta} + \mathbf{z_{iT}'}\mathbf{u}_T + \mathbf{z_{iR}'}\mathbf{u}_R)$$

$$= \Phi(\mathbf{x_i'}\boldsymbol{\beta} + \mathbf{z_i'}\mathbf{u})$$

$$\mathbf{u}_T \sim \mathcal{N}(0, \mathbf{I}\sigma_T^2) \tag{7.43}$$

$$\mathbf{u}_R \sim \mathcal{N}(0, \mathbf{I}\sigma_R^2)$$

$$\mathbf{u} \sim \mathcal{N}(0, \mathbf{D}).$$

In this model, $\mathbf{u}_T$ represents the (random) effects of the donor isolate and $\mathbf{u}_R$ represents the (random) effects of the recipient isolate.

– iii. *Consequences of having random effects*

The unconditional mean is given by

$$E[y_i] = E[E[y_i|\mathbf{u}]]$$

$$= E[\Phi(\mathbf{x_i'}\boldsymbol{\beta} + \mathbf{z_i'}\mathbf{u})].$$

This last quantity can most easily be calculated by appeal to the threshold model. We do not necessarily need to believe that the threshold model holds, but can merely use it as a probabilistic identity.

$$E[y_i] = E[P\{\mathbf{x_i'}\boldsymbol{\beta} + \mathbf{z_i'}\mathbf{u} + \epsilon_i > 0|\mathbf{u}\}]$$

$$= P\{\mathbf{x_i'}\boldsymbol{\beta} + \mathbf{z_i'}\mathbf{u} + \epsilon_i > 0\}$$

$$= P\{-(\mathbf{z_i'}\mathbf{u} + \epsilon_i) < \mathbf{x_i'}\boldsymbol{\beta}\}$$

$$= P\{W < \mathbf{x_i'}\boldsymbol{\beta}\}, \tag{7.44}$$

where $W \sim \mathcal{N}(0, \mathbf{z_i'}\mathbf{D}\mathbf{z}_i + 1)$. The marginal probability is thus

$$E[y_i] = \Phi\left(\frac{\mathbf{x_i'}\boldsymbol{\beta}}{\sqrt{\mathbf{z_i'}\mathbf{D}\mathbf{z}_i + 1}}\right)$$

$$= \Phi(\mathbf{x_i'}\boldsymbol{\beta}^*), \tag{7.45}$$

where β^* is equal to $\beta/\sqrt{z_i'Dz_i + 1}$.

This result is interesting in two ways. First, it is somewhat surprising (and, as it turns out, special) that the form of the relationship of the mean of y to the fixed effects is probit either conditionally or unconditionally. Second, it shows that the marginal coefficients on the probit scale are always *attenuated* as compared to the conditional coefficients. Thus it clearly is important to keep in mind when considering any of these models whether they represent the response conditional on the random effects or are, instead, marginal calculations.

Since y_i is binary, it has a marginal Bernoulli distribution with mean, $E[y_i]$, given by (7.45). Its variance is therefore $E[y_i] (1 - E[y_i])$.

A typical consequence of including random effects is that they induce a correlation between observations sharing the random effects and this model is no exception. From first principles, the covariance of two observations with the same donor and recipient isolates would be given by $cov(y_i, y_j) = E[y_iy_j] - E[y_i]E[y_j]$. The second part of this can be evaluated using (7.45) and the first part calculated as

$$E[y_iy_j] = \int_{-\infty}^{\infty} \Phi(x_i'\beta + \sigma z)\Phi(x_j'\beta + \sigma z)\frac{1}{\sqrt{2\pi}}e^{-z^2/2}\, dz, \qquad (7.46)$$

where $\sigma = \sqrt{z_i'Dz_j}$.

– iv. *Likelihood analysis*

Again, with a conditional specification, the likelihood is most naturally calculated by first writing out the conditional distribution and then integrating out the random factors. The conditional distribution given u is the product of Bernoulli densities:

$$f_{Y|u}(y|u) = \prod_i \Phi(x_i'\beta + z_i'u)^{y_i}[1 - \Phi(x_i'\beta + z_i'u)]^{1-y_i}. \qquad (7.47)$$

The likelihood would then be given by

$$L(\beta, D|y) = \int f_{Y|u}(y|u)f_U(u)\, du, \qquad (7.48)$$

where the integral is of order equal to the dimension of u, which in this example is 259. Furthermore, for the design of this experiment, the likelihood does *not* break down into smaller-dimensional pieces, as it might with longitudinal data. This poses a serious computational problem.

– v.   Results

Given the difficulty of calculating the likelihood, the techniques of Section 14.3c were used to fit the model. A logistic version of (7.47) was fitted using the Monte Carlo Newton-Raphson technique. This was done since the computations were somewhat faster than for (7.47). The maximized value of the likelihood was estimated by importance sampling (Geyer and Thompson, 1992).

The variance components were estimated to be

$$\hat{\sigma}_T^2 \;=\; 1.6$$

$$\hat{\sigma}_R^2 \;=\; 0.5,$$

indicating a small to moderate correlation among observations taken on the same donor isolate and a somewhat smaller correlation among observations taken on the same recipient isolate. A likelihood ratio test of H_0 : all $\gamma_i = 0$ gives a value for $-2\log\Lambda$ of about 160 (since the value is determined by simulation it is not known exactly), which is highly statistically significant when referred to a chi-square distribution with 6 df. This indicates that, unfortunately, transmission is asymmetric: it depends on the value at that locus, not just on whether or not there is a match.

Further analysis shows that the fourth gene (tentatively identified from previous research) does not have an effect on transmission. Neither its direct effect nor its asymmetry effect is statistically significant.

7.8   EXERCISES

E 7.1 Show that, if $u_i \sim \mathcal{N}(0, \sigma_u^2)$ and that each row of $\mathbf{Z}$ has a single entry equal to 1 with all the rest being zero, then $M_u(\mathbf{z}) = \exp\{\sigma_u^2/2\}$.

E 7.2 Prove (7.12).

E 7.3 Show that the log likelihood for (7.15) can be written as (7.16).

E 7.4 Suppose that $Y = \delta X$, where $P\{\delta = 1\} = P\{\delta = 0\} = \frac{1}{2}$ and $X \sim \chi_1^2$ independent of δ. Show that

$$P\{Y > \chi_{1,1-2\alpha}^2\} = \alpha.$$

E 7.5 Derive the log likelihood for the Poisson-gamma model described in Section 7.5d.

E 7.6 Show that for (7.30), where the u_i are treated as fixed, unknown parameters to be estimated, that the MLE of β is given by (7.30). *Hints*: Consider separately pairs in which there are zero, one and two successes and first maximize with respect to the u_i, then β.

E 7.7 Show that $\hat{\beta}$ of (7.30) converges to 2β. *Hint*: Calculate $P\{y_{i1} = 1, y_{i2} = 0\}$ and $P\{y_{i1} = 0, y_{i2} = 1\}$ and hence the expected value of N_{01} and N_{10}.

E 7.8 Prove (7.38).

Chapter 8

MODELS FOR LONGITUDINAL DATA

8.1 INTRODUCTION

Investigators gather repeated measures or longitudinal data in order to study change in a response variable over time as well as to relate these changes to changes in explanatory variables over time. Sections 1.5e and 6.2d briefly describe the general nature of longitudinal data. The basic feature of such data is successive observations on each of a number of subjects (often people or animals). This can be likened to a randomized complete blocks experiment where the subjects, as blocks, are treated as random, and the successive occasions on which observations are taken are akin to treatments. One difference with longitudinal data is that the correlation structure among observations on the same subject is often more complicated than that among treatments in the same block.

This chapter will first consider the case of balanced, normally distributed data, meaning that on each subject there is the same number of observations, to be denoted by n. (It is more difficult to derive exact results for unbalanced data than for balanced data.) We will also discuss models and results for non-normal longitudinal data with a focus on repeated binary responses. For ease of description we refer to the occasions when observations are taken as times; thus each of say, m, subjects provides a datum at n times. For subject i, with $i = 1, 2, \ldots, m$, the datum at time j (for $j = 1, 2, \ldots, n$) is denoted by

y_{ij} and the vector of data for subject i is

$$\mathbf{y}_i = [y_{i1} \ y_{i2} \cdots y_{ij} \cdots y_{in}]' = \left\{ _c \ y_{ij} \right\}_{j=1}^n. \tag{8.1}$$

And the vector of data on all m subjects is

$$\mathbf{y} = \left\{ _c \ \mathbf{y}_i \right\}_{i=1}^m = \left\{ _c \left\{ _c \ y_{ij} \right\}_{j=1}^n \right\}_{i=1}^m. \tag{8.2}$$

8.2   A MODEL FOR BALANCED DATA

a.   Prescription

We first consider the case where the mean and variance of $\mathbf{y}_i$ are the same for each i:

$$\mathrm{E}[\mathbf{y}_i] = \boldsymbol{\mu} = \left\{ _c \ \mu_j \right\}_{j=1}^n \quad \text{so that} \quad \mathrm{E}[\mathbf{y}] = \mathbf{X}\boldsymbol{\mu} \text{ for } \mathbf{X} = \mathbf{1}_m \otimes \mathbf{I}_n.$$

And on taking the $\mathbf{y}_i$s to be uncorrelated with $\mathrm{var}(\mathbf{y}_i) = \mathbf{V}_0 \ \forall \ i$ we have

$$\mathbf{V} = \mathrm{var}(\mathbf{y}) = \mathbf{I}_m \otimes \mathbf{V}_0. \tag{8.3}$$

b.   Estimating the mean

On assuming normality,

$$\mathbf{y} \sim \mathcal{N}(\mathbf{X}\boldsymbol{\mu}, \mathbf{V}),$$

the ML estimator of $\boldsymbol{\mu}$ is $(\mathbf{X}'\mathbf{V}^{-1}\mathbf{X})^{-1}\mathbf{X}'\mathbf{V}^{-1}\mathbf{y}$ from (6.19), whenever $(\mathbf{X}'\mathbf{V}^{-1}\mathbf{X})^{-1}$ exists. This gives

$$\mathrm{ML}(\boldsymbol{\mu}) = \hat{\boldsymbol{\mu}} = [(\mathbf{1}'_m \otimes \mathbf{I}_n)(\mathbf{I}_m \otimes \mathbf{V}_0^{-1})(\mathbf{1}_m \otimes \mathbf{I}_n)]^{-1}(\mathbf{1}'_m \otimes \mathbf{I}_n)(\mathbf{I}_m \otimes \mathbf{V}_0^{-1})\mathbf{y},$$

which reduces (see E 8.1) to

$$\hat{\boldsymbol{\mu}} = \left\{ _c \ \bar{y}_{\cdot j} \right\}_{j=1}^n, \quad \text{i.e.,} \quad \hat{\mu}_j = \bar{y}_{\cdot j}. \tag{8.4}$$

Notice that this result does not involve $\mathbf{V}_0$; thus it holds no matter what $\mathbf{V}_0$ is. And this is important, because in what follows we consider several forms of $\mathbf{V}_0$, but for all of them with $\mathbf{X} = \mathbf{1}_m \otimes \mathbf{I}_n$ the estimator of $\boldsymbol{\mu}$ is as in (8.4).

c.   Estimating $\mathbf{V}_0$

In attributing no structure to $\mathbf{V}_0$, we want to estimate (by ML) its every element. This necessitates differentiating the likelihood with respect to every element of $\mathbf{V}_0$. But since the likelihood

$$L = \frac{\exp\{-\frac{1}{2}[\mathbf{y} - (1 \otimes \mathbf{I})\boldsymbol{\mu}\,]'(\mathbf{I} \otimes \mathbf{V}_0)^{-1}[\mathbf{y} - (1 \otimes \mathbf{I})\boldsymbol{\mu}\,]\}}{(2\pi)^{\frac{mn}{2}}|\mathbf{I} \otimes \mathbf{V}_0|^{\frac{1}{2}}} \tag{8.5}$$

involves $\mathbf{V}_0^{-1}$ that differentiating is somewhat cumbersome. It can be circumvented by writing

$$\mathbf{V}_0^{-1} = \mathbf{W}$$

and then, ignoring the 2π term,

$$l = \log L = \tfrac{1}{2}\log|\mathbf{I} \otimes \mathbf{W}| - \tfrac{1}{2}\Big\{{}_r\, (\mathbf{y}_i - \boldsymbol{\mu})'\Big\}(\mathbf{I} \otimes \mathbf{W})\Big\{{}_c\, \mathbf{y}_i - \boldsymbol{\mu}\,\Big\}. \tag{8.6}$$

We differentiate this with respect to an element w_{jk} of $\mathbf{W}$, for this purpose treating w_{jk} as different from w_{kj} (even though they are equal because $\mathbf{W}$ is symmetric). Then, on recalling that $\partial(\log|\mathbf{A}|)/\partial x = \mathrm{tr}[\mathbf{A}^{-1}(\partial \mathbf{A}/\partial x)]$, and that $\log|\mathbf{I}_m \otimes \mathbf{W}| = m\log|\mathbf{W}|$, and observing that

$$\frac{\partial \mathbf{W}}{\partial w_{jk}} = \mathbf{E}_{jk},$$

where $\mathbf{E}_{ij}$ is a matrix of all zeros except with element (j, k) being one, we have

$$\frac{\partial l}{\partial w_{jk}} = \tfrac{1}{2}m\,\mathrm{tr}(\mathbf{W}^{-1}\mathbf{E}_{jk}) - \tfrac{1}{2}\Big\{{}_r\, (\mathbf{y}_i - \boldsymbol{\mu})'\Big\}(\mathbf{I} \otimes \mathbf{E}_{jk})\Big\{{}_c\, \mathbf{y}_i - \boldsymbol{\mu}\,\Big\}$$

$$= \tfrac{1}{2}m\, v_{0,jk} - \tfrac{1}{2}\sum_{i=1}^{m}(y_{ij} - \mu_j)(y_{ik} - \mu_k), \tag{8.7}$$

where $v_{0,jk}$ is the (j, k)th element of $\mathbf{V}_0$. We already know from (8.4) that $\mathrm{MLE}(\mu_j) = \hat{\mu}_j = \bar{y}_{\cdot j}$. To get the $\mathrm{MLE}(v_{0,jk})$ we equate (8.7) to zero, with μ_j replaced by $\hat{\mu}_j$. Thus

$$\mathrm{MLE}(v_{0,jk}) = \frac{1}{m}\sum_{i=1}^{m}(y_{ij} - \bar{y}_{\cdot j})(y_{ik} - \bar{y}_{\cdot k}).$$

And since this is true for all j, k we have the matrix result

$$\hat{\mathbf{V}}_0 = \frac{1}{m}\left\{\sum_{i=1}^{m}(y_{ij} - \bar{y}_{\cdot j})(y_{ik} - \bar{y}_{\cdot k})\right\}_{j,k=1}^{n}; \tag{8.8}$$

that is, $\hat{\mathbf{V}}_0$ is a Wishart matrix. Since the MLE not constraining $\mathbf{V}_0$ to be symmetric happens to be symmetric, this is also the constrained MLE.

8.3 A MIXED MODEL APPROACH

Having dealt with general $\mathbf{V}_0$, we now consider some special cases, where $\mathbf{V}_0$ is structured in terms of a few (often just two or three) parameters. We do this by specifying a mixed model for the data.

a. Fixed and random effects

A starting point for a model equation for y_{ij} being the datum on subject i taken at time j is

$$E[y_{ij}|u_i] = \alpha_j + u_i, \tag{8.9}$$

where u_i is a random effect for subject i and α_j is now playing the part of μ_j of Section 8.2. And for $\mathbf{y}$ defined in (8.2) and for

$$\mathbf{u}_{m\times1} = \left\{_c u_i\right\}_{i=1}^m \tag{8.10}$$

we write

$$E[\mathbf{y}|\mathbf{u}] = \mathbf{X}\boldsymbol{\beta} + \mathbf{Z}\mathbf{u} \tag{8.11}$$

for

$$\mathbf{X} = \mathbf{1}_m \otimes \mathbf{I}_n, \quad \mathbf{Z} = \mathbf{I}_m \otimes \mathbf{1}_n \quad \text{and} \quad \boldsymbol{\beta} = \left\{_c \alpha_j\right\}_{j=1}^n. \tag{8.12}$$

b. Variances

To calculate a variance-covariance matrix $\mathbf{V}$ for $\mathbf{y}$, we begin by defining

$$\text{var}(\mathbf{y}|\mathbf{u}) = \mathbf{R}. \tag{8.13}$$

Then from (S.3)

$$\begin{aligned}
\mathbf{V} &= \text{var}(E[\mathbf{y}|\mathbf{u}]) + E[\text{var}(\mathbf{y}|\mathbf{u})] \\
&= \text{var}(\mathbf{Z}\mathbf{u}) + E[\mathbf{R}] \\
&= \mathbf{Z}\mathbf{D}\mathbf{Z}' + \mathbf{R}
\end{aligned}$$

for $\mathbf{D}$ defined as

$$\mathbf{D} = \text{var}(\mathbf{u}).$$

Thus

$$\mathbf{V} \;=\; (\mathbf{I}_m \otimes \mathbf{1}_n)(\mathbf{D} \otimes 1)(\mathbf{I}_m \otimes \mathbf{1}'_n) + \mathbf{R} \tag{8.14}$$

$$\;=\; \mathbf{D} \otimes \mathbf{J}_n + \mathbf{R}. \tag{8.15}$$

Using the notation $\text{var}(u_i) = \sigma_u^2 \; \forall \; i$ and assuming that the responses of different subjects are uncorrelated, (8.3), $\mathbf{D}$ is a diagonal matrix having common elements σ_u^2. That is,

$$\mathbf{D} = \text{var}(\mathbf{u}) = \sigma_u^2 \mathbf{I}_m \tag{8.16}$$

and so $\mathbf{V}$ of (8.15) is

$$\mathbf{V} = \mathbf{I}_m \otimes \sigma_u^2 \mathbf{J}_n + \mathbf{R}. \tag{8.17}$$

A tractable form for $\mathbf{R} = \text{var}(\mathbf{y}|\mathbf{u})$ is to take the $\mathbf{y}_i|u_i$ variables as being independent and all having the same variance-covariance matrix $\mathbf{R}_0$, so that $\mathbf{R} = \mathbf{I}_m \otimes \mathbf{R}_0$ so giving

$$\mathbf{V} = \mathbf{I}_m \otimes \sigma_u^2 \mathbf{J}_n + \mathbf{R}_0. \tag{8.18}$$

For balanced data, this is a fairly general form for $\mathbf{V}$. It provides for a uniform variance σ_u^2 of the random effects and for the same variance, $\mathbf{R}_0$, of $\mathbf{y}_i|u_i$ for each subject. With $\sigma_u^2 = 0$ it reduces to $\mathbf{V} = \mathbf{I} \otimes \mathbf{R}_0$, which is the same form as $\mathbf{I} \otimes \mathbf{V}_0$ treated earlier.

8.4   RANDOM INTERCEPT AND SLOPE MODELS

Time or a time-varying covariate x is often an explantory variable of interest in longitudinal studies and the associations between explanatory variables and responses as well as the average response level may vary between subjects. For example, with the Osteoarthritis Initiative data (1.23) of Chapter 1 it is natural to consider models that allow heterogeneity in average level of knee pain as well as heterogeneity in the magnitude of change in pain over time between subjects; that is, models that include both random intercepts and slopes. We can specify such a model as a special case of (8.11). As in (1.23), consider the case where we fit a mixed model with a single covariate and both random intercepts and slopes:

$$y_{ij} = \beta_0 + b_{0i} + (\beta_1 + b_{1i})x_{ij} + e_{ij}, \tag{8.19}$$

where b_{0i} and b_{1i} denote the random intercepts and slopes, respectively. In this case, the matrix $\mathbf{Z}$ of (8.11) is $\mathbf{Z} = \left\{ {}_d \mathbf{X}_i \right\}_{i=1}^m$, where

$$\mathbf{X}_i = \left\{ {}_r (1 \quad x_{ij}) \right\}_{j=1}^n \tag{8.20}$$

and $\mathbf{u} = \left\{ {}_c \mathbf{u}_i \right\}_{i=1}^m$ with $\mathbf{u}_i' = (b_{0i}, b_{1i})$.

a. Variances

If we let $\text{var}(b_{ki}) = \sigma_k^2$, $k = 0, 1$, and $\text{cov}(b_{0i}, b_{1i}) = \sigma_{01}$ then

$$\mathbf{D} = \begin{bmatrix} \sigma_0^2 & \sigma_{01} \\ \sigma_{01} & \sigma_1^2 \end{bmatrix}. \tag{8.21}$$

An important special case of (8.19) has x_{ij} equal to the time of measurement or visit number so that $\beta_1 + b_{1i}$ measures change in the response over time specific to the ith subject while $\beta_0 + b_{0i}$ measures the average response level for the ith subject when time or visit is zero. Non-zero values of σ_{01} indicate that subject-specific rates of change are associated with subject-specific average response levels.

If we let

$$\text{var}(\mathbf{y}_i|\mathbf{b}_i) = \mathbf{R}_i = \sigma_e^2 \mathbf{I}_n, \tag{8.22}$$

it is straightforward (see E 8.6) to calculate the variance-covariance matrix $\mathbf{V}_i$ for $\mathbf{y}_i$,

$$\mathbf{V}_i = \sigma_0^2 \mathbf{1}_n \mathbf{1}_n' + \sigma_{01}(\mathbf{x}_i \mathbf{1}_n' + \mathbf{1}_n \mathbf{x}_i') + \sigma_1^2 \mathbf{x}_i \mathbf{x}_i' + \sigma_e^2 \mathbf{I}_n. \tag{8.23}$$

b. Within-subject correlations

The elements of $\mathbf{V}_i$ in (8.23) are given by

$$V_{ijj} = \sigma_0^2 + 2\sigma_{01}x_{ij} + \sigma_1^2 x_{ij}^2 + \sigma_e^2, \tag{8.24}$$

$$V_{ijk} = \sigma_0^2 + \sigma_{01}(x_{ij} + x_{ik}) + \sigma_1^2 x_{ij}x_{ik}, \tag{8.25}$$

for $j = 1, \ldots, n$, $j \neq k$ and depend on the value of the covariate x_{ij}, unlike random intercept only, compound symmetric, models. The correlation between the ith subject's responses on occasions j and k is

$$\mathrm{corr}(y_{ij}, y_{ik}) =$$

$$\frac{\sigma_0^2 + \sigma_{01}(x_{ij} + x_{ik}) + \sigma_1^2 x_{ij} x_{ik}}{\sqrt{(\sigma_0^2 + 2\sigma_{01}x_{ij} + \sigma_1^2 x_{ij}^2 + \sigma_e^2)(\sigma_0^2 + 2\sigma_{01}x_{ik} + \sigma_1^2 x_{ik}^2 + \sigma_e^2)}}. \tag{8.26}$$

One can show (see E 8.7) that $\mathrm{corr}(y_{ij}, y_{ik})$ of (8.26) is monontonically decreasing in x_{ik} for $x_{ik} > x_{ij}$. That is, the dependence between pairs (j, k) of responses of the ith subject decay with the distance between the associated covariate values. In particular, when x_{it} denotes the time of the tth measurement, the correlation between y_{ij} and y_{it} decreases as the time between the two measurements increases.

Considering various limits of (8.26) illustrates features of random intercept and slope models. As a first case, consider the setting where we fix x_{ij} and let $x_{ik} \to \infty$. Standard calculations using (8.26) (see E 8.8) yield

$$\lim_{x_{ik} \to \infty} \mathrm{corr}(y_{ij}, y_{ik}) = \frac{\sigma_{01} + \sigma_1^2 x_{ij}}{\sqrt{\sigma_1^2(\sigma_0^2 + 2\sigma_{01}x_{ij} + \sigma_1^2 x_{ij}^2 + \sigma_e^2)}}. \tag{8.27}$$

Equation (8.27) shows that responses with very different covariate values are correlated and that the magnitude of the correlation depends on many of the parameters in the variance-covariance matrix of the responses (8.23) and on σ_{01} and σ_1^2 in particular. This contrasts with settings where within-subject correlations follow an autoregressive pattern and correlations decay to zero as observations are farther and farther apart on the covariate (are farther separated in time).

As a second case, consider the setting where we fix the distance between two covariate values $\delta = x_{ik} - x_{ij}$ and let $x_{ij} \to \infty$. Standard calculations using (8.26) (see E 8.9) yield

$$\lim_{x_{ij} \to \infty} \mathrm{corr}(y_{ij}, y_{ik}) = 1. \tag{8.28}$$

Each subject follows an individual trajectory $\beta_0 + b_{0i} + (\beta_1 + b_{1i})x$ and when we consider pairs of responses for the ith subject that are associated with very large covariate values, these responses are completely separated from the responses of another subject and hence perfectly correlated.

8.5 PREDICTING RANDOM EFFECTS

We start from the general expression for $\tilde{\mathbf{u}}^0$, the best linear unbiased predictor of the random effects

$$\tilde{\mathbf{u}}^0 = \mathbf{D}\mathbf{Z}'\mathbf{V}^{-1}(\mathbf{y} - \mathbf{X}\boldsymbol{\beta}^0). \tag{8.29}$$

This result is presented in Section 6.5 and is established in Chapter 13. Clearly this expression assumes that $\mathbf{D} = \text{var}(\mathbf{u})$ and $\mathbf{V} = \text{var}(\mathbf{y})$ are both known. $\mathbf{Z}$ and $\mathbf{X}$ are arbitrary matrices for random effects and covariates, respectively, and for our purpose we take $\boldsymbol{\beta}^0$ to be the maximum likelihood estimator in (6.19),

$$\hat{\boldsymbol{\beta}} = (\mathbf{X}'\mathbf{V}^{-1}\mathbf{X})^{-}\mathbf{X}'\mathbf{V}^{-1}\mathbf{y}. \tag{8.30}$$

Thus (8.29) becomes

$$\tilde{\mathbf{u}}^0 = \mathbf{D}\mathbf{Z}'\mathbf{V}^{-1}(\mathbf{y} - \mathbf{X}\hat{\boldsymbol{\beta}}). \tag{8.31}$$

Simplifications of (8.31) for special cases are as follows.

a. Uncorrelated subjects

We suppose that the model contains only random intercepts. That is, $\mathbf{Z} = \mathbf{I}_m \otimes \mathbf{1}_n$ as in (8.12) Using (8.18), the predictor (8.31) reduces (see Section 8.10a) to

$$\tilde{\mathbf{u}}^0 = \left[\mathbf{I}_m \otimes \frac{\mathbf{1}'\mathbf{R}_0^{-1}}{1/\sigma_u^2 + \mathbf{1}'\mathbf{R}_0^{-1}\mathbf{1}} \right] \left\{_c \, \mathbf{y}_i - \mathbf{X}_i\hat{\boldsymbol{\beta}} \right\}_{i=1}^m,$$

where $\mathbf{X}_i$ is the matrix of covariates for the ith subject as in (8.20). This gives

$$\tilde{u}_i^0 = \frac{\mathbf{1}'\mathbf{R}_0^{-1}(\mathbf{y}_i - \mathbf{X}_i\hat{\boldsymbol{\beta}})}{1/\sigma_u^2 + \mathbf{1}'\mathbf{R}_0^{-1}\mathbf{1}} \tag{8.32}$$

for $\hat{\boldsymbol{\beta}}$ given by (8.30).

This is not particularly tractable unless $\mathbf{R}_0^{-1}$ is analytically manageable; but numerically it will usually offer little difficulty, especially because $\mathbf{R}_0$ has order n, the number of observations on a subject, and this is often not very large.

b.   Uncorrelated between, and within, subjects

We now consider a case where the correlation between subjects, and between observations on each subject, are each taken as zero. This simply involves putting $\mathbf{R}_0 = \sigma^2\mathbf{I}$ in (8.32) which (in Section 8.10b) yields

$$\tilde{u}_i^0 = \frac{n\sigma_u^2}{\sigma^2 + n\sigma_u^2}(\bar{y}_{i\cdot} - \bar{y}_{\cdot\cdot}). \tag{8.33}$$

This is a very familiar estimator, known as a *Stein*, or *shrinkage*, *estimator*. It occurs widely in animal genetics when wanting to calculate the estimated genetic value of animals. With genetic definitions

$$\tau = \text{repeatability} = \frac{\sigma_u^2}{\sigma^2 + \sigma_u^2}$$

and

$$h^2 = \text{heritability} = \frac{4\sigma_u^2}{\sigma^2 + \sigma_u^2}$$

the fraction multiplying $(\bar{y}_{i\cdot} - \bar{y}_{\cdot\cdot})$ then has several forms that are very familiar to animal geneticists:

$$\frac{n\sigma_u^2}{\sigma^2 + n\sigma_u^2} = \frac{\sigma_u^2}{\sigma^2/n + \sigma_u^2} = \frac{n\tau}{1 + (n-1)\tau} = \frac{nh^2}{4 + (n-1)h^2}. \tag{8.34}$$

c.   Uncorrelated between, and autocorrelated within

Another tractable form for $\mathbf{R}_0/\sigma^2$ is a first-order autocorrelation matrix, a 5×5 example of which is

$$\mathbf{A} = \begin{bmatrix} 1 & \rho & \rho^2 & \rho^3 & \rho^4 \\ \rho & 1 & \rho & \rho^2 & \rho^3 \\ \rho^2 & \rho & 1 & \rho & \rho^2 \\ \rho^3 & \rho^2 & \rho & 1 & \rho \\ \rho^4 & \rho^3 & \rho^2 & \rho & 1 \end{bmatrix}$$

with

$$\mathbf{A}^{-1} = \frac{1}{1-\rho^2} \begin{bmatrix} 1 & -\rho & 0 & 0 & 0 \\ -\rho & 1+\rho^2 & -\rho & 0 & 0 \\ 0 & -\rho & 1+\rho^2 & -\rho & 0 \\ 0 & 0 & -\rho & 1+\rho^2 & -\rho \\ 0 & 0 & 0 & -\rho & 1 \end{bmatrix}. \tag{8.35}$$

Using that inverse (generalized to order n) for $\mathbf{R}_0^{-1}\sigma^2$ in (8.32), the value of $\tilde{u}_i^0$ is (see E 8.3)

$$\tilde{u}_i^0 = \frac{\sigma_u^2\left[(1-\rho)n(\bar{y}_{i\cdot} - \bar{y}_{\cdot\cdot}) + \rho(y_{i1} - \bar{y}_{\cdot 1} + y_{in} - \bar{y}_{\cdot n})\right]}{\sigma^2(1+\rho) + \sigma_u^2[n - (n-2)\rho]}. \qquad (8.36)$$

Note that the $\rho(y_{i1} - \bar{y}_{\cdot 1} + y_{in} - \bar{y}_{\cdot n})$ in the numerator represents "end effects" commensurate with the autocorrelation matrix $\mathbf{A}^{-1}$ of (8.35) having first and last diagonal elements different from all other diagonal elements. Also note that for large n, (8.36) reduces to $\bar{y}_{i\cdot} - \bar{y}_{\cdot\cdot}$, as it should.

d. Random intercepts and slopes

Although it is simple to construct the $\mathbf{V}$ matrix for random intercept and slope models such as (8.19), general, closed-form expressions for $\mathbf{V}^{-1}$ do not exist so that we cannot find simplifications of (8.29) for these models. However, it is straightforward to numerically invert $\mathbf{V}$ and predict random intercepts and slopes using (8.29).

8.6 ESTIMATING PARAMETERS

Having dealt with estimating $\boldsymbol{\beta}$ and predicting $\mathbf{u}$ in $\mathbf{X}\boldsymbol{\beta} + \mathbf{Z}\mathbf{u}$ (for $\mathbf{V}$ assumed known) we now consider estimating the parameters that occur in the forms of $\mathbf{V}$ considered in Section 8.5.

a. The general case

The distributional assumption for the variance components model of Chapter 6, as in (6.56), is

$$\mathbf{y} \sim \mathcal{N}\left(\mathbf{X}\boldsymbol{\beta}, \quad \mathbf{V} = \sum_i \mathbf{Z}_i\mathbf{Z}_i'\sigma_i^2\right)$$

and equations (6.60) for ML estimation of the σ^2s are

$$\left\{_c \operatorname{tr}(\mathbf{V}^{-1}\mathbf{Z}_i\mathbf{Z}_i')\right\} = \left\{_c \mathbf{y}'\mathbf{P}\mathbf{Z}_i\mathbf{Z}_i'\mathbf{P}\mathbf{y}\right\}, \qquad (8.37)$$

where the appearance of $\mathbf{Z}_i\mathbf{Z}_i'$ comes about because

$$\mathbf{Z}_i\mathbf{Z}_i' = \frac{\partial\mathbf{V}}{\partial\sigma_i^2}.$$

Also

$$\mathbf{Py} = \left[\mathbf{V}^{-1} - \mathbf{V}^{-1}\mathbf{X}(\mathbf{X'V}^{-1}\mathbf{X})^{-}\mathbf{X'V}^{-1}\right]\mathbf{y} = \mathbf{V}^{-1}(\mathbf{y} - \mathbf{X}\beta^0)$$

for $\beta^0 = (\mathbf{X'V}^{-1}\mathbf{X})^{-}\mathbf{X'V}^{-1}\mathbf{y}$ from (6.27). Hence (8.37) is

$$\left\{_c \operatorname{tr}\left(\mathbf{V}^{-1}\frac{\partial\mathbf{V}}{\partial\sigma_i^2}\right)\right\} = \left\{_c (\mathbf{y} - \mathbf{X}\beta^0)'\mathbf{V}^{-1}\frac{\partial\mathbf{V}}{\partial\sigma_i^2}\mathbf{V}^{-1}(\mathbf{y} - \mathbf{X}\beta^0)\right\}.$$

$$(8.38)$$

As in Section 6.4, we now think of $\mathbf{V}$ being structured, with elements which are functions of just a few scalar parameters: denote one such element as φ playing the part of σ_i^2 in (8.38). Its equations are of the form

$$\operatorname{tr}\left(\mathbf{V}^{-1}\frac{\partial\mathbf{V}}{\partial\varphi}\right) \quad = \quad (\mathbf{y} - \mathbf{X}\beta^0)'\mathbf{V}^{-1}\frac{\partial\mathbf{V}}{\partial\varphi}\mathbf{V}^{-1}(\mathbf{y} - \mathbf{X}\beta^0) \quad (8.39)$$

$$= \quad (\mathbf{y} - \mathbf{X}\beta^0)'(-1)\frac{\partial\mathbf{V}^{-1}}{\partial\varphi}(\mathbf{y} - \mathbf{X}\beta^0). \quad (8.40)$$

We use either (8.39) or (8.40) for special cases of $\mathbf{V} = \mathbf{I} \otimes \mathbf{V}_0$, often using the notation

$$\text{LHS}(\varphi) \quad = \quad \operatorname{tr}\left(\mathbf{V}^{-1}\frac{\partial\mathbf{V}}{\partial\varphi}\right) = m \operatorname{tr}\left(\mathbf{V}_0^{-1}\frac{\partial\mathbf{V}_0}{\partial\varphi}\right) \quad (8.41)$$

and

$$\text{RHS}(\varphi) \quad = \quad \text{right-hand side of (8.39) and (8.40)},$$

so that the estimating equations are then

$$\text{LHS}(\varphi) = \text{RHS}(\varphi). \quad (8.42)$$

for φ taking in turn each parameter in $\mathbf{V}$, for example, ρ_u, σ_u^2 and σ^2 in $\mathbf{V}$ of (8.17) when $\mathbf{R} = \sigma^2\mathbf{I}_n$. We do this for two of the cases of Section 8.5.

b.   Uncorrelated subjects

This has

$$\mathbf{V} = \mathbf{I}_m \otimes \mathbf{V}_0 \quad \text{for} \quad \mathbf{V}_0^{-1} = \sigma_u^2\mathbf{J}_n + \mathbf{R}_0$$

with

$$\mathbf{V}_0^{-1} = \mathbf{R}_0^{-1} - \frac{\mathbf{R}_0^{-1}\mathbf{1}\mathbf{1}'\mathbf{R}_0^{-1}}{1/\sigma_u^2 + \mathbf{1}'\mathbf{R}_0^{-1}\mathbf{1}}.$$

Using this in (8.41) for $\varphi = \sigma_u^2$ gives

$$\text{LHS}(\sigma_u^2) = m \operatorname{tr}\left(\mathbf{V}_0^{-1}\mathbf{J}_n\right) = m\mathbf{1}'\mathbf{V}_0^{-1}\mathbf{1} = \frac{m\mathbf{1}'\mathbf{R}_0^{-1}\mathbf{1}}{1 + \sigma_u^2\mathbf{1}'\mathbf{R}_0^{-1}\mathbf{1}}. \qquad (8.43)$$

Similarly, using (8.40) gives

$$\text{RHS}(\sigma_u^2) = \left\{_r (\mathbf{y}_i - \hat{\boldsymbol{\beta}})'\right\}_{i=1}^m \left(\mathbf{I}_m \otimes \frac{\mathbf{R}_0^{-1}\mathbf{1}\mathbf{1}'\mathbf{R}_0^{-1}}{1 + \sigma_u^2\mathbf{1}'\mathbf{R}_0^{-1}\mathbf{1})^2}\right)\left\{_c \mathbf{y}_i - \hat{\boldsymbol{\beta}}\right\}_{i=1}^m$$

$$= \sum_{i=1}^m \left[\frac{\mathbf{1}'\mathbf{R}_0^{-1}(\mathbf{y}_i - \hat{\boldsymbol{\beta}})}{1 + \sigma_u^2\mathbf{1}'\mathbf{R}_0^{-1}\mathbf{1}}\right]^2$$

and so the ML equation is

$$m\mathbf{1}'\mathbf{R}_0^{-1}\mathbf{1}\left(1 + \dot{\sigma}_u^2\mathbf{1}'\mathbf{R}_0^{-1}\mathbf{1}\right) = \sum_{i=1}^m \left[\mathbf{1}'\mathbf{R}_0^{-1}(\mathbf{y}_i - \hat{\boldsymbol{\beta}})\right]^2. \qquad (8.44)$$

If $\mathbf{R}_0$ is unspecified, with elements functionally independent of σ_u^2 (which has been assumed in deriving the preceding result), then ML estimation of $\mathbf{R}_0$ will be exactly like that of $\mathbf{V}_0$ in Section 8.3c. We therefore consider having $\mathbf{R}_0 = \sigma^2\mathbf{I}_n$.

c. Uncorrelated between, and autocorrelated within, subjects

We here have $\mathbf{V} = \mathbf{I}_m \otimes \mathbf{A}\sigma^2$ for $\mathbf{A}_{n \times n}$ being the n-order form of the 5×5 example in (8.35). Therefore in (8.40) and (8.41) $\mathbf{A}\sigma^2$ plays the part of $\mathbf{V}_0$. So from using (8.35) for order n in $\mathbf{V} = \mathbf{I} \otimes \mathbf{V}_0 = \mathbf{I} \otimes \sigma^2\mathbf{A}$ we get from (8.41) for $\varphi = \sigma^2$

$$\text{LHS}(\sigma^2) = m \operatorname{tr}\left[(\sigma^2\mathbf{A})^{-1}\frac{\partial(\sigma^2\mathbf{A})}{\partial\sigma^2}\right] = m \operatorname{tr}(\mathbf{A}^{-1}\mathbf{A})/\sigma^2 = mn/\sigma^2.$$

And from RHS(φ) of (8.40)

$$\text{RHS}(\sigma^2) = -(\mathbf{y} - \mathbf{X}\hat{\boldsymbol{\beta}})'\left[\mathbf{I}_m \otimes \frac{\partial(\mathbf{A}\sigma^2)^{-1}}{\partial\sigma^2}\right](\mathbf{y} - \mathbf{X}\hat{\boldsymbol{\beta}}) \quad (8.45)$$

$$= \left\{_r (\mathbf{y}_i - \hat{\boldsymbol{\beta}})'\right\}\left[\mathbf{I}_m \otimes \frac{-\mathbf{A}^{-1}}{\sigma^4}\right]\left\{_c \mathbf{y}_i - \hat{\boldsymbol{\beta}}\right\}$$

$$= \frac{1}{\sigma^4}\sum_{i=1}^m (\mathbf{y}_i - \hat{\boldsymbol{\beta}})'\mathbf{A}^{-1}(\mathbf{y}_i - \hat{\boldsymbol{\beta}}). \qquad (8.46)$$

With $\hat{\boldsymbol{\beta}}$ from (8.4), we can write

$$\mathbf{y}_i - \hat{\boldsymbol{\beta}} = \left\{ {}_c\, y_{ij} - \bar{y}_{\cdot j} \right\}_{j=1}^m \equiv \left\{ {}_c\, \delta_{ij} \right\}_{j=1}^m, \qquad (8.47)$$

so defining δ_{ij}. Then, on using (8.35) for $\mathbf{A}^{-1}$ generalized from order 5 to order n, we get

$$\mathrm{RHS}(\sigma^2) = \frac{1}{\sigma^4}\frac{1}{1-\rho^2}\sum_{i=1}^m\left[(1+\rho^2)\sum_{j=1}^n\delta_{ij}^2 - \rho^2(\delta_{i1}^2 + \delta_{in}^2)\right.$$

$$\left. -2\rho\sum_{j=2}^n\delta_{ij}\delta_{i,j-1}\right]. \qquad (8.48)$$

Therefore $\mathrm{LHS}(\sigma^2) = \mathrm{RHS}(\sigma^2)$ gives the estimating equation

$$mn\hat{\sigma}^2(1-\hat{\rho}^2) = (1+\hat{\rho}^2)\sum_{i=1}^m\sum_{j=1}^n\delta_{ij}^2 - \hat{\rho}^2\sum_{i=1}^m(\delta_{i1}^2 + \delta_{in}^2) - 2\hat{\rho}\sum_{i=1}^m\sum_{j=2}^n\delta_{ij}\delta_{i,j-1}.$$
$$(8.49)$$

Now doing the same thing for $\varphi = \rho$ gives

$$\mathrm{LHS}(\rho) = m\,\mathrm{tr}\left[(\sigma^2\mathbf{A})^{-1}\frac{\partial(\sigma^2\mathbf{A})}{\partial\rho}\right] = m\,\mathrm{tr}\left(\mathbf{A}^{-1}\frac{\partial\mathbf{A}}{\partial\rho}\right).$$

On generalizing $\mathbf{A}^{-1}$ and $\mathbf{A}$ of (8.35) to order n it will be found (see E 8.4) that $\mathrm{LHS}(\rho)$ reduces to

$$\mathrm{LHS}(\rho) = \frac{-2m(n-1)\rho}{1-\rho^2}.$$

And $\mathrm{RHS}(\rho)$ will be the same as $\mathrm{RHS}(\sigma^2)$ of (8.45) but with

$$\frac{\partial(\mathbf{A}\sigma^2)^{-1}}{\partial\sigma^2} = \frac{-\mathbf{A}^{-1}}{\sigma^4} \quad \text{replaced by} \quad \frac{1}{\sigma^2}\frac{\partial(\mathbf{A}^{-1})}{\partial\rho}.$$

Therefore from (8.46)

$$\mathrm{RHS}(\rho) = \frac{1}{\sigma^2}\sum_{i=1}^m(\mathbf{y}_i - \hat{\boldsymbol{\beta}})'\frac{\partial\mathbf{A}^{-1}}{\partial\rho}(\mathbf{y}_i - \hat{\boldsymbol{\beta}}).$$

Now in looking at $\mathbf{A}^{-1}$ it will be found that in $\mathbf{A}^{-1}$ the element

$$
\left.
\begin{array}{lcl}
1/(1-\rho^2) & \text{becomes} & 2\rho/(1-\rho^2)^2 \\
-\rho/(1-\rho^2) & \text{becomes} & -(1+\rho^2)/(1-\rho^2)^2 \\
(1+\rho^2)/(1-\rho^2)^2 & \text{becomes} & 4\rho(1-\rho^2)^2
\end{array}
\right\} \text{ in } \frac{\partial \mathbf{A}^{-1}}{\partial \rho}.
$$

Therefore, in making these changes in going from (8.46) to (8.48) for $\mathrm{RHS}(\rho)$ gives

$$
\mathrm{RHS}(\rho) = -\frac{1}{\sigma^2} \frac{1}{(1-\rho^2)^2} \sum_{i=1}^{m} \left[4\rho \sum_{j=1}^{n} \delta_{ij}^2 - 2\rho(\delta_{i1}^2 + \delta_{in}^2) \right.
$$

$$
\left. - 2(1+\rho^2) \sum_{j=2}^{n} \delta_{ij}\delta_{i,j-1} \right].
$$

Then equating this to $\mathrm{LHS}(\rho)$ gives the second estimating equation as

$$
m(n-1)\hat{\rho}\hat{\sigma}^2(1-\hat{\rho}^2) = 2\hat{\rho}\sum_{i=1}^{m}\sum_{j=1}^{n} \delta_{ij}^2 - \hat{\rho}\sum_{i=1}^{m}(\delta_{i1}^2 + \delta_{in}^2)
$$

$$
- (1+\hat{\rho}^2) \sum_{i=1}^{m}\sum_{j=2}^{n} \delta_{ij}\delta_{i,j-1}. \quad (8.50)
$$

Clearly, this and (8.49) have to be solved numerically. They appear to have no algebraic solution for $\hat{\sigma}^2$ and $\hat{\rho}$. And the solutions will be ML estimators only if $1 \leq \hat{\rho} \leq 1$ and $\hat{\sigma}^2 > 0$.

8.7 UNBALANCED DATA

a. Example and model

Suppose a clinical trial consists of $m = 5$ patients observed on $n = 7$ occasions but where some of the patients from time to time fail to visit the clinic. Table 8.1 is an example of this.

We still write $\mathrm{E}[\mathbf{y}|\boldsymbol{\mu}] = \mathbf{X}\boldsymbol{\beta} + \mathbf{Z}\mathbf{u}$ as in (8.11) but now the specification of $\mathbf{X}$ and $\mathbf{Z}$ is more complicated than in (8.12). For Table 8.1 the

Table 8.1: Patients Visiting a Clinic ($\sqrt{}$ indicates a visit)

Patient $i = 1, \ldots, m$	Clinic Visit $j = 1, \ldots, n$						
	1	2	3	4	5	6	7
1	$\sqrt{}$	$\sqrt{}$		$\sqrt{}$		$\sqrt{}$	$\sqrt{}$
2	$\sqrt{}$		$\sqrt{}$			$\sqrt{}$	$\sqrt{}$
3	$\sqrt{}$	$\sqrt{}$	$\sqrt{}$	$\sqrt{}$	$\sqrt{}$		
4		$\sqrt{}$		$\sqrt{}$			$\sqrt{}$
5	$\sqrt{}$			$\sqrt{}$	$\sqrt{}$		$\sqrt{}$

data for patient 1 have

$$
\mathrm{E}[\mathbf{y}_1 | u_1] =
\begin{bmatrix}
\mathrm{E}[y_{11}|u_1] \\
\mathrm{E}[y_{12}|u_1] \\
\mathrm{E}[y_{14}|u_1] \\
\mathrm{E}[y_{16}|u_1] \\
\mathrm{E}[y_{17}|u_1]
\end{bmatrix}
=
\begin{bmatrix}
1 & 0 & 0 & 0 & 0 & 0 & 0 \\
0 & 1 & 0 & 0 & 0 & 0 & 0 \\
0 & 0 & 0 & 1 & 0 & 0 & 0 \\
0 & 0 & 0 & 0 & 0 & 1 & 0 \\
0 & 0 & 0 & 0 & 0 & 0 & 1
\end{bmatrix} \boldsymbol{\beta} +
\begin{bmatrix}
1 \\ 1 \\ 1 \\ 1 \\ 1
\end{bmatrix} u_1 \quad (8.51)
$$

$$
= \mathbf{X}_1 \boldsymbol{\beta} + \mathbf{1}_{n_1} u_1, \tag{8.52}
$$

where n_i is the number of times patient i visits the clinic using $i = 1$ in (8.52).

Assembling $\mathbf{y} = \left\{ {}_c \mathbf{y}_i \right\}_{i=1}^{m}$ as in (8.2) gives

$$
\mathrm{E}[\mathbf{y}|\mathbf{u}] = \left\{ {}_c \mathbf{X}_i \right\}_{i=1}^{m} \boldsymbol{\beta} + \left\{ {}_d \mathbf{1}_{n_i} \right\}_{i=1}^{m} \mathbf{u}, \tag{8.53}
$$

where $\boldsymbol{\beta}$ is $n \times 1$ and $\mathbf{u}$ is $m \times 1$. In comparing (8.51) and (8.52), it is apparent that for patient i, $\mathbf{X}_i$ is $n_i \times n$ with one row for each checkmark for that patient in Table 8.1; and that row of $\mathbf{X}_i$ is null except for a one corresponding to the checkmark. As examples, in the first line of Table 8.1 there are checkmarks for $j = 1$ and $j = 2$. These generate $[1\ 0\ 0\ 0\ 0\ 0\ 0]$ and $[0\ 1\ 0\ 0\ 0\ 0\ 0]$ as rows of $\mathbf{X}_i$.

Using $\mathrm{var}(\mathbf{y}|\mathbf{u}) = \mathbf{R}$, (8.53), and with $\mathbf{D} = \mathrm{var}(\mathbf{u})$ of (8.16), gives

$$
\mathrm{var}(\mathbf{y}) = \mathbf{V} = \left\{ {}_d \mathbf{1}_{n_i} \right\}_{i=1}^{m} \sigma_u^2 [(1 - \rho_u)\mathbf{I}_m + \rho_u \mathbf{J}_m] \left\{ {}_d \mathbf{1}'_{n_i} \right\}_{i=1}^{m} + \mathbf{R}
$$

$$
= \sigma_u^2 (1 - \rho_u) \left\{ {}_d \mathbf{J}_{n_i} \right\}_{i=1}^{m} + \sigma_u^2 \rho_u \mathbf{J}_N + \mathbf{R}, \tag{8.54}
$$

where

$$N = \sum_{i=1}^{m} n_i.$$ (8.55)

We now consider three special cases similar to those dealt with in Section 8.5 for deriving $\tilde{\mathbf{u}}$. Now, though, $\mathbf{X} = \left\{ {}_c \mathbf{X}_i \right\}$ of (8.53) is no longer as simple as the $\mathbf{1}_m \otimes \mathbf{I}_n$ of (8.12) for balanced data, as is evident from the description of $\mathbf{X}_i$ following (8.53). Therefore for each of our three special cases we deal with $\mathbf{V}$, $\mathbf{V}^{-1}$, $\hat{\boldsymbol{\beta}}$ and $\tilde{\mathbf{u}}$.

b.   Uncorrelated subjects

For balanced data, every patient had n observations and one could assume the same variance-covariance matrix, $\mathbf{R}_0$, for all patients and so have $\mathbf{R} = \mathbf{I}_m \otimes \mathbf{R}_0$ as in Section 8.4a. But that is not possible for patient i having n_i data with n_i not being the same for all patients. The nearest counterpart is to have $\mathbf{R}$ be block diagonal, of blocks of order n_i: $\mathbf{R}_0 = \left\{ {}_d \mathbf{R}_i \right\}_{i=1}^{m}$.

Notation

At this point our curly bracket notation mostly involves i ranging from 1 to m, so we cease indicating that range unless context demands it.

– i.   *Matrix* $\mathbf{V}$ *and its inverse*

Then with $\mathbf{R} = \left\{ {}_d \mathbf{R}_i \right\}$ and $\rho_u = 0$, (8.54) gives

$$\mathbf{V} = \left\{ {}_d \mathbf{V}_i \right\} \quad \text{for} \quad \mathbf{V}_i = \sigma_u^2 \mathbf{J}_{n_i} + \mathbf{R}_i$$ (8.56)

and so

$$\mathbf{V}^{-1} = \left\{ {}_d \mathbf{V}_i^{-1} \right\}.$$ (8.57)

– ii.   *Estimating the fixed effects*

With

$$\mathbf{X} = \left\{ {}_c \mathbf{X}_i \right\}$$

from (8.53), we get

$$\mathbf{X}'\mathbf{V}^{-1}\mathbf{X} = \sum_{i=1}^{m} \mathbf{X}_i'\mathbf{V}_i^{-1}\mathbf{X}_i$$

and

$$\mathbf{X}'\mathbf{V}^{-1}\mathbf{y} = \sum_{i=1}^{m} \mathbf{X}_i'\mathbf{V}_i^{-1}\mathbf{y}_i \qquad (8.58)$$

and so

$$\hat{\boldsymbol{\beta}} = \left(\sum_{i=1}^{m} \mathbf{X}_i'\mathbf{V}_i^{-1}\mathbf{X}_i\right)^{-1} \sum_{i=1}^{m} \mathbf{X}_i'\mathbf{V}_i^{-1}\mathbf{y}_i. \qquad (8.59)$$

This has no simplification such as $\hat{\beta}_j = \hat{\mu}_j = \bar{y}_{.j}$ of (8.4) and (8.32) in the balanced data case.

– iii.   *Predicting the random effects*

We have

$$\mathbf{DZ}' = \sigma_u^2 \mathbf{I}_m \left\{ _d \; \mathbf{1}_{n_i}' \right\} = \sigma_u^2 \left\{ _d \; \mathbf{1}_{n_i}' \right\}$$

and so with $\mathbf{V}^{-1}$ of (8.57)

$$\begin{aligned}
\tilde{\mathbf{u}}^0 \;&=\; \mathbf{DZ}'\mathbf{V}^{-1}(\mathbf{y} - \mathbf{X}\hat{\boldsymbol{\beta}}) \\[2mm]
&=\; \sigma_u^2 \left\{ _d \; \mathbf{1}_{n_i}' \right\} \left\{ _d \; \mathbf{V}_i^{-1} \right\} \left\{ _c \; \mathbf{y}_i - \mathbf{X}_i\hat{\boldsymbol{\beta}} \right\} \\[2mm]
&=\; \sigma_u^2 \left\{ _c \; \mathbf{1}_{n_i}' \mathbf{V}^{-1}\mathbf{y}_i - \mathbf{1}_{n_i}' \mathbf{V}_i^{-1}\mathbf{X}_i\hat{\boldsymbol{\beta}} \right\}. \qquad (8.60)
\end{aligned}$$

The nature of $\mathbf{X}_i$ discussed following (8.53), together with $\mathbf{V}_i$ not being the same for all i (not even of the same order), makes $\hat{\boldsymbol{\beta}}$ of (8.59) not very tractable; and so (8.60) is not amenable to further simplification. But for known $\mathbf{V}_i$, both it and (8.59) are reasonable computations.

8.8   MODELS FOR NON-NORMAL RESPONSES

As (7.2) indicates, we can extend longitudinal models for continuous, normally distributed outcomes Y such as (8.11) to non-normal longitudinal data such as repeated binary outcomes by adding random effects to the linear predictor of generalized linear models, (5.5) and (5.6). That is, as in (7.2), we model non-normal y_i as

$$\mathrm{E}[y_i|\mathbf{u}] \;=\; \mu_i$$

$$g(\mu_i) \;=\; \mathbf{x}_i'\boldsymbol{\beta} + \mathbf{z}_i'\mathbf{u}, \qquad (8.61)$$

where $g(\cdot)$ is a known link function , $\mathbf{u}$ are the random effects, $\mathbf{x}_i'$ is the ith row of the model matrix for the fixed effects, $\mathbf{z}_i'$ is the ith row of the model matrix associated with the random effects and $\boldsymbol{\beta}$ is the fixed effects parameter vector. As with continuous response data, $\mathbf{z}_i'$ will often contain time or other covariates whose association with the response varies between subjects. As in (7.3), we complete the specification of the model for non-normal responses by specifying a density $f_\mathbf{U}(\mathbf{u})$ for the random effects.

a. Covariances and correlations

As (7.13) shows, the pattern of within-subject correlations of responses is more complicated for non-normal responses than for normal ones. For example, the correlations induced by simple random intercepts models for non-normal responses depend on the covariates, x, unlike the case for normal responses. Different specifications of $\mathbf{z}_i'$ and $\mathbf{u}$ can generate a range of patterns for $\mathbf{V} = \text{var}(\mathbf{y})$ for non-normal responses but generalized linear mixed models typically determine an expression for $\text{var}(\mathbf{y}|\mathbf{u})$. Hence, data analysts are not free to specify the $\mathbf{R}$ matrix of (8.14) as they could for normally distributed responses.

b. Estimation

While closed-form expressions such as (6.24) exist for maximum likelihood estimators of linear mixed models, the likelihood for generalized linear mixed models (7.14) does not typically simplify to provide such closed-form estimators. We often must numerically maximize the likelihood using iterative techniques, as in Chapter 14, to obtain estimates of regression coefficients and variance components. For standard models that include just random intercepts and slopes, the integrals in the likelihood are all two–dimensional and numerical integration methods are highly accurate. For models involving large numbers of random effects numerical integration methods can be inaccurate but simulation-based techniques such as Markov Chain Monte Carlo (Gilks et al., 1996), in particular, Gibbs sampling (Zeger and Karim, 1991), and Monte Carlo EM (McCulloch, 1997) have proven useful in these settings.

c. Prediction of random effects

Sections 6.5, 6.6, 8.5 and Chapter 13 focus on predictions $\tilde{u}$ of random effects u that minimize the mean square error of prediction. That is, when u is a scalar, $\tilde{u}$ that minimizes

$$E[\tilde{u} - u]^2. \tag{8.62}$$

When $\mathbf{u}$ is a vector, we seek $\tilde{\mathbf{u}}$ that minimize

$$E[(\tilde{\mathbf{u}} - \mathbf{u})'\mathbf{A}(\tilde{\mathbf{u}} - \mathbf{u})], \tag{8.63}$$

where $\mathbf{A}$ is a positive definite symmetric matrix. Section 13.8 shows that

$$\tilde{u}_i = E(u_i|\mathbf{y}_i) \tag{8.64}$$

minimizes (8.62) and

$$\tilde{\mathbf{u}} = BP(\mathbf{u}) = E[\mathbf{u}|\mathbf{y}], \tag{8.65}$$

minimizes (8.63). Although (8.64) and (8.65) often yield closed-form expressions such as (8.32) for linear mixed effects models, they will typically not do so for generalized linear models, such as the mixed-effects logistic model, that do not have an identity link. With non-identity link GLMMs one can calculate the best predictors in (8.64) and (8.65) using numerical integration methods but the calculations required to compute the expectations in (8.64) and (8.65) can be difficult. As an alternative, many statistical packages shift the focus of prediction away from the mean of the distribution of $u \mid \mathbf{y}$ to the mode of this posterior distribution. The mode represents the most likely value for u given the data and is equal to the mean when $u \mid \mathbf{y}$ is normally distributed, as in the linear mixed effects model. Thus, for GLMMs with non-identity links we predict u as the value $\tilde{u}$ that maximizes the posterior mode of the distribution of $u \mid Y$, that is, the value u that maximizes

$$\log\{f(\mathbf{y}_i \mid x_{i1}, \ldots, x_{in_i}, \beta, \mathbf{u}_i)g(\mathbf{u}_i \mid \theta_G)\} . \tag{8.66}$$

In practice, statistical packages often calculate predictions of the random effects u_i in (8.61) as the values $\hat{u}_i$ that maximize

$$\log\{f(\mathbf{y}_i \mid x_{i1}, \ldots, x_{in_i}, \hat{\beta}, \mathbf{u}_i)g(\mathbf{u}_i \mid \hat{\theta}_G)\}, \tag{8.67}$$

where θ_G denotes the parameters of G and $\hat{\beta}$ and $\hat{\theta}_G$ are current estimates. We can find closed-form expressions for predicted random effects in some special cases (see E 8.10, part(b)). However, closed-form expressions for the joint densities in (8.66) and (8.67) typically do not usually exist, so that one must calculate predicted random effects for generalized linear mixed models using iterative methods to calculate modes and numerical integration techniques to evaluate the integrals in the relevant joint densities.

d. Binary responses, random intercepts and slopes

We can generalize model (8.19) for continuous responses to accommodate longitudinal binary responses by including random intercepts and slopes in the linear predictor of binary generalized linear models. One useful model assumes that

$$\text{logit pr}(y_{ij} = 1 | x_{ij}, b_{0i}, b_{1i}) = \beta_0 + b_{0i} + (\beta_1 + b_{1i})x_{ij}, \qquad (8.68)$$

where b_{0i} and b_{1i} denote the random intercepts and slopes, respectively, which follow a bivariate normal distribution with mean vector $(0, 0)'$ and covariance matrix

$$\mathbf{D} = \begin{bmatrix} \sigma_0^2 & \sigma_{01} \\ \sigma_{01} & \sigma_1^2 \end{bmatrix}. \qquad (8.69)$$

The likelihood for this model is

$$L = \prod_{i=1}^{m} \int \prod_{j=1}^{n_i} f(y_{ij} | b_0, b_1, x_{ij}) g(b_0, b_1) db_0 db_1, \qquad (8.70)$$

where f is the conditional density of response y_{ij} under model (8.68)

$$f(y_{ij} | b_0, b_1, x_{ij}) = \frac{e^{y_{ij}(\beta_0 + b_{0i} + (\beta_1 + b_{1i})x_{ij})}}{1 + e^{\beta_0 + b_{0i} + (\beta_1 + b_{1i})x_{ij}}} \qquad (8.71)$$

and g is the bivariate normal density with mean vector $(0, 0)'$ and covariance matrix $\mathbf{D}$. We maximize (8.70) with respect to the model parameters using computational methods as in Section 14.3 to obtains MLEs for β_0, β_1, σ_0^2 σ_1^2 and σ_{01}. We calculate predicted random intercepts and slopes by maximizing (8.67) using the MLEs from (8.70) and the density f given by (8.71).

8.9   A SUMMARY OF RESULTS

The results developed in Sections 8.4 and 8.5 are sequenced by model within each estimation situation. Here we give a summary of those results and those of Section 8.2 sequenced by estimation situation within each model.

Note: Equation numbers in square brackets refer to equations occurring earlier in the chapter.

a.   Balanced data

– i.   *With some generality*

$$\mathbf{y} \sim \mathcal{N}\left[(\mathbf{1}_m \otimes \mathbf{I}_n)\mu, \mathbf{V} = \mathbf{I}_m \otimes \mathbf{V}_0\right],$$

and

$$\mu = \left\{{}_c \mu_j\right\}^n_{j=1}$$

the MLE of μ_j is

$$\hat{\mu}_j = \bar{y}_{\cdot j} \qquad\qquad [8.4]$$

and that of $\mathbf{V}_0$ is

$$\hat{\mathbf{V}}_0 = \frac{1}{m} \left\{ \sum_{i=1}^{m} (y_{ij} - \bar{y}_{\cdot j})(y_{ik} - \bar{y}_{\cdot k}) \right\}^n_{j,k=1}, \qquad\qquad [8.8]$$

a Wishart matrix. The estimator $\hat{\mu}_j = \bar{y}_{\cdot j}$ applies for the special cases of balanced data we now list; it is not repeated, and it is denoted $\hat{\boldsymbol{\beta}}$.

– ii.   *Uncorrelated subjects*

With $\mathbf{R}_0$ defined just prior to (8.18)

$$\tilde{u}_i^0 = \frac{\mathbf{1}'\mathbf{R}_0^{-1}(\mathbf{y}_i - \hat{\boldsymbol{\beta}})}{1/\sigma_u^2 + \mathbf{1}'\mathbf{R}_0^{-1}\mathbf{1}} \qquad\qquad [8.32]$$

$$\hat{\sigma}_u^2 = \frac{\sum_{i=1}^m \mathbf{1}'\mathbf{R}_0^{-1}(\mathbf{y}_i - \hat{\boldsymbol{\beta}}) - m'\mathbf{1}'\mathbf{R}_0^{-1}\mathbf{1}}{m(\mathbf{1}'\mathbf{R}_0^{-1}\mathbf{1})^2}. \qquad\qquad [8.44]$$

– iii.   *Uncorrelated between, and autocorrelated within, subjects*

$$\tilde{u}_i^0 = \frac{\sigma_u^2\left[(1-\rho)n(\bar{y}_{i\cdot} - \bar{y}_{\cdot\cdot}) + \rho(y_{i1} - \bar{y}_{\cdot 1} + y_{in} - \bar{y}_{\cdot n}\right]}{\sigma^2(1+\rho) + \sigma_u^2[n - (n-2)\rho]}. \qquad\qquad [8.36]$$

For $\delta_{ij} \equiv y_{ij} - \bar{y}_{\cdot j}$

$$mn\hat{\sigma}^2(1-\hat{\rho}^2) = (1+\hat{\rho}^2)\sum_i\sum_j \delta_{ij}^2 - \hat{\rho}^2\sum_{i=1}^m(\delta_{i1}^2 + \delta_{in}^2)$$

$$- 2\hat{\rho}\sum_{i=1}^m\sum_{j=2}^n \delta_{ij}\delta_{i,j-1} \qquad [8.49]$$

$$m(n-1)\hat{\rho}\hat{\sigma}^2(1-\hat{\rho}^2) = 2\hat{\rho}\sum_i\sum_j \delta_{ij}^2 - \hat{\rho}\sum_i(\delta_{i1}^2 + \delta_{in}^2)$$

$$- (1+\hat{\rho}^2)\sum_{i=1}^m\sum_{j=2}^n \delta_{ii}\delta_{i,j-1}. \qquad [8.50]$$

b. Unbalanced data

− i. *Uncorrelated subjects*

$$\hat{\beta} = \sum_{i=1}^m(\mathbf{X}_i'\mathbf{V}_i^{-1}\mathbf{X}_i)^{-1}\sum_{i=1}^m \mathbf{V}_i^{-1}\mathbf{y}_i \qquad [8.59]$$

$$\tilde{u}_i^0 = \sigma_u^2\left(\mathbf{1}_{n_i}'\mathbf{V}_i^{-1}\mathbf{y}_i - \mathbf{1}_{n_i}'\mathbf{V}_i^{-1}\mathbf{X}_i\hat{\beta}\right). \qquad [8.60]$$

8.10 APPENDIX

a. For Section 8.4a

In $\mathbf{V} = \mathbf{I}_m \otimes (\sigma_u^2\mathbf{J}_n + \mathbf{R}_0)$ write $\sigma_u^2\mathbf{J}_n = \sigma_u^2\mathbf{1}_n\mathbf{1}_n'$ and use the standard result for any nonsingular $\mathbf{A}$:

$$(\mathbf{A} + \lambda\mathbf{t}\mathbf{t}')^{-1} = \mathbf{A}^{-1} - \frac{\mathbf{A}^{-1}\mathbf{t}\mathbf{t}'\mathbf{A}^{-1}}{1/\lambda + \mathbf{t}'\mathbf{A}^{-1}\mathbf{t}}$$

to get

$$(\mathbf{R}_0 + \sigma_u^2\mathbf{1}\mathbf{1}')^{-1} = \mathbf{R}_0^{-1} - \frac{\mathbf{R}_0^{-1}\mathbf{1}\mathbf{1}'\mathbf{R}_0^{-1}}{1/\sigma_u^2 + \mathbf{1}'\mathbf{R}_0^{-1}\mathbf{1}}.$$

Then, since $\rho_u = 0$ gives $\mathbf{D} = \sigma_u^2\mathbf{I}_m$, with $\mathbf{Z} = \mathbf{I}_m \otimes \mathbf{1}_n$ we get for $\tilde{u}$

$$\mathbf{D}\mathbf{Z}'\mathbf{V}^{-1} = \sigma_u^2\mathbf{I}_m(\mathbf{I}_m \otimes \mathbf{1}_n')\mathbf{V}^{-1} = \sigma_u^2(\mathbf{I}_m \otimes \mathbf{1}_n')\mathbf{V}^{-1}$$

$$= \sigma_u^2 \mathbf{I}_m \otimes \left(\mathbf{1}_n' \left[\mathbf{R}_0^{-1} - \frac{\mathbf{R}_0^{-1} \mathbf{1} \mathbf{1}' \mathbf{R}_0^{-1}}{1/\sigma_u^2 + \mathbf{1}' \mathbf{R}_0^{-1} \mathbf{1}} \right] \right)$$

$$= \mathbf{I}_m \otimes \frac{\mathbf{1}' \mathbf{R}_0^{-1}}{1/\sigma_u^2 + \mathbf{1}' \mathbf{R}_0^{-1} \mathbf{1}} = \left\{ {}_d \frac{\mathbf{1}' \mathbf{R}_0^{-1}}{1/\sigma_u^2 + \mathbf{1}' \mathbf{R}_0^{-1} \mathbf{1}} \right\}_{i=1}^m .$$

$$\tilde{\mathbf{u}}^0 = \left\{ {}_d \frac{\mathbf{1}' \mathbf{R}_0^{-1}}{1/\sigma_u^2 + \mathbf{1}' \mathbf{R}_0^{-1} \mathbf{1}} \right\}_{i=1}^m \left\{ {}_c \mathbf{y}_i - \mathbf{X}_i \hat{\boldsymbol{\beta}} \right\}_{i=1}^m .$$

$$\tilde{u}_i^0 = \frac{\mathbf{1}' \mathbf{R}_0^{-1} (\mathbf{y}_i - \mathbf{X}_i \hat{\boldsymbol{\beta}})}{1/\sigma_u^2 + \mathbf{1}' \mathbf{R}_0^{-1} \mathbf{1}} . \qquad\qquad [8.32]$$

b.   For Section 8.4b

Put $\mathbf{R}_0^{-1} = (1/\sigma^2) \mathbf{I}_n$ in (8.32), to get

$$\tilde{u}_i^0 = \frac{(1/\sigma^2) \left(y_{i\cdot} - \sum_{j=1}^n \bar{y}_{\cdot j} \right)}{1/\sigma_u^2 + n/\sigma^2} = \frac{n\sigma_u^2 (\bar{y}_{i\cdot} - \bar{y}_{\cdot\cdot})}{\sigma^2 + n\sigma_u^2} .$$

8.11   EXERCISES

E 8.1 Reduce $\text{ML}(\mu)$ in Section 8.2b to $\hat{\mu}$ of (8.4).

E 8.2 If $\mathbf{H}^{-1}$ exists and for $\mathbf{X}$ of full column rank, show that $\mathbf{V}\mathbf{X} = \mathbf{X}\mathbf{H}$ leads to $(\mathbf{X}'\mathbf{V}^{-1}\mathbf{X})^{-1}\mathbf{X}'\mathbf{V}^{-1}\mathbf{y} = (\mathbf{X}'\mathbf{X})^{-1}\mathbf{X}'\mathbf{y}$.

E 8.3 In Section 8.5c show that $\mathbf{A}\mathbf{A}^{-1} = \mathbf{I}$ and derive (8.36).

E 8.4 Derive (8.49) and (8.50) from their respective LHS($\cdot$) = RHS($\cdot$) equations.

E 8.5 In E 6.6(c) write

$$\mathbf{y} = \left\{ {}_c \left\{ {}_c \left\{ {}_c y_{ijk} \right\}_{k=1}^r \right\}_{j=1}^n \right\}_{i=1}^m .$$

Then

$$\mathbf{V} = \text{var}(\mathbf{y})$$

$$= (\mathbf{I_m} \otimes \mathbf{J}_n \otimes \mathbf{J}_r)\sigma_a^2 + (\mathbf{I_m} \otimes \mathbf{I}_n \otimes \mathbf{J}_r)\sigma_g^2 + (\mathbf{I_m} \otimes \mathbf{I}_n \otimes \mathbf{I}_r)\sigma^2$$

and

$$\mathbf{X} \;=\; [(\mathbf{1_m} \otimes \mathbf{1}_n \otimes \mathbf{1}_r) \;\; (\mathbf{1_m} \otimes \mathbf{I}_n \otimes \mathbf{I}_r) \;\; (\mathbf{1_m} \otimes \mathbf{1}_n \otimes \mathbf{I}_r)].$$

(a) For $m = 2$, $n = 3$, and $r = 2$ write out $\mathbf{y}$, $\mathbf{V}$, and $\mathbf{X}$.

(b) Find $\mathbf{H}$ such that $\mathbf{VX} = \mathbf{XH}$.

(c) With $r = 1$ (i.e., effectively deleting k from $\mathbf{y}$), what are $\mathbf{V}$ and $\mathbf{X}$ for E 6.6(a)? And what is $\mathbf{H}$ such that $\mathbf{VX} = \mathbf{XH}$?

(d) Repeat (c) for E 6.6(b).

E 8.6 Show that the variance-covariance matrix $\mathbf{V}_i$ for $\mathbf{y}_i$ in the random intercepts and slopes model (8.19) is (8.23).

E 8.7 Show that (8.26) is monotonically decreasing in x_{ik} for $x_{ik} > x_{ij}$.

E 8.8 Derive (8.27).

E 8.9 Derive (8.28).

E 8.10 (a) Derive the likelihood equations for the log link, random intercept model for binary matched pairs

$$\log pr(y_{ij} = 1|u_i) = \beta_0 + \beta x_{ij} + \sigma u_i \quad j = 1, 2$$

with $x_{i1} \equiv 0$, $x_{i2} \equiv 1$, and random intercepts u that follow a Uniform$(-\sqrt{3}, \sqrt{3})$ distribution. Do closed form expressions exist for any of the maximum likelihood estimators of β_0, β or σ?

(b) Calculate predicted random effects for each response pattern (y_{i1}, y_{i2}) for the binary matched pairs model of part (a) using (8.66), assuming known values for β_0, β and σ.

Chapter 9

MARGINAL MODELS

9.1 INTRODUCTION

Instead of starting from the conditional specification as in (7.1), (7.2), and (7.3), we might directly hypothesize a model for the mean of $\mathbf{y}$. As an example, suppose y_{ij} is equal to 1 if the jth child of woman i is born prematurely and is zero otherwise and assume we have a single predictor x_{ij} = number of drinks of alcohol per day. The marginal approach would model the marginal mean of y_{ij} directly by, for example, assuming that a logistic regression model, as in (3.109), fits the data:

$$\text{logit}(\text{E}[y_{ij}]) = \text{logit}(\text{P}\{y_{ij} = 1\}) = \beta_0 + \beta_1 x_{ij}. \tag{9.1}$$

In words, the model would be for logit of the probability of premature birth, averaged over a population of women. Of course, if the model was for correlated data, we would not be able to assume the observations were independent.

On the other hand, our typical conditional approach corresponds to hypothesizing the existence of a random factor for women and specifying a conditional model such as

$$\text{logit}(\text{E}[y_{ij}|\mathbf{u}]) = \beta_0 + \beta_1 x_{ij} + u_i, \tag{9.2}$$

where u_i represents the random woman effect. This corresponds to modeling the conditional probability of a premature birth for each woman.

From a probabilistic perspective, we can calculate the marginal distribution of $\mathbf{y}$ (at least conceptually — it might be computationally difficult) from the distribution of $\mathbf{u}$ and the conditional distribution of

y|u. As we will demonstrate later in this chapter, in general it is not possible to recover the marginal of **u** and the conditional distribution of **y|u** from the marginal distribution of **y**. Furthermore, having a conditionally specified model with a random effect is the basis for deriving predicted values in Chapter 13. These considerations would seem to favor a conditional specification of the model.

However, in some cases, the marginal distribution (or perhaps only the marginal mean) may be adequate for answering questions of interest. Thus, in the alcohol consumption example, a natural question of interest is how much the incidence of premature birth could be reduced by lowering, on average, women's alcohol consumption. In such cases, the potentially difficult problem of specifying the conditional distribution of **y|u** and the marginal distribution of **u** can be avoided. This is an advantage of marginal modeling and is the basis for the generalized estimating equations approach, which is described in Section 9.3.

Distinguishing conditional from marginal models is straightforward probabilistically, but it is often difficult in practice. For example, a researcher might be interested in "the influence of alcohol consumption on premature birth", which would not specify which type of model to build. In our experience, researchers often think about building models in a mechanistic way, which seems more compatible with the conditional approach. Again considering the premature birth example, a researcher might think about the influence of alcohol consumption by trying to understand how alcohol influenced an individual woman's physiology.

The distinction between conditional and marginal models is an important one to keep in mind in practice. The reason is perhaps easiest to see in a binary data, probit-normal model:

$$\mathrm{E}[y_i|\mathbf{u}] = \mathrm{P}\{y_i = 1|\mathbf{u}\} = \Phi(\mathbf{x}_i'\boldsymbol{\beta} + \mathbf{z}_i'\mathbf{u}), \tag{9.3}$$

$$\mathbf{u} \sim \mathcal{N}(\mathbf{0}, \mathbf{D}).$$

It is not hard to show, as in (7.45), that if the conditional mean is given by (9.3), then the unconditional mean is

$$\mathrm{E}[y_i] = \Phi\left(\frac{\mathbf{x}_i'\boldsymbol{\beta}}{\sqrt{\mathbf{z}_i'\mathbf{D}\mathbf{z}_i + 1}}\right) \equiv \Phi\left(\mathbf{x}_i'\boldsymbol{\beta}^*\right), \tag{9.4}$$

so that $\boldsymbol{\beta}^* = \boldsymbol{\beta}/\sqrt{\mathbf{z}_i'\mathbf{D}\mathbf{z}_i + 1}$. Hence $\boldsymbol{\beta}$ represents the magnitude of

Figure 9.1: Probability of success versus a predictor for the marginal and conditional versions of a probit-normal model. Solid line, marginal model; dashed lines, realizations of the conditional model.

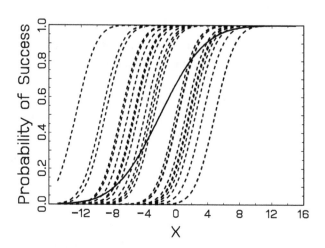

the effect of the predictors on the conditional distribution, while β^* represents the magnitude of the effect on the marginal distribution.

Clearly, β is always larger than β^* in absolute value. Why this is so is perhaps easiest to understand graphically, as shown in Figure 9.1. Each of the realized values of the conditional model [i.e., equation (9.3) plotted for various realized u_i], shown by the dashed lines in the figure, have a large value of β. However, the variance of the random effect is quite large and when all the curves are averaged, the resulting unconditional mean has a much smaller slope and much smaller value for β^* as compared to β.

9.2   EXAMPLES OF MARGINAL REGRESSION MODELS

Many classes of models have been suggested to model the mean of y as a function of predictors. Some approaches, such as Generalized Estimating Equations (described in the next Section), do this without fully specifying the functional form of the joint distribution of the observations. Others explicitly build joint distributions that can be easily parameterized in terms of the marginal mean, $E[y_i|\mathbf{x_i}]$.

In constructing a marginal model we will often be interested in explicitly modeling the marginal means and accommodating associations among the observations. As an illustration, consider modeling a multivariate binary vector $\mathbf{y} = (y_1, y_2, \ldots, y_n)$. Its joint density will be proportional to a function of the y_i, crossproducts of $y_i y_j$, and so forth:

$$f_{\mathbf{Y}} \propto \exp \left\{ \sum_i u_i y_i + \sum_{i<j} u_{ij} y_i y_j + \ldots + u_{12\ldots m} y_1 y_2 \cdots y_n \right\}, \quad (9.5)$$

where, for example,

$$u_{12} = log \left(OR[y_1, y_2 | y_j = 0; j > 3] \right), \quad (9.6)$$

and OR represents the odds ratio (described in Section 3.7a).

This is not, however, a convenient parameterization for describing the marginal means of the y_i and transformation back and forth from the above parameters and the marginal means and odds ratios (for example) can lead to difficulties in restrictions on the parameters. See Liang et al. (1992), Zhao and Prentice (1990), and Ekholm et al. (1995) for more details.

9.3 GENERALIZED ESTIMATING EQUATIONS

We begin by considering the case of a linear model, $E[\mathbf{y}] = \mathbf{X}\beta$ and $var(\mathbf{y}) = \mathbf{V}$. If we define balanced data as the case where $\mathbf{VX} = \mathbf{XH}$ for some $\mathbf{H}$ (some justification of so defining it is given via examples in the exercises) then the ordinary least squares estimator of β is equal to the generalized least squares estimator which is the same as the maximum likelihood estimator. What would happen if the ordinary least squares estimator were used for unbalanced data? For this section, for ease of exposition we will assume that $\mathbf{X}$ is of full column rank, and use the notation $\hat{\beta}_{\mathbf{W}^{-1}}$ to denote the weighted least squares estimator $(\mathbf{X}'\mathbf{WX})^{-1}\mathbf{X}'\mathbf{Wy}$. Thus $\hat{\beta}_{\mathbf{I}}$ denotes the ordinary least squares estimator and $\hat{\beta}_{\mathbf{V}}$ denotes the MLE with a known variance-covariance matrix $\mathbf{V} = var(\mathbf{y})$, i.e., $\hat{\beta}_{\mathbf{V}} = (\mathbf{X}'\mathbf{V}^{-1}\mathbf{X})^{-1}\mathbf{X}'\mathbf{V}^{-1}\mathbf{y}$.

If primary interest lies in estimating β a possible estimator is $\hat{\beta}_{\mathbf{I}}$. How well does it perform? First, note that $\hat{\beta}_{\mathbf{I}}$ is unbiased no matter what the value of $\mathbf{V}$:

$$\begin{aligned} E[\hat{\beta}_{\mathbf{I}}] &= (\mathbf{X}'\mathbf{X})^{-1}\mathbf{X}'E[\mathbf{y}] \\[2mm] &= (\mathbf{X}'\mathbf{X})^{-1}\mathbf{X}'\mathbf{X}\beta \\[2mm] &= \beta. \end{aligned} \quad (9.7)$$

Furthermore, it is straightforward to calculate the variance of $\hat{\beta}_{\mathbf{I}}$:

$$\text{var}(\hat{\beta}_{\mathbf{I}}) = (\mathbf{X'X})^{-1}\mathbf{X'}\text{var}(\mathbf{y})\mathbf{X}(\mathbf{X'X})^{-1}$$

$$= (\mathbf{X'X})^{-1}\mathbf{X'VX}(\mathbf{X'X})^{-1}. \tag{9.8}$$

How does this compare with $\hat{\beta}_{\mathbf{V}}$, which is the optimal estimator for known $\mathbf{V}$? We know that $\hat{\beta}_{\mathbf{V}}$ is also unbiased and has variance equal to $(\mathbf{X'V^{-1}X})^{-1}$, which is smaller than the variance of $\hat{\beta}_{\mathbf{I}}$, but how much smaller? Interestingly, it is often not very much smaller.

For example, consider the extremely simple situation where the mean of all the observations is μ and the data come in m equicorrelated clusters with correlation $\rho > 0$. Further assume that half the clusters are of size n and the other half are of size λn. We thus have

$$\mathbf{y} \sim (\mathbf{1}\mu, \mathbf{V}),$$

where

$$\mathbf{V} = \begin{bmatrix} \mathbf{I}_{m/2} \otimes \mathbf{V}_{0,n} & \mathbf{0} \\ \mathbf{0} & \mathbf{I}_{m/2} \otimes \mathbf{V}_{0,\lambda n} \end{bmatrix} \tag{9.9}$$

and

$$\mathbf{V}_{0,n} = \sigma^2 \left[(1 - \rho)\mathbf{I}_n + \rho\mathbf{J}_n \right]. \tag{9.10}$$

It is straightforward to derive an expression (see E 9.4) for the variance of $\hat{\beta}_{\mathbf{I}}$ and $\hat{\beta}_{\mathbf{V}}$ and to show that the relative efficiency of the ordinary least squares estimator decreases as a function of increasing ρ. Furthermore, the relative efficiency in the worst case, when $\rho = 1$, is given by $2(1 + \lambda^2)/(1 + \lambda)^2$. So if the data are balanced with $\lambda = 1$, the relative efficiency is 1, as expected. If the sample size is 50% larger in one group than in the other, then the worst the variance of ordinary least squares estimator can be is 4% larger than the optimal estimator. Even in the case when the sample size is double ($\lambda = 2$) the variance is only $2(5)/9 = 1.1$ times as large (see E 9.4 and E 9.5).

Certainly it is possible to construct examples where the ordinary least squares estimator does arbitrarily badly in relation to the optimal estimator, but the point of the above calculation is that often the ordinary least squares estimator is quite good for moderate degrees of unbalancedness. If the ordinary least squares estimator performs so well for a wide variety of problems and, furthermore, obviates the need to estimate the variance-covariance structure, then why not use standard regression and regression software whenever the efficiency is high?

The problem is that even though the efficiency is high, the apparent variance can be grossly incorrect. Intuitively, if the observations are positively correlated and we treat them as if they are independent, then the data appear much less variable than they actually are and we can drastically underestimate the variance. This would hold true whether or not the data were balanced.

Again consider a clustered variance scenario with m clusters of size n

$$\mathbf{y}_i \sim \text{ i.i.d. } (\mathbf{1}\beta, \mathbf{V}_{0,n}) \tag{9.11}$$

so that

$$\mathbf{V} = \mathbf{I}_m \otimes \mathbf{V}_{0,n}, \tag{9.12}$$

where $\mathbf{V}_{0,n}$ is defined in (9.10) and the (scalar) β is the mean. The variance that would be estimated from a standard regression or ANOVA program would be

$$
\begin{aligned}
\widehat{\text{var}}(\hat{\beta}_\mathbf{I}) &= \hat{\sigma}^2 (\mathbf{X}'\mathbf{X})^{-1} \\
&= \hat{\sigma}^2 / mn \\
&= \hat{\sigma}^2 / N
\end{aligned}
\tag{9.13}
$$

with

$$\hat{\sigma}^2 = \mathbf{y}[\mathbf{I} - \mathbf{X}(\mathbf{X}'\mathbf{X})^{-1}\mathbf{X}']\mathbf{y}/[N - r(\mathbf{X})] \tag{9.14}$$

and using $N = mn$.

On average $\widehat{\text{var}}(\hat{\beta}_\mathbf{I})$ estimates

$$
\begin{aligned}
\text{E}[\widehat{\text{var}}(\hat{\beta}_\mathbf{I})] &= \text{E}[\hat{\sigma}^2]/N \\
&= \frac{1}{N - r(\mathbf{X})} \text{tr}[\{\mathbf{I} - \mathbf{X}(\mathbf{X}'\mathbf{X})^{-1}\mathbf{X}'\}\mathbf{V}]/N \\
&= \frac{1}{N-1}\text{tr}[\mathbf{V} - \frac{1}{N}\mathbf{1}\mathbf{1}'\mathbf{V}]/N \\
&= \frac{1}{N-1}\sigma^2[N - (1 + \{n - 1\}\rho)]/N
\end{aligned}
$$

$$= \frac{\sigma^2}{N}\left[1 - \frac{(n-1)\rho}{N-1}\right] < \frac{\sigma^2}{N}. \tag{9.15}$$

On the other hand, the true variance of $\hat{\beta}_{\mathbf{I}}$ is

$$\mathrm{var}(\hat{\beta}_{\mathbf{I}}) = (\mathbf{X}'\mathbf{X})^{-1}\mathbf{X}'\mathbf{V}\mathbf{X}(\mathbf{X}'\mathbf{X})^{-1}$$

$$= \frac{1}{N}\mathbf{1}_N'\mathbf{V}\mathbf{1}_N\frac{1}{N}$$

$$= \frac{\sigma^2}{N}[1 + (n-1)\rho] > \frac{\sigma^2}{N}. \tag{9.16}$$

Thus the estimated variance averages less than σ^2/N, the variance if all the observations were independent, while the true variance is larger than σ^2/N; and it can be substantially so. If $\rho = 0.5$, a small to moderate correlation, and if $m = 5$ and $n = 11$, then the underestimation is by a factor of $(1+5)/(1-10[0.5]/54) \doteq 6.6!$

The remedy is relatively straightforward; use $\hat{\beta}_{\mathbf{I}}$ but correctly assess its variance. For example, with model (9.11), we estimate $\hat{\beta} = (\mathbf{1}'\mathbf{1})^{-1}\mathbf{1}'\mathbf{y} = \bar{y}_{..}$. Then

$$\hat{\mathbf{V}}_0 = \frac{1}{m-1}\sum_{i=1}^{m}(\mathbf{y}_i - \mathbf{1}_n\hat{\mu})(\mathbf{y}_i - \mathbf{1}_n\hat{\mu})' \tag{9.17}$$

is a consistent estimator (see E 9.7) of $\mathrm{var}(\mathbf{y}_i)$, no matter what its form. When the mean structure takes a more complicated form, namely $\mathrm{E}[\mathbf{y}_i] = \mathbf{X}_i\beta$, we would generalize accordingly:

$$\hat{\mathbf{V}}_0 = \frac{1}{m-1}\sum_{i=1}^{m}(\mathbf{y}_i - \mathbf{X}_i\hat{\beta})(\mathbf{y}_i - \mathbf{X}_i\hat{\beta})'. \tag{9.18}$$

This is the basic idea behind robust variance estimation which has become a central feature of the approach known as *generalized estimating equations* (GEEs). A primary aspect of GEEs is the specification of a *working* variance-covariance structure. That is, in some cases we might suspect that a specific form of covariance might hold for $\mathbf{y}_i$ of (9.11). Also, the use of the independence assumption can lead to inefficient estimators (Fitzmaurice, 1995). In such cases it is straightforward to make adjustments. Let $\mathbf{W}$ represent the *inverse* of the assumed variance-covariance structure of each of the $\mathbf{y}_i$s:

$$\mathbf{W} = [\mathrm{var}_W(\mathbf{y}_i)]^{-1}, \tag{9.19}$$

where $\text{var}_W(\cdot)$ denotes a temporarily assumed variance-covariance structure. If we believed this structure, our estimator of β would be $\hat{\beta}_{\mathbf{W}^{-1}} = (\mathbf{X}'\mathbf{W}\mathbf{X})^{-1}\mathbf{X}'\mathbf{W}\mathbf{y}$ with variance

$$\text{var}(\hat{\beta}_{\mathbf{W}^{-1}}) = (\mathbf{X}'\mathbf{W}\mathbf{X})^{-1}\mathbf{X}'\mathbf{V}\mathbf{X}(\mathbf{X}'\mathbf{W}\mathbf{X})^{-1}, \qquad (9.20)$$

which is sometimes called the *sandwich variance formula* since $\mathbf{X}'\mathbf{V}\mathbf{X}$ is "sandwiched" between two $(\mathbf{X}'\mathbf{W}\mathbf{X})^{-1}$s. As before, a consistent estimator of $\mathbf{V}$ is formed from the independent "replicates," $\mathbf{y}_i$. For more details see Diggle et al. (1994).

For situations that cannot be described by linear models, the generalized estimating equations approach begins by positing a marginal generalized linear model for the mean of $\mathbf{y}$ as a function of the predictors. For example, for binary data we might hypothesize a logistic regression for the mean:

$$\text{logit}(\text{E}[\mathbf{y}]) = \mathbf{X}\beta.$$

If we used a working assumption of independence of all the elements of $\mathbf{y}$, the ML estimating equations for β would be, from E 5.9,

$$\mathbf{X}'\mathbf{y} = \mathbf{X}'\text{E}[\mathbf{y}]. \qquad (9.21)$$

These are *unbiased estimating equations*, meaning that the difference between the right-hand-side and the left-hand-side is zero, namely

$$\text{E}\left(\mathbf{X}'\mathbf{y} - \mathbf{X}'\text{E}[\mathbf{y}]\right) = \mathbf{0}. \qquad (9.22)$$

It is not surprising and is true under regularity conditions (Heyde, 1997, Sect. 12.2) that solutions to unbiased estimating equations give consistent estimators.

Operationally, the estimator defined as the solution (for β) of (9.21) could be calculated by pretending that all the data were independent and conducting a standard logistic regression analysis. As is true for the linear model, this is often a nearly fully efficient estimator but the standard errors, variance estimates, tests and confidence intervals would often be highly misleading. Again, this can be dealt with by using the *estimator* that naively assumes independence but properly calculating its (large-sample) variance.

For longitudinal data with m subjects and with $\mathbf{y}_i$ denoting the data for the ith subject, we have

$$\mathbf{y} = \left\{ {}_c\, \mathbf{y}_i \right\}_{i=1}^{m} \text{ and } \mathbf{X} = \left\{ {}_c\, \mathbf{X}_i \right\}_{i=1}^{m}$$

and the estimating equation, (9.21), for binary data becomes

$$\sum \mathbf{X}_i \mathbf{y}_i = \sum \mathbf{X}_i \mathrm{E}[\mathbf{y}_i]. \tag{9.23}$$

From this it can be shown (Heyde, 1997, Secs. 4.2 and 12.4) that the large-sample variance of $\hat{\boldsymbol{\beta}}$ is

$$\mathrm{var}_\infty(\hat{\boldsymbol{\beta}}) = (\mathbf{X}'\mathbf{X})^{-1}\mathbf{X}' \, \mathrm{var}(\mathbf{y})\mathbf{X}(\mathbf{X}'\mathbf{X})^{-1} \tag{9.24}$$

$$= \left(\sum \mathbf{X}_i'\mathbf{X}_i\right)^{-1} \left(\sum \mathbf{X}_i' \, \mathrm{var}(\mathbf{y}_i)\mathbf{X}_i\right) \left(\sum \mathbf{X}_i'\mathbf{X}_i\right)^{-1},$$

upon the assumption of independence among the $\mathbf{y}_i$. This can be consistently estimated as $m \to \infty$ with

$$\widehat{\mathrm{var}}_\infty(\hat{\boldsymbol{\beta}}) =$$

$$\left(\sum \mathbf{X}_i'\mathbf{X}_i\right)^{-1} \left(\sum \mathbf{X}_i'(\mathbf{y}_i - \hat{\mathrm{E}}[\mathbf{y}_i])(\mathbf{y}_i - \hat{\mathrm{E}}[\mathbf{y}_i])'\mathbf{X}_i\right) \left(\sum \mathbf{X}_i'\mathbf{X}_i\right)^{-1}, \tag{9.25}$$

where $\hat{\mathrm{E}}[\mathbf{y}_i] = 1/(1 + \exp\{-\mathbf{X}_i\hat{\boldsymbol{\beta}}\})$.

The working assumption of independence may lead to inefficient estimators (Fitzmaurice, 1995) and other working variance-covariance structures can be entertained. In that case, for (9.23) we have

$$\sum \mathbf{X}_i \mathbf{W}_i \mathbf{y}_i = \sum \mathbf{X}_i \mathbf{W}_i \mathrm{E}[\mathbf{y}_i], \tag{9.26}$$

where $\mathbf{W}_i^{-1} = \mathrm{var}_W(\mathbf{y}_i)$ is the working variance for $\mathbf{y}_i$. This engenders corresponding changes in (9.24) and (9.25). See Diggle et al. (1994, Secs. 8.2.3 and 8.4.2) for more details.

a. Models with marginal and conditional interpretations

Since marginal regression models yield straightforward interpretations of regression parameters, but conditional models offer a basis for deriving predicted values, it is natural to seek models that combine the features of both. For example, the probit model, (9.3) and (9.4), gives easily calculated regression parameters that have either marginal or conditional interpretations, and also incorporates random effects. For binary data models, however, logistic regression models are much more commonly used.

– i. *Conditionally-specified, marginal logistic model*

A starting point is to specify a logistic regression equation for the marginal mean using (9.1):

$$y_{ij} \sim \text{Bernoulli}(p_{ij}) \tag{9.27}$$

$$p_{ij} = \text{E}[y_{ij}]$$

$$\text{logit}(\text{E}[y_{ij}]) = \beta_0 + \beta_1 x_{ij}.$$

To this is added a conditional specification of the regression including the random effects:

$$\text{logit}(\text{E}[y_{ij}|u_i]) = \Delta_{ij} + u_i, \tag{9.28}$$

where Δ_{ij} is defined implicitly by (9.27) and (9.28). That is, Δ_{ij} is defined by the requirement that the expected value of the conditional mean is the marginal mean:

$$\text{E}\left[\text{E}[y_{ij}|u_i]\right] = \text{E}[y_{ij}] \tag{9.29}$$

or

$$\int_{-\infty}^{\infty} \frac{1}{1 + \exp[-\{\Delta_{ij} + u_i\}]} g(u_i) du_i = \frac{1}{1 + \exp[-\{\beta_0 + \beta_1 x_{ij}\}]},$$

where $g(\cdot)$ is the density of the random effects distribution. If we add a parametric specification for the random effects distribution, for example, $u_i \sim \mathcal{N}(0, \sigma_u^2)$, then (9.29) can be solved numerically, a likelihood can be defined and inference can proceed using the usual maximum likelihood asymptotics. For details see Heagerty (1999). Though numerically intensive, it offers a compromise between the marginal versus conditional approaches.

– ii. *Marginal and conditional logistic model*

The model in the previous subsection sacrifices easy interpretability of the relationship of the predictor to the conditional mean, since the dependence on the predictors is through the implicitly defined Δ_{ij} from (9.29). Are there models that have the logistic regression form for both the marginal and conditional means, like the probit model in (9.3)?

The answer is yes, as long as the random effects distribution is not constrained to be normal.

For $y_{ij} \sim \text{Bernoulli}(p_{ij})$, Wang and Louis (2003) solved for the requisite random effects distribution, $g(\cdot)$, in order to satisfy the following conditional and marginal regression equations

$$\text{logit}(\text{E}[y_{ij}|u_i]) = \mathbf{x}'_{ij}\boldsymbol{\beta} + u_i$$

$$\text{logit}(\text{E}[y_{ij}]) = \phi\mathbf{x}'_{ij}\boldsymbol{\beta}, \qquad (9.30)$$

where ϕ plays the role of a shrinkage factor relating the conditional and marginal regression coefficients (which cannot be equal due to the same phenomena demonstrated in Figure 9.1). That distribution was named the "bridge" distribution and found to be

$$g(u) = \frac{1}{2\pi} \frac{sin(\phi\pi)}{cosh(\phi u) + cos(\phi\pi)}. \qquad (9.31)$$

As with the previous model, since it is a fully parametric and conditionally specified distribution, it contains random effects for use for prediction, and inferences can be based on likelihood theory.

The examples of the previous two subsections illustrate the point that multiple conditional specifications for the distribution of $\mathbf{y}|\mathbf{u}$ may give rise to the same marginal distribution and regression equation. Hence the conditional model cannot be derived from the marginal specification alone.

9.4   CONTRASTING MARGINAL AND CONDITIONAL MODELS

The discussion above has concentrated on the mean of y and the regression function relating it to the predictors $\mathbf{x}$. Conditionally specified models can provide estimates of the dependence of responses within clusters, such as subjects. Measures such as the intraclass correlation coefficient, $\text{corr}(y_{ij}, y_{ij'})$, depend on the random effects distribution and hence can be estimated after fitting such a model.

Marginal models are useful when correlation among observations must be accommodated, but neither the nature of the clustering, nor the individual clusters themselves are of scientific interest. This may apply if the focus of an investigation is impact on the population, e.g., in a public health question like the alcohol consumption example. On the other hand, it would not be useful for advising an individual on the likely effects of modified behavior.

When clustered-data designs are used for logistical (as opposed to scientific) reasons, as in sample surveys, or for statistical efficiency, and correlations are nuisance parameters in analysis, then there is the advantage of not having to specify the conditional model and mixing distribution for the random effects. For example, GLMMs that mistakenly assume the variance of the random effects distribution is unrelated to the predictor can produce biased estimators (Heagerty and Zeger, 2000; Heagerty and Kurland, 2001). In contrast, when paired with a GEE approach to estimation, estimates of marginal model parameters are consistent, even if the association structure is misspecified.

A drawback of marginal models is that they may not measure covariate effects of primary scientific interest, for example, in longitudinal studies in which explanatory variables change over time within a cluster and interest focusses on the response to such changes. In the alcohol consumption investigation, for example, interest may focus on the effects of a woman reducing her alcohol consumption. In such situations, GLMMs are able to explicitly describe the sources of variation that produce correlated observations, model cluster-specific trajectories of change and derive predicted values of those trajectories. For example, they can be used to model individual trajectories of growth in experimental animals, breeding values of individual bulls, susceptibilities of individual families to inheritable diseases, and resource usage patterns of individual doctors. The costs of this greater model detail are more extensive model assumptions, more complex model fitting, and intensive computations. For a more detailed critique of marginal modeling see Lindsey and Lambert (1998).

9.5   EXERCISES

E 9.1 Prove (9.6).

E 9.2 Write down a model similar to (9.28) but for count data outcomes. Start by hypothesizing that the conditional distribution given u_i is Poisson and then specify the marginal regression:

$$y_{ij}|u_i \sim \text{Poisson}(\exp\{\Delta_{ij} + u_i\})$$

$$u_i \sim \mathcal{N}(0, \sigma_u^2) \tag{9.32}$$

$$\log(\mathrm{E}[y_{ij}]) = \mathbf{x}'_{ij}\boldsymbol{\beta}.$$

Solve for the Δ_{ij} required for this model to be consistent.

E 9.3 Using model (9.3) but with an inverse probit link instead of a logit link, show that the marginal regression equation is approximately

$$\mathrm{logit}\,(\mathrm{E}[y_{ij}|x_{ij}]) = \beta_0^* + \beta_1^* x_i, \qquad (9.33)$$

where $\beta_i^* = \beta_i/\sqrt{1 + 3\sigma_u^2/\pi^2}$. Hint: Assume that the logistic c.d.f., properly scaled, can be approximated with a standard normal c.d.f.

E 9.4 Derive the relative efficiency of $\hat{\boldsymbol{\beta}}_{\mathbf{I}}$ and $\hat{\boldsymbol{\beta}}_{\mathbf{V}}$ as described in the paragraph immediately following (9.10).

E 9.5 Again following the discussion after (9.10), show that the relative efficiency of the ordinary least squares estimator decreases as a function of increasing ρ. Furthermore, show that the relative efficiency in the worst case, when $\rho = 1$, is given by $2(1+\lambda^2)/(1+\lambda)^2$.

E 9.6 Show the calculation of $\mathrm{tr}(\mathbf{V} - \mathbf{11}'\mathbf{V}/N)$ involved in deriving (7.74).

E 9.7 Show that $\hat{\mathbf{V}}_0$ of (9.17) is consistent for $\mathbf{V}_0$ as m tends to ∞.

E 9.8 The characteristic function of the bridge distribution, (9.31), is given by

$$\xi_u(t) = \frac{\sinh(\pi t)}{\phi \sinh(\pi t/\phi)}. \qquad (9.34)$$

By taking the first two derivatives at $t = 0$ show that the mean and variance are 0 and $\pi^2(\phi^{-2} - 1)/3$, respectively. Hence show that the bridge distribution generates the same scaling of the regression coefficients as the logistic/normal of E 9.3.

E 9.9 The shape of a marginal regression can be quite different from the conditional regression equation. For the conditional logistic regression equation below, show that the marginal regression equation approximates a three-part step function as $\beta_1 \to \infty$. Model:

$$\mathrm{logit}\,(\mathrm{E}[y_{ij}|x_{ij}, u_i]) = \beta_0 + \beta_1 x_{ij} + u_i$$

$$u_i = \begin{cases} -\sigma_u \text{ with probability } \frac{1}{2} \\ \sigma_u \text{ with probability } \frac{1}{2} \end{cases}.$$

Chapter 10

MULTIVARIATE MODELS

10.1 INTRODUCTION

It is sometimes desirable to construct models for multivariate outcomes, which may or may not have the same marginal distribution. For example, in a study of an intervention to decrease utilization of health care resources by the homeless, it may be of interest to simultaneously model the cost of care, number of emergency room (ER) visits in a two year period, and death within two years (yes/no).

Multivariate models have several potential advantages. They can assess the joint influence of predictors on all outcomes or assess tradeoffs between outcomes such as increased cost of an intervention with a resultant decrease in ER visits. They can avoid issues of multiple testing and inflation of type I error rates as compared to a series of univariate analyses. They can model the association between outcomes. Finally, multivariate models can, in some situations, be more efficient.

Many of the mixed models studied in previous chapters can be considered simple models for multivariate outcomes. For example, the log diameters in Figure 3.1 can be regarded as multiple outcomes per leaflet. But what about the situation in which the outcomes have different distributions, such as the health resources example above?

Joint multivariate distributions that are sufficiently flexible to accommodate multiple outcome distributions and multiple predictors are uncommon. Exceptions are the multivariate normal distribution and joint distributions built using copulas (Joe, 1997).

In this chapter we consider ways to use mixed models to accommodate multivariate outcomes for both normally and non-normally distributed variables. We start with the well-studied case of multivariate normally distributed outcomes.

10.2   MULTIVARIATE NORMAL OUTCOMES

Consider a simple situation in which r outcomes are collected on each of n subjects (similar to Chapter 8). For subject $i, (i = 1, \ldots, n)$ arrange the r outcomes in a column vector, $\mathbf{y}_i = (y_{i1}, \ldots, y_{ir})'$. Further suppose that $\mathbf{y}_i$ are independent and identically distributed according to a multivariate normal distribution with mean vector $\boldsymbol{\mu}$ and variance-covariance matrix $\boldsymbol{\Sigma}$, so that

$$E[\mathbf{y}_i] \;\;=\;\; \boldsymbol{\mu} \tag{10.1}$$

and

$$\mathrm{var}(\mathbf{y}_i) \;\;=\;\; \boldsymbol{\Sigma}.$$

Let $\mathbf{y}$ be a column vector, of order $nr \times 1$, consisting of the data on all the subjects: $\mathbf{y} = (\mathbf{y}_1' \cdots \mathbf{y}_n')'$. Then it is clear that $\mathbf{y}$ is multivariate normally distributed:

$$\mathbf{y} \sim \mathcal{N}(\mathbf{1}_n \otimes \boldsymbol{\mu}, \mathbf{I}_n \otimes \boldsymbol{\Sigma}). \tag{10.2}$$

How does this connect with the idea of mixed models?

If we write

$$\mathbf{y}_i | \boldsymbol{\delta}_i \;\;=\;\; \boldsymbol{\mu} + \boldsymbol{\delta}_i \tag{10.3}$$

with

$$\boldsymbol{\delta}_i \;\;\sim\;\; \text{i.i.d. } \mathcal{N}(\mathbf{0}, \boldsymbol{\Sigma}),$$

using "random effects" $\boldsymbol{\delta}_i$, then we have identified a mixed model that is equivalent to (10.1).

Further suppose that all r outcomes have the same variance and that the pairwise correlations (e.g., $\mathrm{corr}(y_{i1}, y_{i2})$) are all equal to ρ. Then we can use a standard mixed model to generate the distribution:

$$\mathbf{y}_i | b_i \;\;=\;\; \boldsymbol{\mu} + \mathbf{1}_r b_i + \mathbf{e}_i \tag{10.4}$$

with

$$b_i \sim \text{ i.i.d. } \mathcal{N}(0, \sigma_b^2)$$

and

$$\mathbf{e}_i \sim \text{ i.i.d. } \mathcal{N}(\mathbf{0}, \mathbf{I}_r \sigma_e^2),$$

or, in scalar notation

$$y_{ij}|b_i = \mu_j + b_i + e_{ij} \qquad (10.5)$$

with

$$b_i \sim \text{ i.i.d. } \mathcal{N}(0, \sigma_b^2)$$

and

$$e_{ij} \sim \text{ i.i.d. } \mathcal{N}(0, \sigma_e^2),$$

in each case with the b_i independent of the $e_{ij}, j = 1, \ldots, r$. Alternatively we can specify the model as

$$y_{ij}|b_i \sim \mathcal{N}(\mu_j + b_i, \sigma_e^2) \qquad (10.6)$$

$$b_i \sim \text{ i.i.d. } \mathcal{N}(0, \sigma_b^2).$$

This is a quite parsimonious description of a multivariate distribution, using just a single additional parameter, σ_b^2, compared to the independence case. Of course, the assumptions inherent in (10.4) are quite restrictive and usually would not be met in practice.

A more flexible and possibly more realistic model is the random slopes and intercepts model considered in Chapter 8:

$$y_{ij}|b_{0i}, b_{1i} = (\beta_0 + b_{0i}) + (\beta_1 + b_{1i})x_{ij} + e_{ij}$$

$$\begin{pmatrix} b_{0i} \\ b_{1i} \end{pmatrix} \sim \text{ i.i.d. } \mathcal{N}(\mathbf{0}, \Sigma_b) \qquad (10.7)$$

$$e_{ij} \sim \text{ i.i.d. } \mathcal{N}(0, \sigma_e^2) \text{ independent of the } \mathbf{b_i},$$

which has unconditional (marginal) expectation

$$\text{E}[y_{ij}] = \beta_0 + \beta_1 x_{ij}. \qquad (10.8)$$

This generates a moderately complicated variance covariance structure (8.23) using an additional three parameters (the unique elements of

Σ_b), but the addition of the random effects leaves the marginal expectation of the y_{ij} unchanged. Again, this is a simple description of the multivariate distribution of the collection of observations taken over time on a single "subject."

These examples, (10.3), (10.5) and (10.7), serve to show that mixed models can accommodate some simple "multivariate" scenarios.

Several features of this development are worth noting:

- The incorporation of random effects allows a flexible, but lower dimensional specification of the variance-covariance matrix of $\mathbf{y}$,

- Inclusion of the random effects affects only the variance-covariance matrix of $\mathbf{y}$, not the marginal mean, and

- Both the conditional and marginal distributions are multivariate normal.

How much of this carries forward and is useful for distributions which are not normal? Unfortunately, the multivariate normal formulation, (10.2), does not generalize to non-normal distributions in a straightforward way, so it is logical to consider building multivariate models using random effects.

10.3   NON-NORMALLY DISTRIBUTED OUTCOMES

For example, a natural analog of (10.6), but for binary outcomes would be

$$y_{ij}|b_i \quad \sim \quad \text{Bernoulli}\,(p_{ij})$$

$$\text{logit}\,(p_{ij}) \quad = \quad \mu_j + b_i \qquad\qquad (10.9)$$

$$b_i \quad \sim \quad \text{i.i.d. } \mathcal{N}(0, \sigma_b^2).$$

Unfortunately, the introduction of random effects is not as innocuous as in the multivariate normal setting. We explore this in the following sub-sections.

a.   A multivariate binary model

We consider a model similar to (10.9) but with a probit link:

$$y_{ij}|b_i \quad \sim \quad \text{Bernoulli}\,(p_{ij})$$

$$\Phi^{-1}(p_{ij}) \;=\; \mu_j + b_i \tag{10.10}$$

$$b_i \;\sim\; \text{i.i.d. } \mathcal{N}(0, \sigma_b^2),$$

where $\Phi(\cdot)$ is the standard normal c.d.f. How does this compare to the model without the random effects b_i, that is, one in which all the observations are independent?

– i. *Moments*

The mean, variances and covariances of y_{ij} can be calculated as in (7.45), namely

$$E[y_{ij}] \;=\; \Phi\left(\frac{\mu_j}{\sqrt{1+\sigma_b^2}}\right) \tag{10.11}$$

$$\text{var}(y_{ij}) \;=\; E[y_{ij}](1 - E[y_{ij}]) \tag{10.12}$$

$$\text{cov}(y_{ij}, y_{ik}) \;=\; \int_{-\infty}^{\infty} \Phi(\mu_j + \sigma_b z)\Phi(\mu_k + \sigma_b z)\phi(z)dz - E[y_{ik}]E[y_{ij}]. \tag{10.13}$$

Introduction of the random effects has served the purpose of introducing a positive covariance (see E 10.2), which is a function of σ_b. Unfortunately, the mean, (10.11), variance, (10.12), and covariance, (10.13), all depend on σ_b as well.

So the convenient separation under the multivariate normal model, namely that fixed effects determine the mean while the random effects determine the variance-covariance structure, does not hold for the multivariate probit model.

As in model (10.6), the variance-covariance structure is governed by a single additional parameter, σ_b, which induces an equal correlation on the "probit" scale, i.e., for the terms $\mu_j + b_i$. It does *not* lead to an equal correlation among the y_{ij}, however (see E 10.3).

b. A binary/normal example

We now consider a multivariate distribution for mixed outcome types constructed using mixed models. In particular we develop a model for a binary outcome and a normally distributed outcome measured on the same "subject" by using a random effect to build in a correlation between the two outcomes.

Let y_{1i} denote the binary outcome for subject i and let y_{2i} be the normally distributed outcome. As above we hypothesize a probit model for y_1 and a linear model for y_2. Suppose we have predictors, $x_1, x_2, \ldots, x_p$. For the binary outcome we could construct a probit model as in (10.10):

$$y_{1i} \sim \text{indep. Bernoulli}(\nu_i^B) \tag{10.14}$$

$$\Phi^{-1}(\nu_i^B) = \mathbf{x}_i' \boldsymbol{\gamma}^B.$$

And we can construct a linear regression model for the normally distributed outcome:

$$y_{2i} \sim \text{indep. } \mathcal{N}(\nu_i^N, \tau^2) \tag{10.15}$$

$$\nu_i^N = \mathbf{x}_i' \boldsymbol{\gamma}^N,$$

In these models, the superscripts indicate that the parameters are specific to the binary (B) or normal (N) models, the νs are the means, and the γs are the regression parameters.

While (10.14) and (10.15) would suffice for analyses of each outcome separately, they do not accommodate a correlation between y_{1i} and y_{2i}. A simple device to do so is to introduce a random effect that will be shared by both responses for any particular subject. We modify (10.14) and (10.15) accordingly by modeling the distributions conditional on the random effect, b_i:

$$y_{1i} | b_i \sim \text{indep. Bernoulli}(\mu_i^B)$$

$$\Phi^{-1}(\mu_i^B) = \mathbf{x}_i' \boldsymbol{\beta}^B + b_i$$

$$y_{2i} | b_i \sim \text{indep. } \mathcal{N}(\mu_i^N, \sigma^2) \tag{10.16}$$

$$\mu_i^N = \mathbf{x}_i' \boldsymbol{\beta}^N + \lambda b_i,$$

$$b_i \sim \text{i.i.d. } \mathcal{N}(0, \sigma_b^2),$$

where we make the assumption that, conditional on b_i, the y_{1i} and y_{2i} are independent. The λ multiplying b_i in the equation for μ_i^N is to account for the fact that the linear predictors for y_1 and y_2 are measured on different scales and the random effects will almost certainly have different variances.

– i. *Marginal means*

Using (S.1) and the fact that the mean of b_i is zero, the marginal mean for y_2 from (10.16), averaging over the distribution of the random effects, is simply calculated as:

$$
\begin{aligned}
\mathrm{E}[y_{2i}] &= \mathrm{E}[\mathbf{x}_i'\boldsymbol{\beta}^N + \lambda b_i] \\
&= \mathbf{x}_i'\boldsymbol{\beta}^N + \lambda\mathrm{E}[b_i] \qquad (10.17) \\
&= \mathbf{x}_i'\boldsymbol{\beta}^N.
\end{aligned}
$$

As long as this holds for a range of possible values of x, it implies that the marginal and conditional regression parameters must be the same, for example, $\boldsymbol{\beta}^N = \boldsymbol{\gamma}^N$.

Using calculations as in (10.11) the mean of y_1 is given by

$$
\mathrm{E}[y_{1i}] = \Phi\left(\frac{\mathbf{x}_i'\boldsymbol{\beta}^B}{\sqrt{1+\sigma_b^2}}\right). \qquad (10.18)
$$

Therefore, for the binary outcome, the marginal regression coefficients are not the same as the conditional coefficients and, in particular, are smaller by a factor of $\sqrt{1+\sigma_b^2}$:

$$
\gamma_k^B = \beta_k^B/\sqrt{1+\sigma_b^2}, k = 1,\ldots,p. \qquad (10.19)
$$

– ii. *Marginal variances and covariance*

Responses y_{2i} for different subjects are assumed to be independent. The variance can be calculated using (S.3):

$$
\begin{aligned}
\mathrm{cov}(y_{2i}, y_{2i}) &= \mathrm{E}\left[\mathrm{cov}(y_{2i}, y_{2i}|b_i)\right] + \mathrm{cov}\left(\mathrm{E}[y_{2i}|b_i], \mathrm{E}[y_{2i}|b_i]\right) \\
&= \mathrm{E}\left[\mathrm{var}(y_{2i}|b_i)\right] + \mathrm{cov}\left(\mathrm{E}[y_{2i}|b_i], \mathrm{E}[y_{2i}|b_i]\right) \\
&= \sigma^2 + \mathrm{cov}\left(\mathbf{x}_i'\boldsymbol{\beta}^N + \lambda b_i, \mathbf{x}_i'\boldsymbol{\beta}^N + \lambda b_i\right) \\
&= \sigma^2 + \lambda^2\sigma_b^2, \qquad (10.20)
\end{aligned}
$$

so that τ^2 of (10.15) is given by $\sigma^2 + \lambda^2\sigma_b^2$. The interpretation of this result is that part of the marginal variance in y_{2i}, given by $\lambda^2\sigma_b^2$, is being attributed to variation in the shared random effect.

The y_{1i} are similarly assumed to be independent and, since they have a marginal Bernoulli distribution with mean given by (10.18), the variance is given by

$$
\begin{aligned}
\text{var}(y_{1i}) &= E[y_{1i}]\,(1 - E[y_{1i}]) \\[2mm]
&= \Phi(\mathbf{x}_i'\boldsymbol{\gamma}^B)\left[1 - \Phi(\mathbf{x}_i'\boldsymbol{\gamma}^B)\right]. \qquad (10.21)
\end{aligned}
$$

Using the same identity as in (10.20) we now calculate the covariance:

$$
\begin{aligned}
\text{cov}(y_{1i}, y_{2i}) &= E\left[\text{cov}(y_{1i}, y_{2i}|b_i)\right] + \text{cov}\left(E[y_{1i}|b_i], E[y_{2i}|b_i]\right) \\[2mm]
&= 0 + \text{cov}\left(\Phi(\mathbf{x}_i'\boldsymbol{\beta}^B + b_i), \mathbf{x}_i'\boldsymbol{\beta}^N + \lambda b_i\right) \\[2mm]
&= \text{cov}\left(\Phi(\mathbf{x}_i'\boldsymbol{\beta}^B + b_i), \lambda b_i\right). \qquad (10.22)
\end{aligned}
$$

Letting $\theta_i = \mathbf{x}_i'\boldsymbol{\beta}^B$ we can rewrite (10.22) as

$$
\begin{aligned}
\text{cov}(y_{1i}, y_{2i}) &= \text{cov}\left(\Phi(\theta_i + b_i), \lambda b_i\right) \\[2mm]
&= \lambda \int_{-\infty}^{\infty} \sigma_b z \Phi(\theta_i + \sigma_b z)\phi(z)\,dz, \qquad (10.23)
\end{aligned}
$$

where $\phi(z)$ is the standard normal p.d.f. Although this cannot be evaluated in closed form, it is not too hard to evaluate numerically.

Using (10.23) and the variance calculations, (10.20) and (10.21), the correlation is given by

$$
\text{corr}(y_{1i}, y_{2i}) = \sqrt{\frac{\lambda^2 \sigma_b^2}{\lambda^2 \sigma_b^2 + \sigma^2}}\; \frac{\int_{-\infty}^{\infty} z\Phi(\theta_i + \sigma_b z)\phi(z)\,dz}{\sqrt{\Phi\left(\theta_i/\sqrt{1+\sigma_b^2}\right)\left[1 - \Phi\left(\theta_i/\sqrt{1+\sigma_b^2}\right)\right]}}.
$$

$$(10.24)$$

Table 10.1 gives the correlation, (10.24), for various values of $\theta = \mathbf{x}_i'\boldsymbol{\beta}^B$ and $\rho = (\lambda^2\sigma_b^2)/(\lambda^2\sigma_b^2 + \sigma^2)$, when $\sigma^2 = 1$. We note several features of Table 10.1:

1. When θ is close to 0, the value of the correlation is smaller than but similar to the value of ρ.

2. The value of the correlation also depends on the value of θ.

Table 10.1: Correlations, within a subject, of the binary and normally distributed outcomes for model (10.16) with $\sigma^2 = 1$, for various values of $\theta = x_i'\beta^B$ and $\rho = (\lambda^2\sigma_b^2)/(\lambda^2\sigma_b^2 + \sigma^2)$.

θ	\multicolumn{7}{c}{ρ}						
	0	0.1	0.3	0.5	0.7	0.9	0.99
0	0.00	0.08	0.24	0.40	0.56	0.72	0.79
1	0.00	0.07	0.21	0.36	0.53	0.71	0.79
2	0.00	0.04	0.14	0.27	0.45	0.68	0.78
3	0.00	0.01	0.07	0.16	0.33	0.61	0.78

3. The limiting value of the correlation is *not* 1 as ρ increases or, equivalently, as σ_b^2 increases for a fixed value of σ^2.

In fact, for a fixed value of σ^2 the limit of the correlation as $\sigma_b^2 \to \infty$ is $\sqrt{2/\pi} \doteq 0.798$ (see E 10.4 and E 10.6).

c. A Poisson/Normal Example

We now consider a slightly different model to illustrate that the conditional and marginal distributions need not be the same and other features of shared random effects models. We deal with a joint model with conditional (on the random effects) Poisson and normal distributions, using canonical links for those two distributions (log and identity, respectively):

$$y_{1i}|b_i \sim \text{indep. Poisson}(\mu_i^C)$$

$$\log(\mu_i^C) = x_i'\beta^C + b_i$$

$$y_{2i}|b_i \sim \text{indep. } \mathcal{N}(\mu_i^N, \sigma^2) \qquad (10.25)$$

$$\mu_i^N = x_i'\beta^N + \lambda b_i,$$

$$b_i \sim \text{i.i.d. } \mathcal{N}(0, \sigma_b^2),$$

where the superscript C denotes aspects of the model specific to the count data, y_1.

The marginal distribution of y_{2i} is the same as for model (10.16) so we focus on the distribution of y_{1i}.

– i.   *Marginal means*

The marginal mean for y_1 from (10.25) is again calculated from (S.1):

$$\mathrm{E}[y_{1i}] = \mathrm{E}[\exp\{\mathbf{x}_i'\boldsymbol{\beta}^C + b_i\}]$$

$$= \exp\{\mathbf{x}_i'\boldsymbol{\beta}^C\}\mathrm{E}[e^{b_i}]$$

$$= \exp\{\mathbf{x}_i'\boldsymbol{\beta}^C\}M_{b_i}(1), \tag{10.26}$$

where $M_W(t)$ is the moment generating function of the random variable W evaluated at t. Since $b_i \sim \mathcal{N}(0, \sigma_b^2)$, $M_{b_i}(1)$ is equal to $e^{\sigma_b^2/2}$ and therefore

$$\mathrm{E}[y_{1i}] = \exp\{\mathbf{x}_i'\boldsymbol{\beta}^C + \sigma_b^2/2\}. \tag{10.27}$$

Of note is that the log of the marginal mean is $\mathbf{x}_i'\boldsymbol{\beta}^C + \sigma_b^2/2$, which is the same as the log of the conditional mean, except offset by $\sigma_b^2/2$. In particular the regression coefficient for $\mathbf{x}_i'$, namely $\boldsymbol{\beta}^C$, is the same in both the marginal and conditional models, (10.27) and (10.26) respectively, except for β_0^C, the intercept.

– ii.   *Marginal variance and covariance*

The y_{1i} are assumed to be independent with the variance calculated by the usual formula:

$$\mathrm{cov}(y_{1i}, y_{1i}) = \mathrm{var}(y_{1i}) = \mathrm{E}\left[\mathrm{var}(y_{1i}|b_i)\right] + \mathrm{cov}\left(\mathrm{E}[y_{1i}|b_i], \mathrm{E}[y_{1i}|b_i]\right)$$

$$= \mathrm{E}\left[\mathrm{E}[y_{1i}|b_i]\right] + \mathrm{var}(\exp\{\mathbf{x}_i'\boldsymbol{\beta}^C + b_i\})$$

$$= \mathrm{E}[y_{1i}] + \exp\{2\mathbf{x}_i'\boldsymbol{\beta}^C\}\mathrm{var}(e^{b_i})$$

$$= \mathrm{E}[y_{1i}] + \exp\{2\mathbf{x}_i'\boldsymbol{\beta}^C\}(M_{b_i}(2) - M_{b_i}(1)^2)$$

$$= \mathrm{E}[y_{1i}] + \exp\{2\mathbf{x}_i'\boldsymbol{\beta}^C\}(e^{2\sigma_b^2} - e^{\sigma_b^2})$$

$$= \mathrm{E}[y_{1i}] + \mathrm{E}[y_{1i}]^2(e^{\sigma_b^2} - 1). \tag{10.28}$$

In first line of the calculation above we use the fact that the conditional distribution is Poisson and hence the conditional mean and variance are equal and then use the iterated expectation identity, (S.1).

The covariance is calculated as before:

$$
\begin{aligned}
\text{cov}(y_{1i}, y_{2i}) &= \text{E}\left[\text{cov}(y_{1i}, y_{2i}|b_i)\right] + \text{cov}\left(\text{E}[y_{1i}|b_i], \text{E}[y_{2i}|b_i]\right) \\
&= 0 + \text{cov}\left(\exp\{\mathbf{x}_i'\boldsymbol{\beta}^C + b_i\}, \mathbf{x}_i'\boldsymbol{\beta}^N + \lambda b_i\right) \\
&= \lambda\exp\{\mathbf{x}_i'\boldsymbol{\beta}^C\}\text{cov}\left(e^{b_i}, b_i\right). \quad (10.29)
\end{aligned}
$$

The last term in this calculation can be evaluated as:

$$
\begin{aligned}
\text{cov}\left(e^{b_i}, b_i\right) &= \text{E}[b_i e^{b_i}] \\
&= \text{E}[\sigma_b Z e^{\sigma_b Z}], \text{ with } Z \sim \mathcal{N}(0,1) \\
&= \sigma_b^2 e^{\sigma_b^2/2}. \quad (10.30)
\end{aligned}
$$

Combining (10.29) and (10.30) gives

$$
\begin{aligned}
\text{cov}(y_{1i}, y_{2i}) &= \lambda\exp\{2\mathbf{x}_i'\boldsymbol{\beta}^C\}\sigma_b^2 e^{\sigma_b^2/2} \\
&= \lambda\text{E}[y_{1i}]\sigma_b^2. \quad (10.31)
\end{aligned}
$$

Therefore the correlation between y_{1i} and y_{2i} is

$$
\text{corr}(y_{1i}, y_{2i}) = \frac{\text{E}[y_{1i}]\sigma_b^2}{\sqrt{(\sigma^2/\lambda^2 + \sigma_b^2)(\text{E}[y_{1i}] + \text{E}[y_{1i}]^2\{e^{\sigma_b^2} - 1\})}}. \quad (10.32)
$$

The correlation (10.32) exhibits even more problematic behavior compared to the binary/normal model correlation, given by (10.24). The value of the correlation is *not* a monotonic function of σ_b and, in fact, tends to zero as $\sigma_b \to \infty$. See (E 10.7). The intuitive explanation for this is that the variability of the count outcome, y_1, increases faster than the covariance, driving the correlation to zero.

– iii.   *Marginal distributions*

The ratio of the variance of the marginal distribution of y_{1i} to its mean is easily calculated from (10.28) as

$$
\text{var}(y_{1i})/\text{E}[y_{1i}] = 1 + \text{E}[y_{1i}](e^{\sigma_b^2} - 1). \quad (10.33)
$$

Since the ratio is greater than 1 whenever the random effects have variance greater than 0, the marginal distribution of y_{1i} is not Poisson (which has equal mean and variance) and is always more highly variable than a Poisson distribution. This ratio, or *overdispersion*, which is a characteristic of the marginal distribution of y_{1i}, is thus integrally tied to the correlation between y_{1i} and y_{2i}.

10.4 CORRELATED RANDOM EFFECTS

Previous sections have shown that inclusion of a shared random effect builds a positive correlation between random variables with different distributions. However, the flexibility of the approach does not match that of the multivariate normal distribution, which has the nice feature that the inclusion of random effects modifies only the variance-covariance structure.

A generalization of the shared random effects models of the previous section is to allow separate but correlated random effects for each of the outcomes. For example, consider a generalization of the Poisson/Normal model, (10.32), in which we have a single Poisson outcome, y_1, but repeated measurements over time in the normal outcome, y_2, and we allow separate, but correlated random effects in the two models:

$$y_{1i}|\mathbf{b}_i \quad \sim \quad \text{indep. Poisson}(\mu_i^C)$$

$$\log(\mu_i^C) \quad = \quad \mathbf{x}_i'\boldsymbol{\beta}^C + b_i^C$$

$$y_{2it}|\mathbf{b}_i \quad \sim \quad \text{indep. } \mathcal{N}(\mu_i^N, \sigma^2) \tag{10.34}$$

$$\mu_i^N \quad = \quad \mathbf{x}_i'\boldsymbol{\beta}^N + b_i^N,$$

$$\mathbf{b}_i \quad = \quad \begin{pmatrix} b_i^C \\ b_i^N \end{pmatrix} \sim \text{i.i.d. } \mathcal{N}_2(\mathbf{0}, \boldsymbol{\Sigma}_b).$$

We assumed repeated measurements over time on the continuous outcome, y_{2it} $(t = 1, \ldots, m_i)$ to identify the subject-to-subject variation, $\text{var}(b_i^N)$.

Using the notation

$$\boldsymbol{\Sigma}_b = \begin{pmatrix} \sigma_{bc}^2 & \sigma_{bc}\sigma_{bn}\rho_{cn} \\ \sigma_{bc}\sigma_{bn}\rho_{cn} & \sigma_{bn}^2 \end{pmatrix}, \tag{10.35}$$

the marginal means and variances are little changed:

$$\mathrm{E}[y_{1i}] = \exp\{\mathbf{x}_i'\boldsymbol{\beta}^C + \sigma_{bc}^2/2\}.$$

$$\mathrm{var}(y_{1i}) = \mathrm{E}[y_{1i}] + \mathrm{E}[y_{1i}]^2(e^{\sigma_{bc}^2} - 1)$$

$$\mathrm{E}[y_{2it}] = \mathbf{x}_i'\boldsymbol{\beta}^N \tag{10.36}$$

$$\mathrm{var}(y_{2it}) = \sigma^2 + \sigma_{bn}^2.$$

The covariance calculation is somewhat different:

$$\mathrm{cov}(y_{1i}, y_{2it}) = \mathrm{E}\left[\mathrm{cov}(y_{1i}, y_{2it}|\mathbf{b}_i)\right] + \mathrm{cov}\left(\mathrm{E}[y_{1i}|\mathbf{b}_i], \mathrm{E}[y_{2it}|\mathbf{b}_i]\right)$$

$$= 0 + \mathrm{cov}\left(\exp\{\mathbf{x}_i'\boldsymbol{\beta}^C b_i^C\}, \mathbf{x}_i'\boldsymbol{\beta}^N + b_i^N\right)$$

$$= \exp\{\mathbf{x}_i'\boldsymbol{\beta}^C\}\mathrm{cov}\left(e^{b_i^C}, b_i^N\right). \tag{10.37}$$

After some calculation (E 10.8), the covariance is given by

$$\mathrm{cov}(y_{1i}, y_{2it}) = e^{\beta_0^C + \beta_1^C x_i}\sigma_{bn}\sigma_{bc}\rho_{cn}e^{\sigma_{bc}^2|\rho_{cn}|/2}, \tag{10.38}$$

with a correlation of

$$\mathrm{corr}(y_{1i}, y_{2it}) = \frac{e^{\beta_0^C + \beta_1^C x_i}\sigma_{bn}\sigma_{bc}\rho_{cn}e^{\sigma_{bc}^2|\rho_{cn}|/2}}{\sqrt{(\sigma^2 + \sigma_{bn}^2)(\mathrm{E}[y_{1i}] + \mathrm{E}[y_{1i}]^2(e^{\sigma_{bc}^2} - 1))}}.$$

$$\tag{10.39}$$

The correlation given by (10.39) gives somewhat more flexibility to the model. First and importantly, it allows negative correlations between the two outcomes. Second, it gives more latitude for disentangling the overdispersion, which is governed by σ_{bc}^2, and the covariance, which is additionally governed by ρ_{cn}.

10.5 LIKELIHOOD-BASED ANALYSIS

Inference for joint modeling can proceed using maximum likelihood principles as in previous chapters. That is, we form the likelihood and maximize it, numerically if need be, to form MLEs. Likelihood ratio tests can be used to compare nested models and the information matrix can be used to derive standard errors and Wald tests.

Not surprisingly, for most models, even calculation of the joint likelihood is numerically difficult, similar to (7.16). We illustrate the issues for the Poisson/normal model, (10.25). Conditional on b_i, the y_{1i} and y_{2i} are independent and hence their joint distribution is a product:

$$f_{y_1,y_2}(y_{1i}, y_{2i}|b_i) = f_{y_1}(y_{1i}|b_i) f_{y_2}(y_{2i}|b_i)$$

$$= \frac{(\mu_i^C)^{y_{1i}} e^{-\mu_i^C}}{y_{1i}!} \frac{e^{-(y_{2i}-\mu_i^N)^2/(2\sigma^2)}}{\sqrt{2\pi\sigma^2}}. \quad (10.40)$$

Since observations on different subjects are assumed to be independent, we can first derive the contribution to the likelihood from (y_{1i}, y_{2i}) and then take the product over i. The contribution for an individual subject is the integral of the conditional joint distribution, (10.40):

$$f_{y_1,y_2}(y_1, y_2) = \int_{-\infty}^{\infty} \frac{(\mu^C)^{y_1} e^{-\mu^C}}{y_1!} \frac{e^{-(y_2-\mu^N)^2/(2\sigma^2)}}{\sqrt{2\pi\sigma^2}} f(b) db$$

$$(10.41)$$

$$= \int_{-\infty}^{\infty} \frac{\exp\{y_1(\mathbf{x}'\boldsymbol{\beta}^C + b) + e^{\mathbf{x}'\boldsymbol{\beta}^C+b}\}}{y_1!} \frac{e^{-(y_2-\mathbf{x}'\boldsymbol{\beta}^N-\lambda b)^2/(2\sigma^2)}}{\sqrt{2\pi\sigma^2}} f(b) db.$$

The full likelihood is then formed as the product

$$L(\boldsymbol{\beta}^C, \boldsymbol{\beta}^N, \sigma, \sigma_b, \lambda|\mathbf{y_1}, \mathbf{y_2}) = \prod_{i=1}^{n} f_{y_1,y_2}(y_{1i}, y_{2i}). \quad (10.42)$$

The likelihood, given by (10.41) and (10.42), cannot be evaluated in closed form, but involves only a single dimensional integral. This can be evaluated numerically using methods such as Gauss-Hermite quadrature (see Section 2.6c). To find the MLEs, the numerically-calculated likelihood must be numerically maximized. While computationally challenging, such methods are incorporated in modern statistical software, such as Stata and SAS.

While likelihood ratio tests of the fixed effects parameters follow the usual large sample chi-square distribution, tests of whether the variance components are zero do not, as we observed in Section 2.6. For example testing $H_0 : \sigma_b^2 = 0$ follows a large sample distribution which is a 50:50 mixture of a χ_0^2 and a χ_1^2. For a more detailed analysis see Crainiceanu and Ruppert (2004).

10.6   EXAMPLE: OSTEOARTHRITIS INITIATIVE

The OAI (Osteoarthritis Initiative) was briefly described in Section 1.6f. Two of the outcomes collected were a numeric score measuring the degree of disability due to knee functioning, and the number of days of work missed out of the last 3 months due to knee pain, which is a count variable. We consider modeling the log transformation of the disability score plus 1 (which is closer to approximate normality than the score itself) and the number of days of missed work in the past three months. Using subjects that were diagnosed with osteoarthritis (OA) at enrollment as well as those who might develop OA gave a total of 2,678 subjects for analysis. The model will incorporate three predictors: age, sex, and body mass index (BMI).

We fit a model of the form (10.25), that is, a linear regression of log of the disability score plus 1 (LNDIS) on AGE, SEX, and BMI and a Poisson regression of missed days of work (MISSW) on the same three predictors:

$$\text{MISSW}_i | b_i \quad \sim \quad \text{indep. Poisson}(\mu_i^C)$$

$$\log(\mu_i^C) \quad = \quad \beta_0^C + \beta_1^C \text{AGE}_i + \beta_2^C \text{SEX}_i + \beta_3^C \text{BMI}_i + b_i,$$

$$\text{LNDIS}_i | b_i \quad \sim \quad \text{indep. } \mathcal{N}(\mu_i^N, \sigma^2) \tag{10.43}$$

$$\mu_i^N \quad = \quad \beta_0^N + \beta_1^N \text{AGE}_i + \beta_2^N \text{SEX}_i + \beta_3^N \text{BMI}_i + \lambda b_i,$$

with $b_i \sim$ i.i.d. $\mathcal{N}(0, \sigma_b^2)$, using SAS Proc NLMIXED (SAS Institute, Cary NC).

Table 10.2 gives the value of the negative log likelihood for three comparison models: the model of (10.43), given in the first row of the table, that with no random effects in the second row, and that with the BMI predictor removed from the regression models for $\log(\mu_i^C)$ and μ_i^N in the last row.

The model without the random effects has a much lower value of the log likelihood compared to model (10.25). The asymptotic distribution of the likelihood ratio test is problematic in this case because of the boundary issues of testing $H_0 : \sigma_b^2 = 0$ and because, when $\sigma_b^2 = 0, \lambda$ is not identifiable. However, by any measure, the model without the random effects is much worse.

The model without BMI in either linear predictor enables a straightforward simultaneous test of $H_0 : \beta_3^C = \beta_3^N = 0$. The likelihood ratio

Table 10.2: Models fit and negative log likelihoods for the OAI data.

Predictors	Random effect?	$-2\log L$	Change in $-2\log L$
BMI, SEX, AGE	Yes	4472.9	reference model
BMI, SEX, AGE	No	5467.0	994.1
SEX, AGE	Yes	4647.3	174.4

statistic is 174.4 with an asymptotic χ_2^2 distribution, giving a p-value of approximately $0 = Pr\{\chi_2^2 \geq 174.4\}$. This is an indication of how joint tests of the influence of a predictor (in this case BMI) on multiple outcomes can easily be carried out in the joint modeling context, potentially avoiding multiple testing issues.

These models can be fit using statistical packages such as SAS or Stata, which use adaptive Gauss-Hermite quadrature to approximate the integrals for calculating the log likelihood. The programs then use quasi-Newton algorithms to numerically maximize the likelihood.

10.7   NOTES AND EXTENSIONS

a.   Missing data

An advantage of the joint modeling approach is the ability to accommodate situations in which data are missing in a way predictable by one of the outcomes. For example, suppose that primary interest focusses on y_1 but that data are missing in a way that depends on y_2. In the context of the OAI, let y_2 be knee pain at enrollment in the study and y_1 the occurrence of total knee replacement by year 3. Participants with higher values of pain at baseline may have a higher likelihood of dropping out and therefore not furnishing a value for y_1.

Because the chance that the data are missing may be related to the outcome (it is easy to believe that those with higher pain values may be more likely to have total knee replacements), biased estimators can result. If the missing data process depends on y_2 and the data are analyzed by maximum likelihood methods, then the argument of Laird (1988) can be easily applied to joint models with shared or correlated random effects to show that consistent estimators will result for the parameters of the y_1 model.

b. Efficiency

Especially in the situation where there are missing data, joint analysis of the outcomes can produce more efficient estimators compared to analyzing the outcomes separately. The gains can be significant when the outcome of interest is a "low-information" outcome (like a binary outcome) and has a large percentage of missing data. McCulloch (2008) reports standard errors that can be up to 40% smaller and Gueorguieva and Sanacora (2006) show little improvement for a continuous outcome, but about a 20% decrease in standard errors for analyzing an ordinal outcome in a joint model. Other research has reported little or no gain (Lesaffre and Molenberghs, 2001; Fitzmaurice and Laird, 1997; Gueorguieva and Agresti, 2001) in using joint models.

10.8 EXERCISES

E 10.1 Show that (10.3) is equivalent to (10.1).

E 10.2 Show that the *covariance* in the multivariate probit model, (10.10), is zero when $\sigma_b = 0$, positive when $\sigma_b > 0$, and an increasing function of σ_b. Find the limit of the *correlation* as $\sigma_b \to \infty$.

E 10.3 Show that the correlation of y_{ij} and y_{ik} in the multivariate probit model, (10.10), depends on the values of μ_j and μ_k.

E 10.4 For a fixed value of σ^2 show that the limit (as $\sigma_b \to \infty$) of the correlation, given by (10.24), is $\sqrt{2/\pi}$.

E 10.5 Prove (10.30).

E 10.6 For $\theta_i = 0$ show that (10.24) simplifies to

$$\sqrt{\frac{\lambda^2 \sigma_b^2}{\lambda^2 \sigma_b^2 + \sigma^2}} \sqrt{\frac{2}{\pi}} \sqrt{\frac{\sigma_b^2}{\sigma_b^2 + 1}}.$$

Hence show that, when $\lambda = 1$ and $\sigma = 1$, the correlation is a linear function of ρ.

E 10.7 Consider the correlation for the Poisson/normal model, (10.32) as a function of $\theta = \mathbf{x}_i' \boldsymbol{\beta}^C$ and $\rho = \sigma_b^2 / (\sigma_b^2 + \sigma^2)$. For a fixed value of θ show the correlation tends to zero as $\sigma_b \to \infty$. Also, for a fixed value of ρ, show the correlation tends to a limit as $\theta \to \infty$.

E 10.8 Prove (10.38).

Chapter 11

NONLINEAR MODELS

11.1 INTRODUCTION

This chapter gives brief consideration to nonlinear mixed models, abbreviated NLMMs, mainly to emphasize that GLMMs are a proper subset of NLMMs, which comes with both advantages and disadvantages. We illustrate the ideas mostly in the context of an example.

11.2 EXAMPLE: CORN PHOTOSYNTHESIS

Parker (1995) at Cornell University studied the photosynthetic ability of wild relatives of corn. The main question of interest was to compare two species (an annual and perennial) with respect to photosynthetic physiology. Seeds from two populations of each species were collected and grown in the greenhouse. The experimental design was a randomized complete block design with four blocks and three seeds from each population in each block (for a total of 12 seeds per block). After 24 days, photosynthesis was recorded at nine different light levels from full sunlight to darkness on one individual from each population in each block (N=16). Measurements on the same 16 plants were repeated after 48 days. From these data, photosynthesis versus irradiance response curves reflecting the change in photosynthetic rate with light level were derived.

The traits of interest are the maximum photosynthetic rate, dark respiration, the light compensation point, and the quantum yield. The maximum photosynthetic rate measures the maximum amount of carbon dioxide the plants are able to assimilate in full sunlight, the dark respiration indicates how much carbon dioxide they respire in the dark,

Figure 11.1: **Photosynthetic rate versus light for two corn plants.**

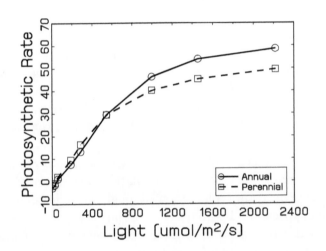

the light compensation point is the light level at which photosynthesis overcomes respiration and carbon assimilation becomes positive, and quantum yield is the efficiency of carbon assimilation at low light levels, or the slope of the light response curve as it crosses the light compensation point.

To elaborate we now describe those ideas from mathematical and graphical viewpoints. Figure 11.1 shows the graph of photosynthetic rate of two representative plants, one annual and one perennial species. The form of the curve typically used to describe this relationship as a function of light, l, is

$$\text{PHOTO}(l) = \beta_1 + \beta_2 e^{-\beta_3 l}. \tag{11.1}$$

This is perhaps the simplest way to write the equation, but not all the parameters β_i are directly of interest. Equation (11.1) has value $\beta_1 + \beta_2$ at $l = 0$, asymptotes at β_1 at $l \to \infty$ (for $\beta_3 > 0$), and crosses the x-axis at $l = -\log(-\beta_1/\beta_2)/\beta_3$. Thus we define

$$\alpha \;=\; \text{asymptote} = \text{maximum photosynthetic rate} = \beta_1$$

$$\delta \;=\; \text{dark respiration rate} = \beta_1 + \beta_2 \tag{11.2}$$

$$\lambda_0 \;=\; \text{light compensation point} = -\log(-\beta_1/\beta_2)/\beta_3$$

and rewrite (11.1) (see E 11.1) as

$$\text{PHOTO}(l) = \alpha - (\alpha - \delta)\left(\frac{\alpha}{\alpha - \delta}\right)^{\frac{l}{\lambda_0}}. \qquad (11.3)$$

This is the equation for a single plant. If we assume that (11.3) represents the mean response as a function of light, this cannot be a GLMM. This is because no function of rate will be linear in the parameters. Hence models like this one which are intrinsically nonlinear in the parameters are not GLMMs. Of course, GLMMs are special cases of nonlinear mixed models; but this example shows that they are not one and the same.

So far, only the effect of light is incorporated into (11.3). What about the effect of plants, blocks, populations and species? If the results in Figure 11.1 are typical then we might consider modeling α as a function of species and perhaps each of the other factors as well. Let i be a subscript representing species, j represent population, k represent block, m represent plant, and t represent replicates. A reasonable model for α itself would be

$$\alpha_{ijkmt} = \mu + \text{SPECIES}_i + \text{POPLN}_j + \text{BLOCK}_k + \text{PLANT}_m, \quad (11.4)$$

where BLOCK and PLANT might be considered random effects. This approach, of modeling the parameters as functions of the other factors, is sensible since they represent (three of) the traits of interest in this study. It is thus easy to assess, for example, the influence of species on the maximum photosynthetic rate, α. To complete the overall model we would need similar submodels for the dependence of δ and λ_0 on the foregoing factors. However, it is certainly also possible to entertain alternate ways of incorporating the factors.

Some of the advantages and disadvantages of GLMMs as compared to NLMMs are clear from this example. To specify the NLMM each of the sub-models for α, δ, and λ_0 must first be specified, then fit and perhaps simplified using the data. For example, the plant effects in each model would need to be considered and separate plant variance components would need to be estimated for each submodel. This would lead to a large number of parameters to be estimated. In contrast, for a GLMM we assume that all model terms enter into the mean of the distribution, simplifying the construction of the model.

Clearly this is a double-edged sword. Although the models are simpler, with fewer parameters to estimate, they may make unreasonably restrictive assumptions. In the photosynthesis example the model is fundamentally nonlinear and a GLMM will not suffice.

11.3 PHARMACOKINETIC MODELS

A common usage of nonlinear mixed models is in pharmacokinetic modeling, that is, models for describing the movement of drugs or other substances through the body. The mean structure for these models is typically derived from a system of differential equations. The differential equations are commonly set up by hypothesizing the existence of two or more *compartments* in the body with differential equations incorporating the rates of flow from one compartment to the next.

For example, suppose a dose D_0 of a drug is administered orally at time $t = 0$. Two compartments might be hypothesized, representing the stomach and the blood system. We model this as a system of differential equations that describe the flows between compartments. Let $S(t)$ be the amount of drug in the stomach at time t and let $B(t)$ be the amount in the bloodstream. We assume that the drug moves from the stomach to the bloodstream at a rate r_{12}, leaves the system from the stomach at rate r_{13}, is reabsorbed from the bloodstream at rate r_{21}, and exits the system from the bloodstream at rate r_{23}. This would give the following set of equations:

$$
\begin{aligned}
\frac{dS(t)}{dt} &= -(r_{12} + r_{13})S(t) + r_{21}B(t) \\[2mm]
\frac{dB(t)}{dt} &= r_{12}S(t) - (r_{21} + r_{23})B(t) \\[2mm]
S(0) &= D_0 \\[2mm]
B(0) &= 0.
\end{aligned}
\tag{11.5}
$$

Since this is a relatively simple system of differential equations, it can be solved explicitly for the amount of drug in the bloodstream at any time t. The solution is of the form

$$
B(t) = \beta(e^{-\lambda_1 t} - e^{-\lambda_2 t}).
\tag{11.6}
$$

Let y_{ij} represent the amount of the drug found in the jth sample taken from the ith person's bloodstream, which occurred at time t_{ij}.

Our model might then be

$$y_{ij}|\beta_i, \lambda_{1i}, \lambda_{2i} \sim \mathcal{N}(\mu_{ij}, \sigma^2)$$

$$\mu_{ij} = \beta_i(e^{-\lambda_{1i}t_{ij}} - e^{-\lambda_{2i}t_{ij}}) \qquad (11.7)$$

$$\begin{pmatrix} \beta_i \\ \lambda_{1i} \\ \lambda_{2i} \end{pmatrix} \sim \text{ i.i.d. } \mathcal{N}(\mathbf{0}, \mathbf{\Sigma}),$$

which allows the parameters to vary from person to person.

Clearly, this is a nonlinear mixed model: The means are nonlinear in the parameters and cannot be linearized with a fixed transformation. And we are allowing the flow rates between the compartments to vary from person to person.

Models that arise from the solution of a differential equation or system of such equations are typically nonlinear in the parameters. When we incorporate random effects to model variation from subject to subject or other forms of correlation we end up with nonlinear mixed models. A much more thorough treatment of these topics can be found in Giltinan and Davidian (1995) and Sheiner et al. (1997)

11.4 COMPUTATIONS FOR NONLINEAR MIXED MODELS

When data are normally distributed and homoscedastic (or can be transformed to be) then the Laplace approximation methods described in Chapter 14 are more successful for NLMMs than they were for GLMMs in general. This is the basis of computing algorithms implemented in S-Plus and SAS for NLMMs. The conceptual basis for them is described in Giltinan and Davidian (1995), Pinheiro and Bates (1995), and Lindstrom and Bates (1990).

11.5 EXERCISES

E 11.1 Prove that (11.1) can be rewritten as (11.3) using the reparameterization given in (11.2).

Chapter 12

DEPARTURES FROM ASSUMPTIONS

12.1 INTRODUCTION

Analyses of clustered and longitudinal data using generalized linear mixed models specify the conditional distribution of the response given random effects as well as the distribution of the random effects. In practice, data analysts often choose particular statistical models for responses and random effects because they are easy to work with mathematically or are simple to fit using available software. For example, it is common to specify a normal distribution for the random effects because it is computationally convenient.

Invariably, to some extent, some or all model assumptions are incorrect. For example, assumptions about the distributions of the response and random effects may be wrong. There may be densities that describe the random effects distribution more accurately, and functions of the covariates that better describe their relationship with the response. When this occurs, we have misspecified the GLMM.

In this chapter we consider misspecifications of GLMMs that come about because of choice of the incorrect link function, omission of important covariates, misclassification of binary outcomes and choice of the wrong distribution for the random effects. We begin by considering incorrect specifications of the conditional distribution of the response given the random effects and then consider wrongly specified random effects distributions. Finally, we discuss methods to diagnose and correct errors using specification tests as well as conditional and nonparametric maximum likelihood.

Model misspecification can produce biased or inefficient estimates of the associations of covariates with the response, invalid variance estimates and less powerful tests of hypotheses. Our main approach to examine effects of model misspecification, such as bias and loss of efficiency, is to calculate the expected value and variance of estimators obtained from the misspecified model with respect a more general model to gauge the impact of our assumptions.

12.2   INCORRECT MODEL FOR RESPONSE

There are a number of ways in which the model for the response, conditional on the random effects, can be misspecified. We first consider the case in which covariates are incorrectly omitted from the model

a.   Omitted covariates

An incomplete understanding of the phenomenon under study or an inability to collect data on all the relevant factors related to the outcome of interest can lead investigators to misspecify statistical models by omitting important covariates. Such omitted covariates may be associated with the covariates included in the model or may be independent of them, as in studies with randomized treatment assignments (see E 12.1). There is a large literature on the effects of omitted covariates in models for independent data (e.g., Gail et al., 1984; Neuhaus and Jewell, 1993) which focuses on settings where the omitted covariates are independent of those included. Results vary by the setting but typically omitting covariates that are independent of those included produces estimates of covariate effects that are smaller in absolute value than the true values, and losses in estimation efficiency.

The literature on the effects of omitted covariates with models for clustered and longitudinal data is less extensive than that for independent data but we present some results here. These results come from settings where the true model for the responses y_{ij}, $i = 1, \ldots, m; j = 1, \ldots, n_i$ is a GLMM with scalar covariates x_{ij} and w_{ij} as well as a random intercept b that follows a distribution G with zero mean. That is, the true model has

$$\mu_{ij} = \mathrm{E}[y_{ij} \mid b_i, x_{ij}, w_{ij}] = h^{-1}(\beta_0 + b_i + \beta_1 x_{ij} + \gamma w_{ij}), \qquad (12.1)$$

where $b_i \sim G$ with $\mathrm{E}[b_i] = 0$ and $\mathrm{var}(b_i) = \sigma_b^2$ and h is a link function.

However, for reasons as listed above, we fit a model that omits w_{ij}. That is, we fit the model

$$\mu_{ij}^* = \mathrm{E}[y_{ij}|\, b_i^*, x_{ij}] = h^{-1}(\beta_0^* + b_i^* + \beta_1^* x_{ij}), \qquad (12.2)$$

where $b_i^* \sim G$ with $\mathrm{E}[b_i^*] = 0$ and $\mathrm{var}(b_i^*) = \sigma_{b*}^2$ and the asterisks indicate that the parameters of (12.2) may differ from those of (12.1).

Our objective is to estimate the within-cluster association of x_{ij} with y_{ij}, adjusted for w_{ij}, i.e. β_1, but equation (12.2) differs from equation (12.1) and estimates $\hat{\beta}_1^*$ may be biased for β_1 or less efficient than $\hat{\beta}_1$ based on fitting (12.1). As with independent data, omitting w_{ij} when it is correlated with x_{ij} will typically yield $\beta_1^* \neq \beta_1$ and biased estimates $\hat{\beta}_1^*$ of β_1. Thus, we will focus on settings where w_{ij} and x_{ij} are independent and consider several special cases to illustrate the effects of omitted covariates. The special cases consider different link functions and omitted covariate types. Note that our development of theory here assumes that both w_{ij} and x_{ij} are random variables standardized to have mean 0 and variance 1 and independent of the random effects b_i.

Case 1: G =normal, $w_{ij} = w_i \sim N(0,1)$

In this setting, w_i is a cluster-level covariate that follows a standard normal distribution. Setting $b_i^* = b_i + \gamma w_i$, we see that equation (12.2) differs from equation (12.1) only in the random effect terms and that we can completely absorb the omitted covariate effect into b_i^*. That is, $\beta_0^* = \beta_0$, $\beta_1^* = \beta_1$, $\mathrm{E}[b_i^*] = 0$, $\mathrm{var}(b_i^*) = \sigma_b^2 + \gamma^2 = \sigma_{b*}^2$ and b_i^* follows a normal distribution. Equation (12.2) correctly specifies the distribution of y_{ij} when we omit w_i and standard estimation methods, such as maximum likelihood, based on (12.2) will yield consistent estimates of β_0 and β_1. Note that this result holds for any distribution G that is closed under convolution.

Case 2: h=identity link, G =normal, $w_{ij} = w_i \sim G_w$, not normal

In this setting, the true model for y_{ij} has

$$\mathrm{E}[y_{ij}|\, b_i, x_{ij}, w_i] = \beta_0 + b_i + \beta_1 x_{ij} + \gamma w_i + e_{ij} = \beta_0 + b_i^* + \beta_1 x_{ij} + e_{ij}, \quad (12.3)$$

where $b_i^* = b_i + \gamma w_i$ with $\mathrm{E}[b_i^*] = 0$ and $\mathrm{var}(b_i^*) = \sigma_b^2 + \gamma^2 = \sigma_{b*}^2$ and where, conditional on b_i, the e_{ij} are independent $(0, \sigma_e^2)$ variates, independent of both b_i and w_i. Averaging over the distribution of b_i^*, the marginal mean of y_{ij} is

$$\mathrm{E}[y_{ij}|\, x_{ij}] = \beta_0 + \beta_1 x_{ij} + e_{ij}^*, \qquad (12.4)$$

where $\mathbf{V} = \mathrm{cov}(\mathbf{e}^*) = \mathrm{cov}(\mathbf{Y}) = \sigma_e^2 \mathbf{I}_n + \sigma_{b^*}^2 \left\{ _d \mathbf{J}_{n_i} \right\}$, and $n = \sum n_i$.

From (12.4), it is clear that under the true model (12.3), the ordinary least squares estimator, $\hat{\boldsymbol{\beta}}_{OLS} = (\mathbf{X}'\mathbf{X})^{-1}\mathbf{X}'\mathbf{y}$, is unbiased for $\boldsymbol{\beta}$ no matter what the distributions of $\mathbf{b}^*$ and $\mathbf{e}$ are. We can view the estimator $\hat{\boldsymbol{\beta}}_{OLS}$ as the maximum likelihood estimator arising from a likelihood which is based on the assumption that $\sigma_{b^*}^2 = 0$ and that the errors $\mathbf{e}$ follow a normal distribution. As is well known, the usual estimate of the standard error of $\hat{\boldsymbol{\beta}}_{OLS}$ is incorrect. A valid estimate of $\mathrm{var}(\hat{\boldsymbol{\beta}}_{OLS})$ is, however, available (e.g., Binder, 1983; Liang and Zeger, 1986)

Although $\hat{\boldsymbol{\beta}}_{OLS}$ is unbiased, it can be inefficient. The generalized least squares estimator, $\hat{\boldsymbol{\beta}}_{GLS} = (\mathbf{X}'\mathbf{V}^{-1}\mathbf{X})^{-1}\mathbf{X}'\mathbf{V}^{-1}\mathbf{y}$, is more efficient and is unbiased no matter what the distributions of $\mathbf{b}^*$ and $\mathbf{e}$ are. We can think of $\hat{\boldsymbol{\beta}}_{GLS}$ as the maximum likelihood estimator from a likelihood which is based on the assumption that the random terms $\mathbf{b}^*$ and $\mathbf{e}$ follow independent normal distributions. A valid estimate of $\mathrm{var}(\hat{\boldsymbol{\beta}}_{GLS})$ can be derived from the information matrix of the likelihood based on the linear mixed model (12.3) with the assumption that both $\mathbf{b}^*$ and $\mathbf{e}$ are normally distributed. The estimator is consistent for $\mathrm{var}(\hat{\boldsymbol{\beta}}_{GLS})$ provided the correlation structure of the errors $\mathbf{e}^*$ are correctly specified by (12.4) and the intraclass correlation coefficient $\rho = \sigma_{b^*}^2/(\sigma_e^2 + \sigma_{b^*}^2)$ is consistently estimated. Note that correct specification of the random effects distribution is not required. Finally, estimates of $\mathrm{var}(\hat{\boldsymbol{\beta}}_{GLS})$ which are robust to misspecification of the intracluster correlation, $\mathrm{cov}(\mathbf{e}^*)$, are available (e.g., Liang and Zeger, 1986).

If the distribution of the cluster-level omitted covariate, G_w differs from the distribution G of the random effects b_i of equation (12.1), then the distribution of b_i^* is the convolution of G and G_w, H say, and this distribution is typically not equal to G nor even in the same family. However, with identity link models, standard methods such as ordinary and generalized least squares yield unbiased estimates of covariate effects with omitted cluster-level variables.

Case 3: h=**logistic link**, G =**normal**, $w_{ij} = w_i \sim G_w$, not normal

As in Case 2, we can view omitting w as misspecifying the distribution of the random effects $\mathbf{b}^*$ of a mixed-effects logistic model; we assume that the random effects $\mathbf{b}^*$ follow a distribution G when they actually follow the convolution of G and G_w, a different distribution. Thus we can assess the effects of omitting a covariate by applying results on the effects of misspecifying the random effects distribution of mixed-effects logistic models. For example, Neuhaus et al. (1992) presented approximations and simulation results to show that misspecifying the random effects distribution results in little bias in estimated covariate effects. Applied to the setting here, these results indicate that omitting cluster-level covariates from mixed-effects logistic models will yield approximately unbiased estimates of β in equation (12.1). Section 12.3a presents some additional results on the effects of misspecifying the distribution of the random effects.

Case 4: h=any link, G =normal, $w_{ij} \sim N(0,1)$

Conditional on b_i, this setting is identical to the independent response situation investigated by Gail et al. (1984) and Neuhaus and Jewell (1993) and results from these papers apply here. For example, Neuhaus and Jewell (1993) show that the presence and direction of bias due to omitted covariates depends on geometric properties of the link function h. In particular, they show that bias due to omitted covariates depends on the curvature of $H(t) = -1/h'(t)$. If $H(t)$ is constant or linear then there is no bias. Links with this feature include the identity and log. Link functions for which H is convex, which include the logistic, probit and complementary log-log (see E 12.2), will yield attenuated estimates of β_1 with omitted covariates. In fact, any link function based on an inverse cumulative distribution function, $h(\mu) = \mathcal{H}^{-1}(\mu)$, with a log concave density function has H convex. Examples include the normal, exponential, gamma and Weibull with shape parameter greater than one. Link functions for which H is concave will yield estimates which are biased away from the null. The power family of links, $h(\mu) = (\mu^a - 1)/a$, for $0 < a < 1$ are links with this property.

b. Misspecified link functions

Standard analysis using GLMMs specifies a link function h but there may exist another function h^* that more accurately describes the relationship between the covariates and the expected value of the response. In such settings we can view the choice of h over h^* as a misspecification of the link function. We describe one such setting in Section 12.2c.

Although several papers have examined the effects of link misspecification on generalized linear models, the literature on the effects for GLMMs is not extensive. We present some results here for generalized linear models to indicate the effects we might expect to see for GLMMs and examine a special case in Section 12.2c.

Czado and Santner (1992) investigate the effect of link misspecification with binary regression models and show that incorrect assumptions about the link function produce biased estimates of covariate effects and losses of efficiency. Li and Duan (1989) show that the estimated slope $\hat{\beta}_1^*$ from a model with a misspecified link function typically consistently estimates the true parameter β_1 up to a scale factor. That is, $E[\hat{\beta}_1^*] = c\beta_1$, for a constant c. Neuhaus (1999) shows that misspecified link functions arise naturally in studies with misclassified binary responses and uses the Li and Duan (1989) result to motivate an expression for bias. Finally, Pregibon (1980) develops a goodness-of-link test that several computer packages now implement.

c.   Misclassified binary outcomes

Methods to analyze clustered and longitudinal binary data assume that one measures the binary response without error but in practice this may not be the case. For example, Shiboski et al. (1999) conducted a longitudinal study to assess the magnitudes of the associations of within-subject changes in health behaviors and immune function with the appearance of an oral lesion known as hairy leukoplakia. The study involved several repeated physical and oral examinations of each subject in the study. The investigators who performed the oral examinations and gathered the study data were not oral medicine specialists but rather physician's assistants trained to diagnose oral lesions. Hairy leukoplakia lesions are sometimes difficult to diagnose and distinguish from other oral lesions and the diagnoses of the physician's assistants were less accurate than those of the experts. This led to misclassified study responses. Misclassified binary responses also arise when the response indicates the presence or absence of a medical condition identified through an imperfect diagnostic test.

To develop theory related to misclassified binary responses we assume that we have true clustered or longitudinal binary responses y_{ij} along with p-dimensional covariates $\mathbf{x}_{ij}$, where i indexes cluster and j unit within cluster, and that we want to fit a binary mixed-effects model to assess the association of $\mathbf{x}_{ij}$ with $E[y_{ij}]$ as well as to estimate within-cluster dependence of the response y_{ij}. In particular, we assume that y_{ij} follows a GLMM with link function h and that the random effects follow a distribution G with mean zero. That is, $P\{y_{ij} = 1 \mid b_i, x_{ij}\} = h^{-1}(z_{ij}b_i + \mathbf{x}'_{ij}\boldsymbol{\beta})$, where the z_{ij} make up the model matrix for the random intercepts.

We do not observe y_{ij} but rather an error-corrupted version w_{ij}. In theory, the probability of misclassification could depend on all the true responses and covariates in a cluster and all other observed responses as well. However, in this chapter we assume that the probability of misclassification of the ij^{th} response only depends on the true response of that unit. This assumption would be reasonable in settings where the response of the ij^{th} unit arises from a diagnostic test or procedure whose result does not depend on the results of any other unit. Specifically, we assume that

$$P\{w_{ij}|w_{ik}, y_{ij}, y_{ik}, \mathbf{x}_{ij}, \mathbf{x}_{ik}\} = P\{w_{ij}|y_{ij}\} \text{ for all } k \neq j.$$

Let

$$P\{w_{ij} = 0|y_{ij} = 1, x_{ij}\} = \gamma_1 \quad (12.5)$$

$$\text{and } P\{w_{ij} = 1|y_{ij} = 0, x_{ij}\} = \gamma_0 . \quad (12.6)$$

We assume that $\gamma_1 + \gamma_0 < 1$ since values of γ_1 and γ_0 larger than one half indicate that the procedure producing the observed classification w_{ij} performs worse than chance.

Simple probability calculations show that the true model for the observed response, w, has

$$P\{w_{ij} = 1|x_{ij}, \mathbf{b}\} = (1 - \gamma_1 - \gamma_0) \, P\{y_{ij} = 1|\mathbf{b}, \mathbf{x}\} + \gamma_0$$

$$= (1 - \gamma_1 - \gamma_0) \, h^{-1}(\eta_x + b_i) + \gamma_0. \quad (12.7)$$

Note that $P\{w_{ij} = 1|\mathbf{x}\}$ depends on $\mathbf{x}$ only through η_x. Simple algebra (see E 12.3) yields

$$\eta_x = h \left(\frac{P\{w_{ij} = 1|\mathbf{x}\} - \gamma_0}{1 - \gamma_1 - \gamma_0} \right) = h^* \left(P\{w_{ij} = 1|\mathbf{x}\} \right) = g^*(\mu_w), \quad (12.8)$$

where $\mu_w = P\{w_{ij} = 1|\mathbf{x}\}$. Since h is an increasing link function, h^* is differentiable with

$$\frac{\partial h^*}{\partial \mu_w} = h'\{(\mu_w - \gamma_0)/(1 - \gamma_1 - \gamma_0)\}/(1 - \gamma_1 - \gamma_0). \quad (12.9)$$

The function h^* will be increasing and therefore a link function when $\gamma_1 + \gamma_0 < 1$.

We see that when $\mathbf{y}$ follows a GLMM with link function h and random effects distribution G, $\mathbf{w}$ will also follow a GLMM with modified link function (12.8) and the same random effects distribution G. Thus, the likelihood for the observed responses will have the same integral form as the likelihood for the true responses $\mathbf{y}$ but will depend on the link h^* instead of h. Maximization of the likelihood for the observed responses will involve its derivatives which depend on the derivative of the link h^*, (12.8). Equation (12.8) shows that we can calculate the derivatives of h^* as simple modifications of the derivatives of h. Alternatively, one can consistently estimate all model parameters by analyzing the observed responses $\mathbf{y}$ using any GLMM routine that allows user-defined link functions.

GLMM analyses that ignore misclassified responses misspecify the link function, and thus the model, and produce biased estimates of covariate effects and other model parameters. Neuhaus (2002) developed approximations and conducted simulation studies to examine the magnitude of the bias due to ignoring misclassified responses. One simulation study generated true clustered binary responses according to a mixed-effects logistic model with a standard normal random intercept and one covariate, x, which also followed a standard normal distribution. The simulation generated the covariate independently within clusters and generated the observed responses by randomly changing true responses $y = 1$ to observations $w = 0$ with probability $\gamma_1 = 0.1$ and randomly changing the responses $y = 0$ to $w = 1$ with probability $\gamma_0 = 0.1$. The simulation ignored misclassification and fit standard mixed-effects logistic models to the observed response data by maximum likelihood using a Newton-Raphson algorithm and Gaussian quadrature to evaluate the integrals in the likelihood. The simulation yielded highly biased estimates of all model parameters. For example, average bias in the estimated slope parameter was -34%, nearly 4.0 standard errors below the true value.

d. Informative cluster sizes

Standard analyses of clustered and longitudinal data using GLMMs assume that the expected value of the response is independent of cluster size but example data show that this is frequently not the case. For instance, in studies of surgical interventions, surgery volume and outcomes may both be related to the skill level of the surgeons. Cluster sizes that are related to the outcome are called response-dependent or informative and this section examines the effect of ignoring this dependence on estimates of parameters of interest. We investigate the effects of ignoring informative cluster sizes by assuming that cluster sizes n_i are random variables and that the responses follow a model that features correlation between cluster sizes and responses but we fit a model that ignores this correlation.

We begin our investigation with the case of no covariates. Specifically, we assume that we have clustered responses y_{ij}, $i = 1, \ldots, m$, $j = 1, \ldots, n_i$, where

$$y_{ij} = \beta_0 + b_i + e_{ij}, \tag{12.10}$$

with random variables b_i and e_{ij} that each have zero means and are independent for all i and j. The objective is to estimate $\mathrm{E}[y_{ij}] = \beta_0$ and the standard estimator is the sample mean $\bar{y} = \sum_{i=1}^m \sum_{j=1}^{n_i} y_{ij} / \sum_{i=1}^m n_i$. As in Benhin et al. (2005), we let the random cluster sizes $n_i = h(b_i)$ for some function h so that the shared random effect b_i induces correlation between responses and cluster sizes In this case, it is straightforward to show (see E 12.4) that

$$\bar{y} \xrightarrow{pr} \beta_0 + \frac{\mathrm{E}[h(b_i)b_i]}{\mathrm{E}[h(b_i)]}, \tag{12.11}$$

so that the sample mean is a biased estimator of β_0 with a bias that depends on the correlation between y_{ij} and n_i (through b_i) as well as the expected cluster size.

We can find closed-form expressions for the bias term in (12.11) in some special cases. One such case has $b_i \sim N(0, \sigma_b^2)$, $e_{ij} \sim N(0, \sigma_e^2)$, with **b** independent of **e** and n_i conditional on b_i being given by $1 + h(b_i)$ where $h(b_i)$ follows a Poisson distribution with expected value $e^{\gamma_0 + \gamma_1 b_i}$. Straightforward calculations (see E 12.5) show that

$$\mathrm{E}[\bar{y}] = \beta_0 + \frac{\gamma_1 \sigma_b^2 \exp(\gamma_0 + \gamma_1^2 \sigma_b^2 / 2)}{1 + \exp(\gamma_0 + \gamma_1^2 \sigma_b^2 / 2)} \tag{12.12}$$

Note that the bias term in (12.12) is 0 when $\gamma_1 = 0$, that is, when the random effects and cluster sizes are uncorrelated.

We can extend results for means and intercepts to examine the effect of ignoring informative cluster sizes on slope estimators by adding covariate effects to (12.10) as in

$$y_{ij} = \beta_0 + b_i + \beta_1 x_{ij} + e_{ij}. \tag{12.13}$$

Since Case 2 of Section 12.2a shows that the ordinary and generalized least squares estimators estimate the same quantity, we can examine the effect of ignoring informative cluster sizes on covariate effect estimators by calculating the expected value of $\hat{\boldsymbol{\beta}}_{OLS} = (\mathbf{X}'\mathbf{X})^{-1}\mathbf{X}'\mathbf{y}$. Calculations similar to those that produced (12.11) (see E 12.6) yield

$$\hat{\boldsymbol{\beta}}'_{OLS} \xrightarrow{pr} (\beta_0^* \quad \beta_1), \tag{12.14}$$

where $\beta_0^* = \beta_0 + \mathrm{E}[n_i b_i]/\mathrm{E}[h(b_i)]$, indicating different behavior of intercept and slope estimators. Ignoring informative cluster sizes yield biased estimates of intercepts and means but unbiased estimates of slopes and regression coefficients. This is important since scientific interest often focuses on covariate effect estimates from regression models and not on the intercepts.

We can assess the effect of ignoring informative cluster sizes on the parameter estimates of any GLMM in a similar fashion. That is, we assume that the true model for responses y and cluster sizes n has

$$\mathrm{E}[y_{ij}|\, b_i, x_{ij}] \;=\; h_y^{-1}(\beta_0 + \sigma_b b_i + \beta_1 x_{ij}) \tag{12.15}$$

$$\mathrm{E}[n_i|\, b_i, x_{ij}] \;=\; h_n^{-1}(\lambda_0 + \lambda_1 b_i), \tag{12.16}$$

where h_y and h_n are link functions for responses and cluster sizes, respectively, and $b_i \sim G$ with $\mathrm{E}[b_i] = 0$ and $\mathrm{var}(b_i) = 1$.

Since closed-form expressions for the estimates of the parameters of (12.15) for models with non-identity links typically do not exist, we will need to modify the approaches that produced (12.11) and (12.14) or use simulation studies to examine the bias that may result from ignoring informative cluster sizes. Below, we report the results of such a simulation study for a mixed-effects logistic model.

The simulations generated clustered binary responses from mixed-effects logistic models where the random intercepts b_i were correlated with cluster size. Specifically, we generated the random intercepts from a normal distribution with mean 0 and variance σ_b^2, independently in different clusters. Given the random effect for a cluster, we then generated a cluster size as 2 plus a Poisson $\{\lambda(b)\}$ random variable with the log of the mean given by $\log\{\lambda(b)\} = \gamma_0 + \gamma_1 b$. We added 2 to the Poisson variate to preclude cluster sizes of 0 or 1. The simulations used $\gamma_0 = \gamma_1 = 1.0$. Simple calculations show that the expected cluster size was $\mathrm{E}[n] = \exp(\gamma_0 + \sigma_b^2 \gamma_1^2/2) + 2$. Given the cluster size n_i, we then generated n_i binary, within-cluster covariates x_{ij} from a

Table 12.1: Observed means and standard deviations of the regression coefficients of standard mixed-effects logistic regression models fit to simulated clustered data where the random intercepts were associated with cluster sizes.

	β_0	β_1	$\log \sigma_b$
True values	1.0	1.0	0.0
Simulation	-0.752	1.002	0.001
	(0.192)	(0.195)	(0.162)

Bernoulli distribution with parameter $p_b = 0.5$. Finally, given b_i, n_i, and $x_{i1}, \dots, x_{in_i}$, we generated the binary responses $y_{i1}, \dots, y_{in_i}$ as n_i conditionally independent draws from a mixed-effects logistic model $\text{logit} (P\{y_{ij} = 1 \mid b_i, x_{ij}\}) = \beta_0 + b_i + \beta_1 x_{ij}$ with $\beta_0 = -1$, $\beta_1 = 1$ and $\sigma_b^2 = 1$.

To these data we fit standard mixed-effects logistic model that ignored the correlation between responses and cluster sizes using Proc NLMIXED in SAS. To accommodate the constraint that variance component estimates must be non-negative, we reparametrized the variance component of the mixed-effects models in terms of $\log \sigma_b$ and report the results on this scale. Note that the value $\sigma_b^2 = 1$ corresponds to $\log \sigma_b = 0$.

Table 12.1 presents the the average values of the parameter estimates along with standard deviations of these estimates from the simulations. As with the linear mixed-effects model, the simulation findings in Table 12.1 indicate that ignoring correlations between cluster sizes and responses results in biased estimates of the intercept, β_0, of mixed-effects logistic models, but, more importantly, unbiased estimates of the slope, β_1, typically the parameter of scientific interest. The simulations also show no bias in the estimate of the variance of the random effects.

12.3 INCORRECT RANDOM EFFECTS DISTRIBUTION

This section examines the effects of misspecification of the distribution of random effects. We first consider the case where the data analyst assumes the incorrect family of distributions for the random effects and then consider two cases where one incorrectly models the relationship between the random effects and covariates.

a.   Incorrect distributional family

Suppose that the true model for the responses y_{ij}, $i = 1, \ldots, n_i; j = 1, \ldots, m$ is a GLMM with a single scalar covariate x_{ij} as well as a random intercept b that follows a distribution G_T of known form with zero mean and unit variance and T stands for true. That is, the true model has

$$\mu_{ij} = \mathrm{E}[y_{ij} \mid b_i, x_{ij}] = h^{-1}(\beta_0 + \sigma_b b_i + \beta_1 x_{ij}), \qquad (12.17)$$

where $b_i \sim G_T$ with $\mathrm{E}[b_i] = 0$ and $\mathrm{var}(b_i) = 1$ and h is a link function. Suppose further that the assumed random effects distribution, G_F, where F stands for fitted, is also of known form but different from G_T. We will typically take G_F to be the standard normal distribution since this is the most popular assumed random effects distribution in practice. Our objective is to assess the effects of incorrectly assuming G_F to be the distribution of the random effects. In particular, we want to assess the magnitude of bias in estimated covariate effects due to misspecification of the random effects distribution.

– i.   *Linear mixed effects model*

We begin our assessment of the consequences of misspecifying the random effects distribution with the linear mixed effects model. Under this model the continuous outcome variable y_{ij} has

$$\mathrm{E}[y_{ij} \mid b_i] = \beta_0 + \sigma_b b_i + \beta_1 x_{ij} + e_{ij}, \qquad (12.18)$$

where, conditional on b_i, the e_{ij} are independent $(0, \sigma_e^2)$ variates, and the random effects b_i are independent $(0, 1)$ variates distributed according to G_T. We separate the standard deviation of the random effects, σ_b, from the standardized random effects in (12.18) to allow a more straightforward connection to other work on misspecified models that requires optimization with respect to the parameters of the fitted model.

Rather than fitting model (12.18) with true distribution G_T, we consider the setting where the data analyst assumes that the random effects follow a distribution G_F. The form of this setting is equivalent to that of Case 2 in Section 12.2a and (12.4) so that the results of that section apply here. In particular, under (12.18),

$$\text{cov}(y_{i1}, \ldots, y_{in_i}) = \mathbf{V} = \sigma_e^2 \mathbf{I} + \sigma_b^2 \mathbf{J}, \qquad (12.19)$$

independent of the shape of the distribution of the random effects and marginally the responses y_{ij} follow the model

$$y_{ij} = \beta_0 + \beta_1 x_{ij} + e_{ij}^*. \qquad (12.20)$$

As a result, the ordinary least squares estimator, $\hat{\boldsymbol{\beta}}_{OLS}$, is unbiased for β no matter what the distributions of $\mathbf{b}$ and $\mathbf{e}$ are. The estimator $\hat{\boldsymbol{\beta}}_{OLS}$ can be thought of as the maximum likelihood estimator arising from a likelihood which is based on the assumption that $\sigma_b^2 = 0$ and that the errors $\mathbf{e}$ follow a normal distribution. As is well known, the usual estimate of the standard error of $\hat{\boldsymbol{\beta}}_{OLS}$ is incorrect. A valid estimate of $\text{var}(\hat{\boldsymbol{\beta}}_{OLS})$ is, however, available (e.g., Binder, 1983).

Although $\hat{\boldsymbol{\beta}}_{OLS}$ is unbiased, it can be inefficient. A more efficient estimator is given by the generalized least squares estimator $\hat{\boldsymbol{\beta}}_{GLS}$ (S.23). The estimator $\hat{\boldsymbol{\beta}}_{GLS}$ is unbiased no matter what the distributions of $\mathbf{b}$ and $\mathbf{e}$ are and can be thought of as the maximum likelihood estimator from a likelihood which is based on the assumption that the random terms $\mathbf{b}$ and $\mathbf{e}$ follow independent normal distributions. A valid estimate of $\text{var}(\hat{\boldsymbol{\beta}}_{GLS})$ can be derived from the information matrix of the likelihood based on the linear mixed model (12.18) with the assumption that both $\mathbf{b}$ and $\mathbf{e}$ are normally distributed. The estimator is consistent for $\text{var}(\hat{\boldsymbol{\beta}}_{GLS})$ provided the correlation structure of the errors $\mathbf{e}^*$ are correctly specified by (12.19) and the intraclass correlation coefficient $\rho = \sigma_b^2/(\sigma_e^2 + \sigma_b^2)$ is consistently estimated. Note that correct specification of the random effects distribution is not required. Finally, estimates of $\text{var}(\hat{\boldsymbol{\beta}}_{GLS})$ which are robust to misspecification of the intracluster correlation, $\text{cov}(\mathbf{e}^*)$, are available (e.g., Liang and Zeger, 1986).

In summary, misspecification of the random effects distribution does not affect the estimates of the regression coefficients of (12.18); the expectations of $\hat{\boldsymbol{\beta}}_{OLS}$ and $\hat{\boldsymbol{\beta}}_{GLS}$ do not depend on the distribution of $\mathbf{b}$ and $\mathbf{e}$. Standard error estimates of these estimates are affected by misspecification of the random effects. That is, regression coefficient estimates obtained under misspecification of the random effects may be inefficient.

For the remainder of this section, we will examine the extent to which these properties of robustness to errors in distributional assumptions extend to other generalized linear mixed models.

– ii. *General approach*

Our general approach is to assume that under the correct model, observations y_{ij} follow a model with true random effects distribution G_T but that we fit a model by maximizing a "likelihood" that we construct under the assumption that the random effects follow a distribution G_F with $G_F \neq G_T$. That is, the fitted model misspecifies the distribution of the random effects. There are many possible forms of misspecification, including : 1) G_F has incorrect distributional shape; 2) G_F incorrectly models the random effects to be uncorrelated with the covariates; and 3) G_F incorrectly models random effects variances to be constant over the different covariate values. In this section we focus on models that contain only random intercepts.

We use and adapt results from the theory of inference with misspecified models (e.g., White, 1994) to examine the effects of misspecifying the random effects distribution. Specifically, we assume that under the true model, the density of the responses in the ith cluster, $y_{i1}, \ldots, y_{in_i}$ is

$$f_T(y_{i1}, \ldots, y_{in_i} | \mathbf{x}_{i1}, \ldots, \mathbf{x}_{in_i}; \xi) = \int \prod_{j=1}^{n_i} f(y_{ij} | \mathbf{b}, \mathbf{x}_{ij}; \beta) \; dG_T(\mathbf{b}; \boldsymbol{\theta}),$$

(12.21)

where f is the conditional density corresponding to a generalized linear model, β is a p-dimensional vector of regression coefficients corresponding to $\mathbf{x}_{ij}$, a p-dimensional vector of covariates, $\boldsymbol{\theta}$ is a vector of parameters of the random effects distribution G_T, $\xi = (\beta, \boldsymbol{\theta})$ and T denotes true. Note that the true distribution of the random effects $G_T(\mathbf{b})$ may depend on the $\mathbf{x}_{ij}$. We will focus on clustered and longitudinal data settings where we can describe the response correlation structure using random intercepts. Since we are interested in settings where random effects may be correlated with covariates and we would like to obtain results that apply to a wide range of covariate configurations, we will often assume that the covariates x are random.

We investigate the setting where the true model is of form (12.21) but we fit a model that incorrectly assumes the random effects distribution to be G_F and the density of the responses in the ith cluster, $y_{i1}, \ldots, y_{in_i}$ to be

$$f_F(y_{i1}, \ldots, y_{in_i} | \mathbf{x}_{i1}, \ldots, \mathbf{x}_{in_i}; \xi^*) = \int \prod_{j=1}^{n_i} f(y_{ij} | \mathbf{b}, \mathbf{x}_{ij}; \beta^*) \; dG_F(\mathbf{b}; \boldsymbol{\theta}^*),$$

(12.22)

where $\boldsymbol{\beta}^*$ and $\boldsymbol{\theta}^*$ are the regression and random effects distribution parameters, respectively, of the misspecified model. Let $\boldsymbol{\xi}^* = (\boldsymbol{\beta}^*, \boldsymbol{\theta}^*)$, where the asterisks indicate that the parameters of (12.22) may differ from those of (12.21) and F indicates the *false* distribution.

Following Neuhaus et al. (1992) and Neuhaus (1998), we examine the bias in parameter estimates due to model misspecification by calculating the expected value estimators obtained from the misspecified model with respect to the true, underlying density of the responses. The work of White (1994) shows that the "maximum likelihood" estimator, $\hat{\boldsymbol{\xi}}^*$, under the false model (12.22) converges to the value $\boldsymbol{\xi}^*$ that minimizes the Kullback-Leibler divergence (Kullback, 1959) between the true and misspecified models. That is, $\boldsymbol{\xi}^*$ minimizes

$$E_x E_{y|x} \left[\log \left\{ f_T(\mathbf{y}|\boldsymbol{\xi}, \mathbf{x}) / f_F(\mathbf{y}|\boldsymbol{\xi}^*, \mathbf{x}) \right\} \right], \qquad (12.23)$$

where one takes the expectation with respect to the true model. White (1994) further shows that $\hat{\boldsymbol{\xi}}^*$ has an asymptotic normal distribution with var($\hat{\boldsymbol{\xi}}^*$) given by a matrix product of the form

$$\text{var}(\hat{\boldsymbol{\xi}}^*) = \mathbf{A}^{-1}(\boldsymbol{\xi}^*) \mathbf{B}(\boldsymbol{\xi}^*) \mathbf{A}^{-1}(\boldsymbol{\xi}^*), \qquad (12.24)$$

where $\mathbf{A}(\boldsymbol{\xi}^*) = \left\{ E_x E_{y|x} \left[\partial^2 \log f_F(\mathbf{y}|\boldsymbol{\xi}^*, \mathbf{x}) / \partial \xi_i^* \, \partial \xi_j^* \right] \right\}$, and

$$\mathbf{B}(\boldsymbol{\xi}^*) = \left\{ E_x E_{y|x} \left[\partial \log f_F(\mathbf{y}|\boldsymbol{\xi}^*, \mathbf{x}) / \partial \xi_i^* \partial \log f_F(\mathbf{y}|\boldsymbol{\xi}^*, \mathbf{x}) / \partial \xi_j^* \right] \right\}$$

and expectations are with respect to the true model (12.21) with f_T. The matrix $\mathbf{A}$ involves the information from the misspecified likelihood, while the matrix $\mathbf{B}$ involves the true variance-covariance structure of the responses. Under the correct model specification, that is $f_F = f_T$, we have $\mathbf{A} = \mathbf{B}$.

Differentiation of (12.23) with respect to $\boldsymbol{\xi}^* = (\xi_1^*, \dots, \xi_q^*)$ yields, respectively, the simultaneous equations

$$E_x \int_y \lambda(\mathbf{y}|\mathbf{x}) \frac{\partial}{\partial \xi_k^*} \left\{ \int \prod_{j=1}^{n_i} f(y_{ij}|\mathbf{b}, \mathbf{x}_{ij}; \boldsymbol{\beta}^*) \, dG_F(\mathbf{b}; \boldsymbol{\theta}^*) \right\} d\mathbf{y} = 0,$$
$$(12.25)$$

for $k = 1, \dots, q$, where $\lambda(\mathbf{y}|\mathbf{x}) = f_T(\mathbf{y}|\mathbf{x}, \boldsymbol{\xi}) / f_F(\mathbf{y}|\mathbf{x}, \boldsymbol{\xi}^*)$ and we interpret $\int_y$ as a sum for discrete random variables.

Note that $\boldsymbol{\xi}^*$ that yield $\lambda(\mathbf{y}|\mathbf{x}) = 1$ for all $\mathbf{x}$ would solve the system (12.25) since in this case the system would be exactly the system of equations obtained when the true mixture model has b_i distributed as G_F. For given conditional density f and mixture distributions G_T and G_F, one can solve the system of equations and compare the solution to the true underlying values. As the next section shows, for some combinations of f, G_T and G_F we can analytically calculate the integrals in (12.25) and solve the system. Such analytic solutions develop intuition about expected effects of misspecification in other settings.

– iii.   *Binary matched pairs*

Consider the case of binary matched pairs, log link models and assumed normally distributed random intercepts

$$\log\left(P\{y_{ij} = 1|b_i\}\right) = \beta_0 + \beta_1 x_{ij} + \sigma_b b_i \quad j = 1, 2 \qquad (12.26)$$

with $x_{i1} \equiv 0$, $x_{i2} \equiv 1$, and random intercepts b that follow a true distribution G_T with $\mathrm{E}[b_i] = 0$, $\mathrm{var}(b_i) = 1$. Such models yield estimates of relative risks, measures of association of great interest to epidemiologists (Lu and Tilley, 2001), although logit link models are more popular and enjoy important advantages. With binary matched pairs, there are three free response pair probabilities and, as noted above, a solution to (12.25) matches these response probabilities under the true and misspecified models. The three relevant true probabilities are

$$P\{y_{i1} = 1, y_{i2} = 1\} = \int (e^{\beta_0 + \sigma_b b})(e^{\beta_0 + \beta_1 + \sigma_b b}) \, dG_T(b)$$

$$P\{y_{i1} = 1, y_{i2} = 0\} = \int (e^{\beta_0 + \sigma_b b})(1 - e^{\beta_0 + \beta_1 + \sigma_b b}) \, dG_T(b)$$

$$P\{y_{i1} = 0, y_{i2} = 1\} = \int (1 - e^{\beta_0 + \sigma_b b})(e^{\beta_0 + \beta_1 + \sigma_b b}) \, dG_T(b),$$

which we can express in terms of the moment generating function $m_{G_T}(.)$ of the distribution G_T as

$$P\{y_{i1} = 1, y_{i2} = 1\} = e^{2\beta_0 + \beta_1} m_{G_T}(2\sigma_b)$$

$$P\{y_{i1} = 1, y_{i2} = 0\} = e^{\beta_0} m_{G_T}(\sigma_b) - e^{2\beta_0 + \beta_1} m_{G_T}(2\sigma_b) \quad (12.27)$$

$$P\{y_{i1} = 0, y_{i2} = 1\} = e^{\beta_0 + \beta_1} m_{G_T}(\sigma_b) - e^{2\beta_0 + \beta_1} m_{G_T}(2\sigma_b).$$

The fitted model assumes normally distributed random intercepts and easy calculations (see E 12.7) show that the relevant fitted response probabilities are

$$P\{y_{i1} = 1, y_{i2} = 1\} = e^{2\beta_0^* + \beta_1^* + 2\sigma_{b*}^2}$$

$$P\{y_{i1} = 1, y_{i2} = 0\} = e^{\beta_0^* + .5\sigma_{b*}^2} - e^{2\beta_0^* + \beta_1^* + 2\sigma_{b*}^2} \qquad (12.28)$$

$$P\{y_{i1} = 0, y_{i2} = 1\} = e^{\beta_0^* + \beta_1^* + .5\sigma_{b*}^2} - e^{2\beta_0^* + \beta_1^* + 2\sigma_{b*}^2}.$$

A solution $\xi^* = (\beta_0^*, \beta_1^*, \sigma_{b*})$ to (12.25) equates equations (12.28) to equations (12.27). A solution with equal regression parameters, $\beta_1^* = \beta_1$, makes the third equation in each set redundant and leaves us to find $(\beta_0^*, \sigma_{b*}^2)$ such that

$$e^{2\beta_0^* + 2\sigma_{b*}^2} = e^{2\beta_0} m_G(2\sigma_b)$$

$$e^{\beta_0^* + .5\sigma_{b*}^2} - e^{2\beta_0^* + 2\sigma_{b*}^2} = e^{\beta_0} m_G(\sigma_b) - e^{2\beta_0} m_G(2\sigma_b).$$

Straightforward calculations show that

$$\beta_0^* = \beta_0 + \log\left\{\frac{[m_G(\sigma_b)]^2}{\sqrt{m_G(2\sigma_b)}}\right\} \text{ and } \sigma_{b*}^2 = \log\left\{\frac{m_G(2\sigma_b)}{[m_G(\sigma_b)]^2}\right\} \qquad (12.29)$$

solve the system. We can obtain explicit expressions for $(\beta_0^*, \sigma_{b*}^2)$ for any true distributions G that have closed form expressions for their moment generating function. For example, the calculations are straightforward for the uniform, gamma, logistic and double exponential distributions. Thus, for matched binary pairs with a log link model, fitting models that assume that the random intercepts are normally distributed when they actually follow a different distribution such as the uniform, gamma, logistic or double exponential, yields a consistent estimate of the regression coefficient of interest. However, estimates of the intercept, β_0, and variance component, σ_b, are biased with misspecified random effects.

Additional results exist for matched pairs binary data and logit link models. In this case, the model of interest has the same right-hand side as (12.26) but the logit link instead of the log;

$$\text{logit}\,(P\{y_{ij} = 1|b_i\}) = \beta_0 + \beta_1 x_{ij} + \sigma_b b_i \quad j = 1, 2. \qquad (12.30)$$

It is useful to consider the set, $\mathcal{S}$, of all probabilities $P\{y_{i1}, y_{i2}\}$ that model (12.30) can generate for varying values of β_0, σ_b and G_T. This is a two dimensional set since $P\{y_{i1} = 0, y_{i2} = 1\} = e^{\beta_1} P\{y_{i1} = 1, y_{i2} = 0\}$, and the four cell probabilities sum to one. Neuhaus et al. (1994) show that if the assumed random effect distribution, G_F can generate all the probabilities in $\mathcal{S}$, then estimates $\hat{\beta}_1^*$ will be consistent for β_1. The authors present sufficient conditions that G_F must satisfy to generate all the probabilities in $\mathcal{S}$ and conjecture that any continuous distribution, and the standard normal in particular, would satisfy the conditions. Neuhaus et al. (1994) also show that when G_F satisfies the sufficient conditions for consistent estimation, that $\hat{\beta}_1^*$ is the standard conditional likelihood estimate (7.35) for binary matched pairs data. Thus, as with log link models, fitting logit link models to binary matched pairs data under the assumption that the random intercepts are normally distributed yields a consistent estimate of the regression coefficient of interest even when this assumption is incorrect.

– iv. *Logit link, arbitrary covariates*

In this section we assume that we have a single, arbitrary covariate, X and that under the true model, the responses, Y, follow a mixed-effects logistic model of form

$$\operatorname{logit} P\{y_{ij} = 1 | b_i\} = \beta_0 + \beta_1 x_{ij} + \sigma_b b_i \qquad (12.31)$$

for $j = 1, \ldots, n_i$, where $b_i \sim G_T$ and x_{ij} is a single, arbitrary covariate which, without loss of generality, we can assume has expectation zero. We fit a model of form (12.31) but assume that $b_i \sim G_F$ with $G_F \neq G_T$. Differentiation of the Kullback-Leibler divergence in this case yields

$$\mathrm{E}_x \sum_y \lambda(\mathbf{y}|\mathbf{x}) \int \left\{ \sum_i (y_i - p_i^*) \prod_j (p_j^*)^{y_j} (q_j^*)^{1-y_j} \right\} dG_F(b) \;=\; 0$$

$$(12.32)$$

$$\mathrm{E}_x \sum_y \lambda(\mathbf{y}|\mathbf{x}) \int \left\{ \sum_i x_i (y_i - p_i^*) \prod_j (p_j^*)^{y_j} (q_j^*)^{1-y_j} \right\} dG_F(b) \;=\; 0$$

$$(12.33)$$

$$E_x \sum_y \lambda(\mathbf{y}|\mathbf{x}) \int \left\{ b \sum_i (y_i - p_i^*) \prod_j (p_j^*)^{y_j} (q_j^*)^{1-y_j} \right\} dG_F(b) = 0,$$

$$(12.34)$$

where $\lambda(\mathbf{y}|\mathbf{x}) = P_{G_T}\{\mathbf{y}|\mathbf{x}\}/P_{G_F}\{\mathbf{y}|\mathbf{x}\}$ and logit $p_j = \beta_0 + \beta_1 x_j + \sigma_b b$, logit $p_j^* = \beta_0^* + \beta_1^* x_j + \sigma_{b^*} b$, $q_j = 1 - p_j$ and $q_j^* = 1 - p_j^*$. Again, values $(\beta_0^*, \beta_1^*, \sigma_{b^*})$ that yield $\lambda(\mathbf{y}|\mathbf{x}) = 1$ for all $\mathbf{x}$ would solve (12.32)–(12.34) since in this case the system would be exactly the system of equations obtained when the true model has b_i distributed as G_F.

We will typically be unable to find an analytic, closed-form solution to the system (12.32)–(12.34) since neither analytic expressions for the integrals nor closed-form expressions for minimizing values will typically exist. However, given specified distributions G_T and G_F, we can numerically solve (12.32)–(12.34) using numerical integration and minimization methods.

Neuhaus et al. (1992) show that $\beta_1^* = 0$ when $\beta_1 = 0$. This is easy to see since $\lambda(\mathbf{y}|\mathbf{x})$, p_i^* and q_i^* are all functions that are independent of $\mathbf{x}$ when $\beta_1^* = \beta_1 = 0$ and since $E[x_{ij}] = 0$, it follows that the left hand side of (12.32) is zero for all $\beta_0, \sigma_b, \beta_0^*, \sigma_{b^*}$. Equations (12.32) and (12.34) determine β_0^*, σ_{b^*} in terms of β_0, σ_b. Thus, when $\beta_1 = 0$, estimates $\hat{\beta}_1^*$ based on a likelihood with a misspecified random effects distribution consistently estimate zero.

Neuhaus et al. (1992) use the result of consistent estimation when $\beta_1 = 0$ as the basis for an approximate solution of the system (12.32)–(12.34). Expanding (12.34) in a Taylor series about $\beta_1^* = \beta_1 = 0$ and approximating $E[p^r q^{n-r}]$ in powers of $E[pq]$, $E[p]$ and $E[q]$, they derive the following approximate solution to the system (12.32)–(12.34):

$$\beta_0[1 - \rho_{G_T}(0)] + \text{logit}\,(E_{G_T}[p_0]) = \beta_0^*[1 - \rho_{G_F}(0)] + \text{logit}\,(E_{G_F}[p_0])$$

$$(12.35)$$

$$\beta_1[1 - \rho_{G_T}(\beta_0)] = \beta_1^*[1 - \rho_{G_F}(\beta_0^*)] \qquad (12.36)$$

$$\rho_{G_T}(\beta_0) = \rho_{G_F}(\beta_0^*), \qquad (12.37)$$

where $\rho_{G_T}(\beta_0) = \text{corr}(y_{ij}, Y_{ij'}|\beta_1 = 0; \beta_0, \sigma_b)$ assuming that $b \sim G_T$ while $\rho_{G_F}(\beta_0^*) = \text{corr}(y_{ij}, y_{ij'}|\beta_1^* = 0; \beta_0^*, \sigma_{b^*})$ assuming that $b \sim G_F$. The approximate solution (12.35)–(12.37) indicates that there will be little bias in $\hat{\beta}_1^*$ since the intracluster correlation induced by the mis-specified model is approximately equal to the true intracluster correlation.

Neuhaus et al. (1992) conducted a set of simulation studies that show the approximate solution (12.35)–(12.37) closely corresponds to observed values and indicate little bias in estimates of regression coefficients obtained from likelihoods based on misspecified random effects distributions. We conducted some additional simulation studies and present the results here.

In these simulations, we generated clustered binary data according to the following mixed-effects model with two covariates

$$\text{logit}\,(P\{y_{ij} = 1|b_i, x_{ij}, z_i\}) = \beta_0 + \sigma_b b_i + \beta_1 x_{ij} + \beta_2 z_i, \qquad (12.38)$$

using several random effect distributions with varying degrees of skewness and kurtosis. To these data, we fit mixed-effects logistic models of form (12.38) which assume that the random effects distribution is normal. Each generated dataset consisted of 100 clusters of size 5. We used five different random effects distributions for the b_i, each standardized to have mean 0 and variance 1: gamma distributions with scale parameter 1 and shape parameters 0.5 and 16, t distributions with 3 and 5 degrees of freedom and a normal distribution. Thus, except for the last case, the fitted model misspecified the random effects distribution. The intercept β_0 was -2, while the variance of the random effects, σ_b^2 was 4. The true regression coefficients were $\beta_1=0.5$ and $\beta_2=1$. We generated the covariates x and z as independent standard normal random variables. Table 12.2 reports the results.

As suggested by the approximate solution, Table 12.2 indicates there was little bias in the estimation of the regression coefficients β_1 and β_2, even when the random effects density was highly skewed ($\Gamma[.5]$). The simulation standard errors of the estimated regression coefficients are all less than 3 percent of the true regression coefficient values. Also, as expected, there is little bias in the estimation of the intercept β_0 for the symmetric random effects distributions and large bias for the highly skewed $\Gamma(.5)$. The estimates of the standard deviation of the random effects, σ_b, exhibit large biases.

b. Correlation of covariates and random effects

This section considers the setting where the random effects in a GLMM are correlated with one of the covariates. Ignoring this correlation rep-

Table 12.2: Logistic-normal models fit to mixed-effects data generated from different mixture models with $\sigma_b = 2$ and fixed cluster size $=5$.

a. Bias of estimators (percent)

Mixture	β_0 Observed	$\log(\sigma_b)$ Observed	β_1 Observed	β_2 Observed
Normal	1	-3	2	3
t_5	3	-12	3	3
t_3	3	-33	2	1
$\Gamma(16,1)$	8	8	4	4
$\Gamma(0.5,1)$	30	19	4	8

b. Bias of estimated variances (percent)

Mixture	β_0 Observed	$\log(\sigma_b)$ Observed	β_1 Observed	β_2 Observed
Normal	2	-1	-1	3
t_5	-2	-8	2	-7
t_3	2	-11	0	1
$\Gamma(16,1)$	2	-4	0	-2
$\Gamma(0.5,1)$	6	-10	0	0

resents a misspecification of the random effects distribution and we will see that this misspecification can produce seriously biased estimates of the regression parameters of a GLMM, unlike the misspecification settings in Section 12.3a, where the random effects and covariates are uncorrelated.

We begin our investigation with a simple linear mixed model without an intercept

$$y_{ij} = b_i + \beta x_{ij} + e_{ij} \tag{12.39}$$

for $i = 1, \ldots, m; j = 1, \ldots, n$ where $x_{ij} \sim (0, \sigma_x^2)$, $b_i \sim (0, \sigma_b^2)$, $e_{ij} \sim (0, \sigma_e^2)$, b_i is independent of e_{ij}, x_{ij} is independent of e_{ij}, but b_i is associated with x_{ij}.

As in Section 12.2d, we examine the behavior of the ordinary least squares estimator

$$\hat{\beta}_{OLS} = \left(\sum_{i,j} x_{ij}^2 \right)^{-1} \left(\sum_{i,j} x_{ij} y_{ij} \right), \quad (12.40)$$

which is unbiased and consistent as the number of clusters increases under the usual linear mixed model in which the covariates and random effects are uncorrelated as in Chapter 6. However, under model (12.39), it is straightforward to show that (see E 12.9) as $m \to \infty$

$$\hat{\beta}_{OLS} \xrightarrow{pr} \beta + \frac{\sigma_{xb}}{\sigma_x^2}, \quad (12.41)$$

where σ_{xb} is the covariance between b_i and x_{ij} and so $\hat{\beta}_{OLS}$ is biased whenever $\sigma_{xb} \neq 0$. This result makes intuitive sense. Incorrectly assuming independence between covariates and random effects is analogous to mis-modelling covariates in a linear regression model and it is well known that such a misspecification produces biased regression parameter estimates.

When b_i and e_{ij} both follow normal distributions with known σ_b^2 and σ_e^2, the optimal estimator for model (12.39) is the weighted least squares estimator,

$$\hat{\beta}_{GLS} = \left(\sum_i \mathbf{x}_i' \mathbf{V}_i^{-1} \mathbf{x}_i' \right)^{-1} \left(\sum_i \mathbf{x}_i' \mathbf{V}_i^{-1} \mathbf{y}_i \right), \quad (12.42)$$

where $\mathbf{V}_i = \sigma_e^2 \mathbf{I} + \sigma_b^2 \mathbf{J}$, and $\mathbf{x}_i$ and $\mathbf{y}_i$ are the covariate matrix and response vector for the ith cluster. Again it is straightforward (see E 12.10) to derive the probability limit under equal cluster sizes:

$$\hat{\beta}_{GLS} \xrightarrow{pr} \beta + \frac{\sigma_{xb}}{\sigma_x^2} \left(\frac{\sigma_e^2}{\sigma_e^2 + (n-1)\sigma_b^2} \right). \quad (12.43)$$

Like (12.41), (12.43) shows that ignoring correlations between random effects and covariates leads to biased estimates of β. However, unlike the ordinary least squares estimator, the bias of the weighted least squares estimator depends on the cluster size, n. In particular, the bias in $\hat{\beta}_{GLS}$ decreases as n increases.

Simple expressions for bias such as (12.41) and (12.43) do not exist for models such as mixed-effects logistic models since closed-form estimators such as (12.40) and (12.42) do not exist for these models.

However, as in Neuhaus and McCulloch (2006), one can assess the bias resulting from ignoring correlations between random effects and covariates using simulation studies.

Neuhaus and McCulloch (2006) conducted simulation studies using three different binary mixed-effects models: the logistic, probit and complementary log-log models and we summarize the results here. The simulations generated clustered binary responses from generalized linear mixed models where the random intercepts b were correlated with the distribution of the covariate and assessed the effect of ignoring this correlation on parameter estimates. Specifically, the simulations generated the random intercepts from a normal distribution with a mean of zero and variance σ_b^2, independently in different clusters. Five binary, within-cluster covariates, x_{ij}, were then generated from a Bernoulli distribution with parameter p_b where $\text{logit}(p_b) = 1.0 + b$. Finally, given b_i and $x_{i1}, \ldots, x_{i5}$, they generated the binary responses $y_{i1}, \ldots, y_{i5}$ as $n = 5$ conditionally independent draws from a generalized linear mixed model with link function g,

$$g\left(\text{P}\{y_{ij} = 1 \mid b_i, x_{ij}\}\right) = \beta_0 + b_i + \beta_1 x_{ij},$$

where g is the link function for the logistic, probit and complementary log-log models. Each simulation generated 1000 data sets, each with $m = 100$ clusters. The mixed-effects model parameter values were $\beta_0 = -1$, $\beta_1 = 1$ and $\sigma_b^2 = 1$.

To these data the authors fit standard binary mixed-effects models that ignored correlations between covariates and random effects. Table 12.3 presents the results. As with the linear mixed model, Table 12.3 shows that ignoring correlations between random effects and covariates can produce highly biased estimates of regression parameters and other parameters of binary mixed-effects models. The large biases $\hat{\beta}_1$ in Table 12.3 contrast with small biases in Table 12.2 where model misspecification did not involve x.

c. Covariate-dependent random effects variance

Heagerty and Kurland (2001) considered settings where the variance of the random effects depends on covariates but the fitted model ignores this feature. Specifically, these authors focused on settings where the true model for clustered binary responses was a mixed-effects logistic

Table 12.3: Observed means and standard deviations of the regression coefficients obtained by fitting standard binary mixed-effects models that ignored correlations between the random effects and the covariate. True values: $\beta_0 = -1$, $\beta_1 = 1$, $\sigma_b^2 = 1$ (log $\sigma_b = 0$).

Link function	β_0	β_1	log σ_b
Logistic	-1.38	1.55	-0.14
	(0.18)	(0.20)	(0.16)
Probit	-1.26	1.37	-0.11
	(0.14)	(0.15)	(0.11)
Complementary	-1.29	1.41	-0.11
log-log	(0.15)	(0.16)	(0.12)

model of form

$$\text{logit}\left(P\{y_{ij} = 1 | x_{i,1}, x_{ij,2}, b_i\}\right) = \beta_0 + \beta_1 x_{i,1} + \beta_2 x_{ij,2} + \beta_3 x_{i,1} x_{ij,2} + b_i,$$
$$(12.44)$$

where $x_{i,1}$ is a binary, cluster-level covariate, $x_{ij,2}$ varies within clusters and $\text{var}(b_i | x_{i,1} = k) = \sigma_k^2$ for $k = 0, 1$ with $\sigma_0^2 \neq \sigma_1^2$. The fitted model was also of form (12.44), but misspecified the distribution of b by assuming $\sigma_0^2 = \sigma_1^2$.

Heagerty and Kurland (2001) assess the magnitude of the bias resulting from incorrectly assuming $\sigma_0^2 = \sigma_1^2$ using the work of White (1994), as in Section 12.3a-ii, and numerical methods to calculate the parameter values that minimize the Kullback-Leibler divergence between the true and fitted models. These calculations indicate that incorrect assumptions about the variance of the random effects can produce highly biased estimates of all parameters in model, (12.44), and that bias increased as the discrepancy between σ_0^2 and σ_1^2 increased. The biases were largest for $\hat{\beta}_1$ and $\hat{\beta}_3$, the regression coefficients associated with $x_{i,1}$ and smallest for $\hat{\beta}_2$, regression coefficient of the within-cluster covariate. Heagerty and Kurland (2001) note that one could consistently estimate β_2 using a conditional likelihood approach that eliminates the b_i from the likelihood rather than integrate them out. We discuss this approach further in Section 12.4c.

12.4 DIAGNOSING MISSPECIFICATION

a. Conditional likelihood methods

This section considers canonical link models, such as the linear mixed effects model for normal responses and mixed-effects logistic model for binary responses, and models whose only random effects are random intercepts, b_i. Rather than integrating the random effects out of the model as in (12.21), we can estimate the effects of within-cluster covariates of canonical link models such as (12.31) using a conditional likelihood that eliminates the random effects from the model. The usual conditional likelihood approach treats the random intercepts as fixed parameters and conditions them away using sufficient statistics for the cluster-specific intercepts. Sartori and Severini (2004) show that a similar interpretation is possible using the integrated, marginal likelihood.

For canonical link models such as identity and logistic link models the sufficient statistics are $\sum_{j=1}^{n_i} y_{ij}$ and the conditional likelihood has terms

$$ f\left(y_{i1}, \ldots, y_{in_i} | x_{i1}, \ldots, x_{in_i}, n_i, b_i, \sum_{j=1}^{n_i} y_{ij} \right). \tag{12.45} $$

For example, the conditional likelihood for a logistic model has terms

$$ \frac{\exp \sum_{j=1}^{n_i} y_{ij} x_{ij} \beta_1}{\sum_L \exp \sum_{k \in L} x_{ik} \beta_1} \tag{12.46} $$

where L ranges over all the $\binom{n_i}{S_i}$ possible subsets of S_i elements from n_i and $\sum_{j=1}^{n_i} y_{ij} = S_i$.

As another example, the contribution from the ith cluster to the conditional likelihood for the linear-normal response model is

$$ f\left(y_i | x_i, b_i, \sum_j y_{ij} \right) = \frac{1}{\sqrt{n_i}} \left\{ \frac{1}{\sqrt{2\pi}\sigma_e} \right\}^{n_i - 1} e^{-\frac{\sum_{j=1}^{n_i} [y_{ij} - \bar{y}_i - \beta_1 (x_{ij} - \bar{x}_{i.})]^2}{2\sigma_e^2}}. $$

$$ \tag{12.47} $$

Calculation of the conditional likelihood based on (12.45),(12.46) or (12.47) shows that it depends on the covariates x_{ij} only through the differences $(x_{ij} - x_{in_i})$ or equivalently through the deviations $(x_{ij} - \bar{x}_{i.})$, where $\bar{x}_{i.} = (1/n_i) \sum_{j=1}^{n_i} x_{ij}$. This suggests a strong connection between conditional likelihood estimates and estimates from GLMMs that model covariate effects in terms of these differences or deviations.

Conditional likelihood approaches can provide consistent estimates of the effects of within-cluster covariates in settings where the random effects are correlated with covariates. To see this, consider the case where the mean of the covariates is dependent on some function, v, of the random effect b_i :

$$x_{ij} = v(b_i) + \eta_{ij}, \text{ with } \eta_{ij} \text{ independent of } b_i. \tag{12.48}$$

Under this model note that $x_{ij} - \bar{x}_{i\cdot} = \eta_{ij} - \bar{\eta}_{i\cdot}$, which is independent of b_i. Next we can rewrite the linear predictor as

$$
\begin{aligned}
g(\mathrm{E}[y_{ij} \mid b_i, x_{ij}]) &= \beta_0 + b_i + \beta_1 x_{ij} \\[2mm]
&= \beta_0 + b_i + \beta_1 \bar{x}_{i\cdot} + \beta_1(x_{ij} - \bar{x}_{i\cdot}) \\[2mm]
&= \beta_0 + b_i + \beta_1[v(b_i) + \bar{\eta}_{i\cdot}] + \beta_1(x_{ij} - \bar{x}_{i\cdot}) \\[2mm]
&= \beta_0 + (b_i + \beta_1[v(b_i) + \bar{\eta}_{i\cdot}]) + \beta_1(x_{ij} - \bar{x}_{i\cdot}) \\[2mm]
&\equiv \beta_0 + d_i + \beta_1(x_{ij} - \bar{x}_{i\cdot}), \tag{12.49}
\end{aligned}
$$

where $x_{ij} - \bar{x}_{i\cdot}$ is uncorrelated with $d_i = b_i + \beta_1[v(b_i) + \bar{\eta}_{i\cdot}]$, since $x_{ij} - \bar{x}_{i\cdot}$ is independent of b_i and uncorrelated with $\bar{\eta}_{i\cdot}$. We can view d_i as a new random effect and eliminate it from the likelihood using standard conditional likelihood methods. This conditional likelihood approach makes no assumptions about the distribution of the d_i, and will give consistent estimates of β_1, under standard regularity conditions (Andersen, 1970), when the predictor is $x_{ij} - \bar{x}_{i\cdot}$.

Since conditional likelihood approaches require no assumptions about the distribution of the random effects, we can use discrepancies between conditional likelihood estimators and estimators from other approaches to identify possible model misspecifications. For example, (12.49) shows that conditional likelihood approaches provide consistent estimates of covariate effects when the covariates are correlated with random effects. Table 12.3 shows that a standard mixed-effects logistic model that ignores such correlations would yield highly biased covariate effect estimates which should greatly differ from conditional likelihood estimates. Tchetgen and Coull (2006) formalize this diagnostic approach as a specification test which we discuss further in Section 12.4c.

b.   Between/within cluster covariate decompositions

The fact that the conditional likelihoods (12.45), (12.46) and (12.47) depend on the covariates x_{ij} only through the deviations $(x_{ij} - \bar{x}_{i\cdot})$ suggests a strong connection between conditional likelihood estimates and estimates from GLMMs that model covariate effects in terms of these deviations. Such models replace $\beta_1 x_{ij}$ in the linear predictor by

$$\beta_B \bar{x}_i \ + \ \beta_W (x_{ij} - \bar{x}_{i\cdot}) \, , \tag{12.50}$$

so that β_B measures the change in $\mathrm{E}[y_{ij}]$ associated with between-cluster differences in covariate means while β_W measures the change in $\mathrm{E}[y_{ij}]$ associated with within-cluster covariate differences. The connection between covariate decomposition and conditional likelihood approaches suggests that GLMMs that include separate between- and within-cluster covariate components can provide conditional likelihood-like inference even for non-canonical link models that do not support conditional likelihood methods.

We examine this issue further by considering additional consequences of expression (12.49). First, suppose that the mean of x_{ij} is a linear function of b_i, namely $v(b_i) = \delta_0 + \delta_1 b_i$ then

$$d_i = b_i + \beta_1 \delta_0 + \beta_1 \delta_1 b_i + \beta_1 \bar{\eta}_{i\cdot} = \beta_1 \delta_0 + (1 + \beta_1 \delta_1) b_i + \beta_1 \bar{\eta}_{i\cdot} \, . \tag{12.51}$$

If, further, $\bar{\eta}_{i\cdot}$ has a normal distribution then the distribution of the d_i is normal. If the assumed random effects distribution in a naive analysis assuming no correlation between the covariates and random effects is also normal then the distribution of the d_i is *not* misspecified and consistent estimators will result when using the predictor $x_{ij} - \bar{x}_{i\cdot}$. Note that this result is not dependent on having a canonical link. Second, suppose that the function v transforms the b_i in such a way that the distribution of the d_i is close to normal, and if the assumed random effects distribution is normal, then the effect of the correlation of the random effects and covariates should be very mild when using the predictor $x_{ij} - \bar{x}_{i\cdot}$.

Neuhaus and McCulloch (2006) further examine the performance of conditional likelihood and covariate decomposition approaches by including them in the simulations described in Section 12.3b and Table 12.3. That is, to clustered binary responses generated from mixed-effects logistic, probit and complementary log-log regression models with random intercepts correlated with the covariate, the authors fit

Table 12.4: Observed means and standard deviations of the regression coefficients obtained by fitting binary mixed-effects models which decompose the covariate into between- and within-cluster components, (12.50), to clustered binary data with correlations between the random effects and the covariate. True values: $\beta_0 = -1$, $\beta_1 = 1$, $\sigma_b^2 = 1$ (log $\sigma_b = 0$).

Link function	β_0	β_W	β_B	log σ_b
Logistic	−2.85	1.01	3.63	−0.36
	(0.31)	(0.20)	(0.41)	(0.20)
Probit	−2.88	1.02	3.66	−0.33
	(0.29)	(0.15)	(0.38)	(0.12)
Complementary	−2.84	1.00	3.61	−0.34
log-log	(0.28)	(0.16)	(0.36)	(0.13)

mixed-effects models that decomposed the covariate into between- and within-cluster components (12.50). In the mixed-effects logistic simulations the authors also fit the standard conditional likelihood approach (12.46). The conditional likelihood approach provided highly accurate estimates of the covariate effect; the mean of the conditional likelihood estimates was 1.01 with a standard deviation of 0.21 while the true value β_1 was 1. Table 12.4 presents the results from fitting the binary mixed-effects models with the covariate decomposed into between- and within-cluster components (12.50).

Although the means of the between-cluster estimators $\hat{\beta}_B$ are highly biased, the means of the within-cluster estimators $\hat{\beta}_W$ have essentially no bias for all three links. As with the conditional likelihood approach, decomposing the covariate into between- and within-cluster components corrects the bias of standard mixed-model approaches and provides unbiased estimates of covariates in settings with correlation between the random effects and covariates. Thus, GLMMs with separate between- and within-cluster covariate effects can provide conditional likelihood-like analysis for models with non-canonical links.

c. Specification tests

Specification tests are formal procedures to compare a consistent estimator $\hat{\theta}_1$ to another estimator $\hat{\theta}_2$ which is consistent and fully efficient under correct model specification (Hausman, 1978). The null hypothesis of the test is that the model that produces $\hat{\theta}_2$ is correctly specified and large discrepancies between $\hat{\theta}_1$ and $\hat{\theta}_2$ lead to small significance probabilities, suggesting model misspecification.

Tchetgen and Coull (2006) adapted the work of Hausman (1978) to develop specification tests to examine the appropriateness of the assumed random effects distribution in generalized linear models with canonical links. The consistent estimator $\hat{\theta}_1$ is the conditional likelihood estimator $\hat{\beta}_{CML}$, while $\hat{\theta}_2$ is the estimated regression coefficient, $\hat{\beta}_G$, obtained from fitting a canonical link generalized linear model that assumed the distribution of the random intercepts to be G. The null hypothesis is $H_0 : G$ is correctly specified. Since conditional likelihood estimators only estimate the effects of covariates that vary within clusters, the components of $\hat{\beta}_G$ also only involve such covariates.

The test statistic involves $\hat{\delta} = \hat{\beta}_{CML} - \hat{\beta}_G$ along with an estimate $\hat{\Sigma}_\delta$ of its asymptotic variance Σ_δ. Under H_0, $D = m\hat{\delta}'\hat{\Sigma}_\delta^{-1}\hat{\delta}$ is asymptotically χ_q where q is dimension of $\hat{\beta}_{CML}$.

d. Nonparametric maximum likelihood

Motivated by the potential consequences of misspecifying the distribution of the random effects, G, several authors have proposed leaving G unspecified and jointly estimating the regression parameters of models such as (12.21) and the nonparametric distribution G. Methods to implement this approach have been proposed by Laird (1978), Follman and Lambert (1989), Lesperance and Kalbfleisch (1992), Lindsay and Grego (1991), Butler and Louis (1997) and Wang (2007). This approach provides consistent estimation of the effects of all covariates and G under conditions of identifiability (Kiefer and Wolfowitz, 1956). However, unless there is a great deal of data, G cannot be at all precisely estimated. In practice, estimates of G tend to be highly discrete with only a few support points. In addition, general properties of estimation within this approach are not known. Nonetheless, the work cited above suggests that this approach yields valid and fairly efficient estimates of the regression coefficients.

One can use the fit of a semi-parametric mixed-effects model as a diagnostic to detect model misspecification. Large discrepancies between features of a semi-parametric and parametrically-specified mixed-effects model, such as large differences between the estimated regression parameters of the two approaches, indicate a problem with the specification of the parametric model. Rather than fit a completely unspecified distribution for the random effects, one could fit a parametric distribution that provides more flexibility than the typically fitted normal distribution but does not require the estimation of a potentially large number of parameters. For example, Piepho and McCulloch (2004) suggest use of the Johnson family of distributions for the random effects distribution, Zhang and Davidian (2001) fit a smooth density, and Magder and Zeger (1996) suggest mixtures of normal densities. One could also model the random effects distribution using the Tukey *gh* (Hoaglin, 1985) family. While these more flexible approaches have a number of advantages, they also have disadvantages. These include:

1. The distributions are usually highly parameterized, which can be statistically inefficient when they are not needed and often lead to numerical difficulties in fitting the distributions.

2. The flexible form often makes it hard to test for model improvement. For example, in the finite mixtures of normals approach, it is a long-standing and difficult statistical problem to test for the need for and decide on the order of mixtures of distributions.

3. The more flexible distributions often fail to capture in a simple way things like familial aggregation of disease; simpler models often describe features of the random effects distribution much more compactly.

4. The more flexible distributions are often more complicated to fit, requiring custom software which may not have all the features of commercial software, such as the availability of best predicted values.

5. The flexibility of the distribution may preclude fitting more complicated models with multiple and/or crossed random effects.

6. The more flexible fits often fail to diagnose the nature of the deviation from the assumed model in a readily interpretable way.

12.5   A SUMMARY OF RESULTS

The results in this chapter indicate that the magnitude of the effects of departures from standard GLMM assumptions varies by the type and severity of the misspecification. In particular, misspecifications unrelated to the covariate of interest have little effect. For example, misspecifying the shape of the random effects distribution produces little bias in the regression coefficients of interest. Similarly, informative cluster sizes or omitting covariates uncorrelated with those included in linear mixed-effects models yields no appreciable bias. However, misspecifications involving the covariate of interest can produce significant bias. Specifically, incorrectly assuming that the random effects are independent of the covariates yields highly biased regression coefficients and unknown direction of the bias. We can correct this bias with canonical link GLMMs using conditional likelihood methods or using GLMMs that partition covariates into between- and within-cluster components, for any GLMM.

Several misspecifications produce attenuated estimates of covariate effects, that is, estimates of covariate effects that are smaller in absolute value. Knowing the direction of the bias provides a bound for the true covariate effect. For example, standard analysis of clustered binary responses subject to misclassification can produce attenuated covariate effect estimates. Thus, in such a setting, we can infer that the true covariate effect is larger than the one we observed.

We can diagnose model misspecification by comparing the fits of standard models with alternative models with include more flexible specifications of covariate effects and random effects distributions. In particular, a useful diagnostic is to fit GLMMs that partition covariates into between- and within-cluster components. Large discrepancies between the coefficient of the within-cluster covariate component and the corresponding coefficient of the standard model, as well as large discrepancies between the coefficients of the between- and within-cluster components suggest misspecification.

12.6 EXERCISES

E 12.1 Consider an experiment in which there are two groups of subjects, randomized to one of two treatments. Prove that the covariate representing treatment assignment is independent of all other covariates (omitted or not).

E 12.2 Show that the function $H(t) = -1/h'(t)$ of Case 4 in Section 12.2a is convex for the logistic, probit and complementary log-log links.

E 12.3 Show that the function g^* of (12.8) satisfies that equation and is a link function.

E 12.4 Derive equation (12.11).

E 12.5 Derive equation (12.12).

E 12.6 Derive equation (12.14).

E 12.7 Show that equations (12.27) reduce to (12.28) for normally distributed random effects.

E 12.8 (a) Derive the expressions for β_0^* and σ_{b*}^2 in (12.29) under the assumption that $\beta_1^* = \beta_1$.

(b) Show that no solution equating (12.28) to equations (12.27) exists assuming $\beta_0^* = \beta_0$ or $\sigma_{b*}^2 = \sigma_b^2$.

E 12.9 Derive equation (12.41).

E 12.10 Derive equation (12.43).

Chapter 13

PREDICTION

13.1   INTRODUCTION

Earlier chapters contain results for estimation known as predicting random effects. Section 1.2 describes what is meant by a random effect; in being a random variable, it has mean and second moments, properties of which are shown in Section 1.4 for traditional LMMs. Some real-life examples of random effects are described in Section 1.5. How one decides whether effects are fixed or random is discussed in Section 1.6, a decision tree in Section 1.7 helps with doing this, and Section 1.7c has a brief discussion of predicting random effects. It is this brevity which we now expand upon. In doing so we provide underlying methodology for expressions given earlier for predicted values of random effects in a variety of models. These predictors can be found in equations (2.56), (2.90), (2.103), (3.92), and (6.42), and in Section 6.6a.

We begin by presenting three different but interrelated methods of prediction. In doing so we make numerous references to Searle et al. (1992). Hereinafter, for the sake of brevity, it is referred to as VC (from its title: *Variance Components*). Its Chapter 7 has the extensive mathematical detail supporting much of what we present here.

We deal with the general case of $\mathbf{y}$ being available data, in the model for which $\mathbf{u}$ is a vector of random effects. First and second moments of $\mathbf{u}$ and $\mathbf{y}$ are defined by

$$
\begin{bmatrix} \mathbf{u} \\ \mathbf{y} \end{bmatrix} \sim \left(\begin{bmatrix} \boldsymbol{\mu}_u \\ \boldsymbol{\mu}_y \end{bmatrix} \begin{bmatrix} \mathbf{D} & \mathbf{C} \\ \mathbf{C'} & \mathbf{V} \end{bmatrix} \right) \tag{13.1}
$$

so that $E[\mathbf{u}] = \boldsymbol{\mu}_u$, $E[\mathbf{y}] = \boldsymbol{\mu}_y$ and

$$\mathbf{D} = \mathrm{var}(\mathbf{u}), \quad \mathbf{C} = \mathrm{cov}(\mathbf{u}, \mathbf{y}') \quad \text{and} \quad \mathbf{V} = \mathrm{var}(\mathbf{y}). \tag{13.2}$$

13.2   BEST PREDICTION (BP)

When $f(u, \mathbf{y})$ is the joint density function of $\mathbf{y}$ and scalar u then, with the predictor of u being denoted by $\tilde{u}$, the mean squared error of prediction is

$$E[\tilde{u} - u]^2 = \int \int (\tilde{u} - u)^2 f(u, \mathbf{y}) \, d\mathbf{y} \, du. \tag{13.3}$$

A generalization of this to a vector of random variables $\mathbf{u}$ is

$$E[(\tilde{\mathbf{u}} - \mathbf{u})' \mathbf{A} (\tilde{\mathbf{u}} - \mathbf{u})] = \int \int (\tilde{\mathbf{u}} - \mathbf{u})' \mathbf{A} (\tilde{\mathbf{u}} - \mathbf{u}) f(\mathbf{u}, \mathbf{y}) d\mathbf{y} \, d\mathbf{u}, \tag{13.4}$$

where $\mathbf{A}$ is a positive definite symmetric matrix. Clearly, for $\mathbf{A}$ being scalar and unity, (13.4) is identical to (13.3).

a.   The best predictor

Our criterion for deriving a predictor is minimum mean square, i.e., we minimize (13.4). The result is what we call the *best predictor*. Note that "best" here means minimum mean squared error of prediction, which is different from a common meaning of "best" being minimum variance. Because variance is variability around a fixed value and because u in (13.3) is a random variable, (13.3) is not the definition of the variance of u. Thus, for estimating a parameter we use the criterion of minimum variance unbiased, while for predicting the realized value of a random variable we use the criterion of minimum mean square error. Thus, as shown in VC p.262, from minimizing (13.4) we get

$$\text{best predictor: } \tilde{\mathbf{u}} = \mathrm{BP}(\mathbf{u}) = E[\mathbf{u}|\mathbf{y}], \tag{13.5}$$

that is, the best predictor of $\mathbf{u}$ is the conditional mean of $\mathbf{u}$ given $\mathbf{y}$. Details are given in Section 13.8.

   Noteworthy features of this result are: (i) it holds for all probability density functions $f(\mathbf{u}, \mathbf{y})$ and (ii) it does not depend on the positive definite symmetric matrix $\mathbf{A}$.

b. Mean and variance properties

First and second moments of the best predictor are important. They are discussed in Cochran (1951) and in Rao (1965, pp. 79 and 220–222) for the case of scalar u. First, the best predictor is unbiased for sampling over $\mathbf{y}$:

$$E_\mathbf{y}[\tilde{\mathbf{u}}] = E_\mathbf{y}\left[E_{\mathbf{u}|\mathbf{y}}[\mathbf{u}|\mathbf{y}]\right] = E[\mathbf{u}], \qquad (13.6)$$

as detailed in Section S.1 of VC. Note that the meaning of the unbiasedness here is that the expected value of the predictor equals that of the random variable for which it is a predictor. This differs from the usual meaning of unbiasedness when estimating a parameter. In that case unbiasedness means that the expected value of (estimator minus parameter) is zero; for example, $E[\hat{\beta} - \beta] = \mathbf{0}$, where β is a constant. With prediction, unbiasedness means that the expected value of (predictor minus random variable) is zero; for example, $E[\tilde{\mathbf{u}} - \mathbf{u}] = \mathbf{0}$ where $\mathbf{u}$ is a random variable. The former gives $E[\hat{\beta}] = \beta$, whereas the latter gives $E[\tilde{\mathbf{u}}] = E[\mathbf{u}]$.

Second, prediction errors $\tilde{\mathbf{u}} - \mathbf{u}$ have a variance-covariance matrix that is the mean value, over sampling on $\mathbf{y}$, of that of $\mathbf{u}|\mathbf{y}$:

$$\mathrm{var}(\tilde{\mathbf{u}} - \mathbf{u}) = E_\mathbf{y}\left[\mathrm{var}(\mathbf{u}|\mathbf{y})\right]. \qquad (13.7)$$

Also,

$$\mathrm{cov}(\tilde{\mathbf{u}}, \mathbf{u}') = \mathrm{var}(\tilde{\mathbf{u}}) \qquad \text{and} \qquad \mathrm{cov}(\tilde{\mathbf{u}}, \mathbf{y}') = \mathrm{cov}(\mathbf{u}, \mathbf{y}'). \qquad (13.8)$$

c. A correlation property

For scalar u there are two further properties of interest. The first is that the correlation between u and any predictor of it that is a function of $\mathbf{y}$ is maximum for the best predictor, that maximum value being

$$\rho(\tilde{u}, u) = \sigma_{\tilde{u}}/\sigma_u. \qquad (13.9)$$

A proof of (13.9) following that of Rao (1973, pp. 265–266) is given in VC, p. 263.

d. Maximizing a mean

A second property of interest concerns the mean of a selected upper fraction of a population of random effects. Making this selection on the basis of values of $\tilde{u}$ ensures that

$$\text{for that selected fraction, } E[u] \text{ is maximized.} \tag{13.10}$$

This, too, has a proof available in VC at pp. 264–265.

e.   Normality

It is to be emphasized that $\tilde{u} = E[u|y]$ is a random variable, being a function of y and unknown parameters. Thus the problem of estimating the best predictor $\tilde{u}$ remains, and demands some knowledge of the joint density $f(u|y)$. Should this be normal,

$$\begin{bmatrix} u \\ y \end{bmatrix} \sim \mathcal{N}\left(\begin{bmatrix} \mu_u \\ \mu_y \end{bmatrix}, \begin{bmatrix} D & C \\ C' & V \end{bmatrix} \right), \tag{13.11}$$

using Section S.2b gives

$$\tilde{u} = E[u|y] = \mu_u + CV^{-1}(y - \mu_y). \tag{13.12}$$

Properties (13.7) through (13.10) of $\tilde{u}$ still hold. In (13.7) we now have from (13.11) that $\text{var}(u|y) = D - CV^{-1}C'$, so that in (13.7)

$$\text{var}(\tilde{u} - u) = D - CV^{-1}C'. \tag{13.13}$$

And using (13.12) in (13.8) gives

$$\text{cov}(\tilde{u}, u') = \text{var}(\tilde{u}) = CV^{-1}C', \quad \text{and hence} \quad \rho(\tilde{u}_i, u_i) = \sqrt{\frac{c_i'V^{-1}c_i}{\sigma_{u_i}^2}},$$
$$\tag{13.14}$$

where c_i' is the ith row of C.

An estimation problem is clearly visible in these results. The predictor is given in (13.12) but it and its succeeding properties cannot be estimated without having values for, or estimating, the four parameters μ_u, μ_y, C and V.

13.3   BEST LINEAR PREDICTION (BLP)

a.   BLP(u)

The best predictor (13.5) is not necessarily linear in y. Suppose attention is now confined to predictors of u that *are* linear in y, of the form

$$\tilde{u} = a + By \tag{13.15}$$

for some vector $\mathbf{a}$ and matrix $\mathbf{B}$. Minimizing (13.4) for $\tilde{\mathbf{u}}$ of (13.15), in order to obtain the best linear predictor, leads (without any assumption of normality) to

$$\text{BLP}(\mathbf{u}) = \tilde{\mathbf{u}} = \boldsymbol{\mu}_u + \mathbf{C}\mathbf{V}^{-1}(\mathbf{y} - \boldsymbol{\mu}_y), \qquad (13.16)$$

where $\boldsymbol{\mu}_u$, $\boldsymbol{\mu}_y$, $\mathbf{C}$ and $\mathbf{V}$ are as defined in (13.11) but without assuming normality as there. Not only do (13.5) and (13.16) demand no assumption of normality, but additionally important is the fact that they also apply no matter what form $\boldsymbol{\mu}_u$ and $\boldsymbol{\mu}_y$ have. Equations (13.5) and (13.16) apply for all forms of those means.

An immediate observation on (13.16) is that it is identical to (13.12). This shows that the best linear predictor (13.16), derivation of which demands no knowledge of the form of $f(\mathbf{u}, \mathbf{y})$, is *identical* to the best predictor under normality, (13.12). Properties (13.13) and (13.14) therefore apply equally to (13.16) as to (13.12). And, of course, $\text{BLP}(\mathbf{u})$ is unbiased, in the sense described following (13.6), namely that $\text{E}[\tilde{\mathbf{u}}] = \text{E}[\mathbf{u}]$. Problems of estimation of the unknown parameters in (13.16) still remain.

b. Example

To illustrate (13.16) we use the beta-binomial model of Section 2.6b wherein we have

$$\text{E}[y_{ij}|p_i] = p_i$$

$$y_{ij}|p_i \sim \text{Bernoulli}(p_i)$$

and

$$\text{E}[p_i|\bar{y}_{i\cdot}] = \frac{\alpha}{\alpha + \beta}$$

$$\text{var}(p_i|\bar{y}_{i\cdot}) = \frac{\alpha\beta}{(\alpha + \beta)^2(\alpha + \beta + 1)};$$

also

$$\text{E}[y_{ij}] = \frac{\alpha}{\alpha + \beta}$$

and

$$\text{var}(y_{ij}) = \frac{\alpha\beta}{(\alpha + \beta)^2}.$$

Therefore using (13.16)

$$\text{BLP}(p_i) = \text{E}[p_i] + \text{cov}(\bar{y}_{i\cdot}, p_i) \left[\text{var}(\bar{y}_{i\cdot})\right]^{-1} \left[\bar{y}_{i\cdot} - \text{E}[\bar{y}_{i\cdot}]\right].$$

To derive $\text{cov}(\bar{y}_{i\cdot}, p_i)$ we adapt (2.75) as follows:

$$\begin{aligned}
\text{cov}(\bar{y}_{i\cdot}, p_i) &= \text{cov}\left(\text{E}[\bar{y}_{i\cdot}|p_i], \text{E}[p_i|p_i]\right) + \text{E}\left[\text{cov}(\bar{y}_{i\cdot}, p_i|p_i)\right] \\
&= \text{cov}(p_i, p_i) + 0 \\
&= \text{var}(p_i).
\end{aligned}$$

Therefore,

$$\begin{aligned}
\text{BLP}(p_i) &= \frac{\alpha}{\alpha + \beta} \\
&\quad + \frac{\alpha\beta}{(\alpha+\beta)^2(\alpha+\beta+1)} \left[\frac{\alpha\beta}{(\alpha+\beta)^2}\right]^{-1} \left(\bar{y}_{i\cdot} - \frac{\alpha}{\alpha+\beta}\right) \\
&= \frac{\alpha + \bar{y}_{i\cdot}}{\alpha + \beta + 1}.
\end{aligned}$$

c.   Derivation

Since we want a predictor $\tilde{\mathbf{u}}$ to be linear (in $\mathbf{y}$) we take $\tilde{\mathbf{u}} = \mathbf{a} + \mathbf{By}$ and proceed to derive $\mathbf{a}$ and $\mathbf{B}$ so that $\tilde{\mathbf{u}} = \mathbf{a} + \mathbf{By}$ is best, meaning that it has minimum mean squared error of prediction. Thus we want to minimize the left-hand side of (13.4), which now gets written as

$$\begin{aligned}
q &= \text{E}\left[(\mathbf{a} + \mathbf{By} - \mathbf{u})'\mathbf{A}(\mathbf{a} + \mathbf{By} - \mathbf{u})\right] \qquad\qquad (13.17) \\
&= \text{E}\left[\mathbf{a}'\mathbf{A}\mathbf{a} + 2\mathbf{a}'\mathbf{A}(\mathbf{By} - \mathbf{u}) + (\mathbf{By} - \mathbf{u})'\mathbf{A}(\mathbf{By} - \mathbf{u})\right].
\end{aligned}$$

Equating $\partial q/\partial \mathbf{a}$ to $\mathbf{0}$ gives

$$2\mathbf{A}(\mathbf{a} + \text{E}[\mathbf{By} - \mathbf{u}]) = \mathbf{0}$$

and so

$$\mathbf{a} = -\text{E}[\mathbf{By} - \mathbf{u}] = -(\mathbf{B}\boldsymbol{\mu}_y - \boldsymbol{\mu}_u).$$

Substituting **a** into (13.18) gives

$$q = -E[\mathbf{By} - \mathbf{u}]'\mathbf{A}E[\mathbf{By} - \mathbf{u}] + E\left[(\mathbf{By} - \mathbf{u})'\mathbf{A}(\mathbf{By} - \mathbf{u})\right]$$

$$= \mathrm{tr}\{\mathbf{A}\,\mathrm{var}(\mathbf{By} - \mathbf{u})\}, \qquad \text{based on the normality,}$$

$$= \mathrm{tr}\{\mathbf{A}(\mathbf{BVB}' + \mathbf{D} - \mathbf{BC}' - \mathbf{CB}')\}, \tag{13.18}$$

from (13.11) and the end of Section S.2. We wish to minimize this with respect to **B**. To do this, ignore **A** and **D** (because they do not involve **B**), and define $\mathbf{b}_i'$ and $\mathbf{c}_j'$ as the ith and jth rows of **B** and **C**, respectively. Then the (i, j)th element of $\mathbf{BVB}' - \mathbf{BC}' - \mathbf{CB}'$ is

$$\varphi_{ij} = \mathbf{b}_i'\mathbf{Vb}_j - \mathbf{b}_i'\mathbf{c}_j - \mathbf{c}_i'\mathbf{b}_j.$$

Thus to minimize this element with respect to $\mathbf{b}_i'$ and $\mathbf{b}_j'$

$$\frac{\partial \varphi_{ij}}{\partial \mathbf{b}_i} = \mathbf{Vb}_j - \mathbf{c}_j \qquad \text{and} \qquad \frac{\partial \varphi_{ij}}{\partial \mathbf{b}_j} = \mathbf{Vb}_i - \mathbf{c}_i.$$

Therefore we take the minimizing form of **B** as $\mathbf{VB}' = \mathbf{C}'$ and so $\mathbf{B} = \mathbf{CV}^{-1}$. Thus

$$\mathrm{BLP}(\mathbf{u}) = \mathbf{a} + \mathbf{By} = \boldsymbol{\mu}_u + \mathbf{B}(\mathbf{y} - \boldsymbol{\mu}_y)$$

$$= \boldsymbol{\mu}_u + \mathbf{CV}^{-1}(\mathbf{y} - \boldsymbol{\mu}_y). \tag{13.19}$$

d. Ranking

In establishing, as observed in (13.10), that selection on the basis of the best predictor $\tilde{u}$ maximizes $E[u]$ of the selected proportion of the population, Cochran's (1951) development implicitly relies on each scalar $\tilde{u}$ having the same variance and being derived from a **y** that is independent of other **y**s. Sampling is over repeated samples of u (scalar) and **y**. However, these conditions are not met for the elements of $\tilde{\mathbf{u}}$ derived in (13.12). Each such element is derived from the whole vector **y**, their variances are not equal, and the elements of **y** used in one element of $\tilde{\mathbf{u}}$ are not necessarily independent of those used for another element of $\tilde{\mathbf{u}}$. Maximizing the probability of correctly ranking individuals on the basis of elements in $\tilde{\mathbf{u}}$ is therefore not assured. In place of this there is a property about pairwise ranking.

Having predicted the (unobservable) realized values of the random variables in the data, a salient problem that is often of great importance is this: How does the ranking on predicted values compare with the ranking on the true (realized but unobservable) values? Henderson (1963) and Searle (1974) show, under certain conditions (including normality), that the probability that predictors of u_i and u_j have the same pairwise ranking as u_i and u_j is maximized when those predictors are elements of BLP($\mathbf{u}$) of (13.19); that is, the probability $P\{\tilde{u}_i - \tilde{u}_j \gtrless 0 | u_i - u_j \gtrless 0\}$ is maximized. Portnoy (1982) extends this to a special components-of-variance model, for which he shows that ranking all the u_is of $\mathbf{u}$ in the same order as the $\tilde{u}_j$ (the best linear predictors) rank themselves does maximize the probability of ranking the u_is correctly. He does, however, go on to show that in models more general than variance components models, there can be predictors that lead to higher values of this probability than do the best linear predictors, which are elements of the vector BLP($\mathbf{u}$) $= \boldsymbol{\mu}_u + \mathbf{CV}^{-1}(\mathbf{y} - \boldsymbol{\mu}_y)$.

13.4   LINEAR MIXED MODEL PREDICTION (BLUP)

The preceding discussion is concerned with the prediction of random variables. Through maximizing the probability of correct ranking, the predictors are appropriate values upon which to base selection; for example, in genetics, selecting the animals with highest predictions to be parents of the next generation. Consideration is now given to linear mixed model prediction, corresponding to mixed models in which some factors are fixed and others are random.

a.   BLUE($\mathbf{X}\boldsymbol{\beta}$)

In Section 6.3 we derived at equation (6.20)

$$\text{ML}(\mathbf{X}\boldsymbol{\beta}) = \mathbf{X}\boldsymbol{\beta}^0 = \mathbf{X}(\mathbf{X}'\mathbf{V}^{-1}\mathbf{X})^-\mathbf{X}'\mathbf{V}^{-1}\mathbf{y}. \qquad (13.20)$$

By concentrating attention on estimating $\mathbf{X}\boldsymbol{\beta}$ rather than $\boldsymbol{\beta}$ we achieved the invariance of $\mathbf{X}\boldsymbol{\beta}^0$ to $(\mathbf{X}'\mathbf{V}^{-1}\mathbf{X})^-$ compared to the non-invariance of $\boldsymbol{\beta}^0$. Deriving ML($\mathbf{X}\boldsymbol{\beta}$) of (13.20) relied upon assuming normality of the data vector $\mathbf{y}$. But that same estimator $\mathbf{X}\boldsymbol{\beta}^0$ can also be derived, without requiring normality, as the best linear unbiased estimator (BLUE) of $\mathbf{X}\boldsymbol{\beta}$:

$$\text{BLUE}(\mathbf{X}\boldsymbol{\beta}) = \mathbf{X}\boldsymbol{\beta}^0. \qquad (13.21)$$

Best in this context means that of all linear (in $\mathbf{y}$) unbiased estimators of $\mathbf{X}\boldsymbol{\beta}$, the "best", that is, the BLUE($\mathbf{X}\boldsymbol{\beta}$), is the one with the smallest variance. This is established by taking the estimator to have the form $\boldsymbol{\lambda}'\mathbf{y}$ and deriving $\boldsymbol{\lambda}$ so that $\boldsymbol{\lambda}'\mathbf{y}$ is both unbiased for $\mathbf{t}'\mathbf{X}\boldsymbol{\beta}$ (for given $\mathbf{t}'$) and also so that var($\boldsymbol{\lambda}'\mathbf{y}$) is minimized. With $2\mathbf{m}$ being a vector of Lagrange multipliers, this leads to wanting

$$\mathbf{X}'\boldsymbol{\lambda} = \mathbf{X}'\mathbf{t} \tag{13.22}$$

and needing to minimize

$$\boldsymbol{\lambda}'\mathbf{V}\boldsymbol{\lambda} + 2\mathbf{m}'(\mathbf{X}'\boldsymbol{\lambda} - \mathbf{X}'\mathbf{t}) \tag{13.23}$$

with respect to $\boldsymbol{\lambda}$. The resulting equations for $\boldsymbol{\lambda}$ and $\mathbf{m}$ (the equation for $\mathbf{m}$ being of little interest) are

$$\mathbf{V}\boldsymbol{\lambda} + \mathbf{X}\mathbf{m} = \mathbf{0} \tag{13.24}$$

$$\mathbf{X}'\boldsymbol{\lambda} = \mathbf{X}'\mathbf{t} \tag{13.25}$$

or, in matrix form

$$\begin{bmatrix} \mathbf{V} & \mathbf{X} \\ \mathbf{X}' & \mathbf{0} \end{bmatrix} \begin{bmatrix} \boldsymbol{\lambda} \\ \mathbf{m} \end{bmatrix} = \begin{bmatrix} \mathbf{0} \\ (\mathbf{t}'\mathbf{X})' \end{bmatrix}, \tag{13.26}$$

an example of equation (1) of Hayes and Haslett (1999). The solution for $\boldsymbol{\lambda}$ is what leads to BLUE($\mathbf{X}\boldsymbol{\beta}$) $= \mathbf{X}\boldsymbol{\beta}^0$ of (13.21).

b. BLUP($\mathbf{t}'\mathbf{X}\boldsymbol{\beta} + \mathbf{s}'\mathbf{u}$)

The counterpart of BLUE($\mathbf{t}'\mathbf{X}\boldsymbol{\beta}$) in an LM or GLM is BLUP($\mathbf{t}'\mathbf{X}\boldsymbol{\beta} + \mathbf{s}'\mathbf{u}$) in an LMM or GLMM where $\mathbf{u}$ is a vector of random effects and $\mathbf{t}'$ and $\mathbf{s}'$ are known vectors. And BLUP is *best linear unbiased predictor*.

Akin to the derivation of BLUE, we seek $\boldsymbol{\lambda}$ for $\boldsymbol{\lambda}'\mathbf{y}$ to be unbiased for $\mathbf{t}'\mathbf{X}\boldsymbol{\beta} + \mathbf{s}'\mathbf{u}$; and in taking E($\mathbf{u}$) $= \mathbf{0}$, which is customary, this unbiasedness leads, just as in (13.22), to

$$\mathbf{X}'\boldsymbol{\lambda} = \mathbf{X}'\mathbf{t}.$$

We additionally seek $\boldsymbol{\lambda}$ so as to minimize the variance of the prediction error of $[\boldsymbol{\lambda}'\mathbf{y} - (\mathbf{t}'\mathbf{X}\boldsymbol{\beta} + \mathbf{s}'\mathbf{u})]$. This is done by minimizing

$$\begin{aligned} \theta &= \text{var}\left[\boldsymbol{\lambda}'\mathbf{y} - (\mathbf{t}'\mathbf{X}\boldsymbol{\beta} + \mathbf{s}'\mathbf{u})\right] + 2\mathbf{m}'(\mathbf{X}'\boldsymbol{\lambda} - \mathbf{X}'\mathbf{t}) \\ &= \boldsymbol{\lambda}'\mathbf{V}\boldsymbol{\lambda} + \mathbf{s}'\mathbf{D}\mathbf{s} - 2\boldsymbol{\lambda}'\mathbf{C}\mathbf{s} + 2\mathbf{m}'(\mathbf{X}'\boldsymbol{\lambda} - \mathbf{X}'\mathbf{t}) \end{aligned}$$

with respect to λ and $\mathbf{m}$. This gives equations

$$\begin{bmatrix} \mathbf{V} & \mathbf{X} \\ \mathbf{X}' & \mathbf{0} \end{bmatrix} \begin{bmatrix} \lambda \\ \mathbf{m} \end{bmatrix} = \begin{bmatrix} \mathbf{Cs} \\ (\mathbf{t}'\mathbf{X})' \end{bmatrix}. \tag{13.27}$$

Solving for λ yields $\lambda'\mathbf{y}$ as

$$\mathrm{BLUP}(\mathbf{t}'\mathbf{X}\beta + \mathbf{s}'\mathbf{u}) = \mathbf{t}'\mathbf{X}\beta^0 + \mathbf{s}'\mathbf{C}'\mathbf{V}^{-1}(\mathbf{y} - \mathbf{X}\beta^0), \tag{13.28}$$

where $\mathbf{X}\beta^0$ is the same $\mathrm{BLUE}(\mathbf{X}\beta)$ of (13.21).

Special cases of (13.28) are (i) for $\mathbf{s} = \mathbf{0}$ and $\mathbf{t}'$ taking successive rows of $\mathbf{I}$, then $\mathrm{BLUP}(\mathbf{X}\beta) = \mathbf{X}\beta^0$; and (ii) for $\mathbf{t}' = \mathbf{0}$ and $\mathbf{s}'$ being successive rows of $\mathbf{I}$, $\mathrm{BLUP}(\mathbf{u}) = \mathbf{C}'\mathbf{V}^{-1}(\mathbf{y} - \mathbf{X}\beta^0)$. And a very familiar special case is that of $\mathbf{C} = \mathbf{ZD}$, giving $\mathrm{BLUP}(\mathbf{u}) = \mathbf{DZ}'\mathbf{V}^{-1}(\mathbf{y} - \mathbf{X}\beta^0)$ of (6.43).

c.   Two variances

The variance and covariances of some eight variants of (13.28) are given in VC p.272. We show just two of them here.

Let

$$w \;=\; \mathbf{t}'\mathbf{X}\beta + \mathbf{s}'\mathbf{u}$$

with

$$\tilde{w} \;=\; \mathbf{t}'\mathbf{X}\beta^0 + \mathbf{s}'\mathbf{C}'\mathbf{V}^{-1}(\mathbf{y} - \mathbf{X}\beta^0).$$

Then for

$$\mathbf{P} \;=\; \mathbf{V}^{-1} - \mathbf{V}^{-1}\mathbf{X}(\mathbf{X}'\mathbf{V}^{-1}\mathbf{X})^{-}\mathbf{X}'\mathbf{V}^{-1} \tag{13.29}$$

$$\mathrm{var}(\tilde{w}) \;=\; \mathbf{t}'\mathbf{X}(\mathbf{X}'\mathbf{V}^{-1}\mathbf{X})^{-}\mathbf{X}'\mathbf{t} + \mathbf{s}'\mathbf{CPC}'\mathbf{s} \tag{13.30}$$

$$\mathrm{var}(\tilde{w} - w) \;=\; \mathrm{var}(\tilde{w}) - 2\mathbf{t}'\mathbf{X}(\mathbf{X}'\mathbf{V}^{-1}\mathbf{X})^{-}\mathbf{X}'\mathbf{V}^{-1}\mathbf{C}'\mathbf{s}. \tag{13.31}$$

d.   Other derivations

Although Henderson had developed and used (13.28) in dairy cow selection programs prior to publication in Henderson (1963), in the statistical literature an early version can be found in Goldberger (1962). His equation (3.12) has $\mathbf{X}$ of full column rank, and with his $\Omega = \mathbf{V}$, $\mathbf{x}'_* = \mathbf{t}'\mathbf{X}$ and with his $\mathbf{w}' = \mathbf{s}'\mathbf{C}$, his (3.12) is a special case of (13.28).

Of the numerous ways for deriving (13.28), five are detailed in VC, pp. 271–275. One is primarily algebraic as in the derivation of (13.18); another simplifies $E[\mathbf{u}|\mathbf{y}_c]$ for $\mathbf{y}_c$ being $\mathbf{y}$ corrected for fixed effects $\boldsymbol{\beta}$; a third starts with assuming that $\mathbf{w}$ is linear in $\mathbf{y}$, such as $\mathbf{a} + \mathbf{By}$, another procedure is based on partitioning $\mathbf{y}$ as $(\mathbf{y} - \mathbf{X}\boldsymbol{\beta}^0)$ and $\mathbf{X}\boldsymbol{\beta}^0$; and finally there is a Bayes method.

13.5   REQUIRED ASSUMPTIONS

It is interesting to note that BP, BLP and BLUP do not all require the same assumptions. In some general sense BP requires more assumptions than BLP which in turn requires more than BLUP. For $BP(\mathbf{u}) = E[\mathbf{u}|\mathbf{y}]$ one needs to know the distribution of $\mathbf{u}|\mathbf{y}$. But BLP demands knowing only first and second moments of $\mathbf{u}$ and $\mathbf{y}$. BLUP requires knowing only $\mathbf{V} = \text{var}(\mathbf{y})$ and $\mathbf{C} = \text{cov}(\mathbf{u}, \mathbf{y}')$, but first moments are not needed, with fixed effects $\boldsymbol{\beta}$ being estimated through using $\mathbf{X}\boldsymbol{\beta}^0 = \mathbf{X}(\mathbf{X}'\mathbf{V}^{-1}\mathbf{X})^-\mathbf{X}'\mathbf{V}^{-1}\mathbf{y}$. In not one of BP, BLP or BLUP is normality needed.

13.6   ESTIMATED BEST PREDICTION

An expression for $BP(\mathbf{u})$ more usable in practice than $E[\mathbf{u}|\mathbf{y}]$ demands knowing the distribution of $\mathbf{y}$ and $\mathbf{u}$. And even then, numerical values are needed for the parameters of that distribution. This need is clearly demonstrated by considering

$$BLP(\mathbf{u}) = \boldsymbol{\mu}_u + \mathbf{CV}^{-1}(\mathbf{y} - \boldsymbol{\mu}_y). \tag{13.32}$$

In most practical situations in which one wants to use (13.32), numerical values for $\boldsymbol{\mu}_u$, $\mathbf{C}$, $\mathbf{V}$ and $\boldsymbol{\mu}_y$ will not be available. So in order to use (13.32) estimates of these parameters need to be found. Often this entails deriving such estimates from the data being used to obtain estimates of (13.32).

Just as one can simultaneously and optimally estimate μ and σ^2 from data $\mathbf{x} \sim \mathcal{N}(\mu\mathbf{1}, \sigma^2\mathbf{I})$, one would ideally like to optimally estimate $\tilde{\mathbf{u}}$ of (13.32) along with $\boldsymbol{\mu}_u$, $\mathbf{C}$, $\mathbf{V}$ and $\boldsymbol{\mu}_y$. But this is seldom (if ever) feasible. The usual procedure, therefore, is to estimate $\boldsymbol{\mu}_u$, $\mathbf{C}$, $\mathbf{V}$ and $\boldsymbol{\mu}_y$ and replace those parameters in (13.32) by their estimates. No matter what estimation methods are used for the parameters, denote the resulting estimates by $\hat{\boldsymbol{\mu}}_u$, $\hat{\mathbf{C}}$, $\hat{\mathbf{V}}$ and $\hat{\boldsymbol{\mu}}_y$. Then in denoting $BLP(\mathbf{u})$

as $\tilde{\mathbf{u}}$,

$$\text{BLP}(\mathbf{u}) \equiv \tilde{\mathbf{u}} = \boldsymbol{\mu}_u + \mathbf{C}\mathbf{V}^{-1}(\mathbf{y} - \boldsymbol{\mu}_y)$$

we can have a calculated value

$$\hat{\mathbf{u}} = \hat{\boldsymbol{\mu}}_u + \hat{\mathbf{C}}\hat{\mathbf{V}}^{-1}(\mathbf{y} - \hat{\boldsymbol{\mu}}_y)$$

as an estimated BLP($\mathbf{u}$). Similarly for

$$\text{BLUP}(\mathbf{u}) \quad = \quad \mathbf{u}^0 = \boldsymbol{\mu}_u + \mathbf{C}\mathbf{V}^{-1}(\mathbf{y} - \mathbf{X}\boldsymbol{\beta}^0),$$

$$\text{EBLUP}(\mathbf{u}) \quad = \quad \hat{\mathbf{u}}^0 = \hat{\boldsymbol{\mu}}_u + \hat{\mathbf{C}}\hat{\mathbf{V}}^{-1}(\mathbf{y} - \mathbf{X}\hat{\boldsymbol{\beta}}^0) \qquad (13.33)$$

is an estimated BLUP, with

$$\mathbf{X}\hat{\boldsymbol{\beta}}^0 = \mathbf{X}(\mathbf{X}'\hat{\mathbf{V}}^{-1}\mathbf{X})^{-}\mathbf{X}'\hat{\mathbf{V}}^{-1}\mathbf{y}.$$

Note that these estimated predictors have been derived from a purely practical viewpoint: Estimate parameters of the distribution of $\mathbf{u}$ and $\mathbf{y}$ and in the predictors simply replace the parameters by those estimates. In doing this no statistical rationale such as minimum variance has been invoked. The estimated predictors have just been set up in what seems like an "obvious" manner. Nevertheless, $\hat{\mathbf{u}}^0$ and $\mathbf{X}\hat{\boldsymbol{\beta}}^0$ are (with $\hat{\mathbf{V}}$ being the MLE of $\mathbf{V}$) ML estimators of $\mathbf{u}^0$ and $\mathbf{X}\boldsymbol{\beta}^0$, respectively. Their properties, such as mean and variance (let alone distribution) are largely intractable. This is so if for no other reason that, if the estimated parameters are based on $\mathbf{y}$, then the parameter estimates are correlated, not only with each other but also with $\mathbf{y}$. And these correlations will have to be taken into account when seeking moments of the estimated predicted values. And doing that is not easy. The best that has been done in the research literature so far is various attempts at developing approximations. Some of those results are dealt with in Chapter 6. Analogous results for generalized linear models, or nonlinear models would be even more complicated than the results of Chapter 6. Kackar and Harville (1984) and Prasad and Rao (1990) consider some of these complications for the estimated BLUP($\mathbf{u}$), namely $\hat{\mathbf{u}}^0$ of (13.33).

13.7   HENDERSON'S MIXED MODEL EQUATIONS

a.   Origin

A set of equations developed by Henderson in Henderson et al. (1959), which simultaneously yield BLUE($\mathbf{X}\boldsymbol{\beta}$) and BLUP($\mathbf{u}$) in the LMM with model equation $E(\mathbf{y}|\mathbf{u}) = \mathbf{X}\beta + \mathbf{Zu}$, have come to be known as the *mixed model equations* (MMEs). For the joint density of $\mathbf{u}$ and $\mathbf{y}$ being normal, as in (13.11), with $\mathbf{C} = \mathbf{DZ'}$, $\text{var}(\mathbf{y}|\mathbf{u}) = \mathbf{R}$, and $\mathbf{V} = \mathbf{ZDZ'} + \mathbf{R}$, this density is

$$f(\mathbf{y}, \mathbf{u}) = f(\mathbf{y}|\mathbf{u})f(\mathbf{u}) \tag{13.34}$$

$$= \frac{\exp\{-\frac{1}{2}[(\mathbf{y} - \mathbf{X}\beta - \mathbf{Zu})'\mathbf{R}^{-1}(\mathbf{y} - \mathbf{X}\beta - \mathbf{Zu}) + \mathbf{u}'\mathbf{D}^{-1}\mathbf{u}]\}}{(2\pi)^{\frac{1}{2}(N+q)}|\mathbf{R}|^{\frac{1}{2}}|\mathbf{D}|^{\frac{1}{2}}},$$

where q is the number of columns in $\mathbf{Z}$ (i.e., the number of random effects in the model for the data).

Henderson's approach was to maximize (13.34) with respect to β and $\mathbf{u}$, which results in

$$\begin{bmatrix} \mathbf{X}'\mathbf{R}^{-1}\mathbf{X} & \mathbf{X}'\mathbf{R}^{-1}\mathbf{Z} \\ \mathbf{Z}'\mathbf{R}^{-1}\mathbf{X} & \mathbf{Z}'\mathbf{R}^{-1}\mathbf{Z} + \mathbf{D}^{-1} \end{bmatrix} \begin{bmatrix} \tilde{\beta} \\ \tilde{\mathbf{u}} \end{bmatrix} = \begin{bmatrix} \mathbf{X}'\mathbf{R}^{-1}\mathbf{y} \\ \mathbf{Z}'\mathbf{R}^{-1}\mathbf{y} \end{bmatrix}. \tag{13.35}$$

These are the MMEs. Their form is worthy of note: Without the $\mathbf{D}^{-1}$ in the lower right-hand submatrix of the matrix on the left, they would be the ML equations for the model treated as if $\mathbf{u}$ represented fixed effects, rather than random effects.

b. Solutions

After a minor amount of algebra (see E 13.6) it will be found that the solutions to (13.35) are

$$\tilde{\beta} = \beta^0 = (\mathbf{X}'\mathbf{V}^{-1}\mathbf{X})^{-}\mathbf{X}'\mathbf{V}^{-1}\mathbf{y}$$

and

$$\tilde{\mathbf{u}} = \mathbf{u}^0 = \text{BLUP}(\mathbf{u}) = \mathbf{DZ}'\mathbf{V}^{-1}(\mathbf{y} - \mathbf{X}\beta^0).$$

The MMEs not only represent a procedure for calculating a β^0 and $\tilde{\mathbf{u}}$, but are also computationally more economical than the ML equations that lead to $\mathbf{X}\beta^0$. Those equations required inversion of $\mathbf{V}$ of order N. But the MMEs need inversion of a matrix of order only $p + q$, the total number of levels of fixed and random effects in the data. And this number is usually much smaller than N, the number of observations. True, the MMEs do require inversion of both $\mathbf{R}$ and $\mathbf{D}$, but for variance components linear models these are often diagonal, which makes these inversions easy.

c.   Use in ML estimation of variance components

An interesting feature of the MMEs is that parts of them can be used for setting up iterative procedures for calculating ML and REML estimates of variance components in variance components models. Derivation of these iterative procedures is shown in great detail in VC, pp. 277–285. Unfortunately the detailed algebra for these derivations is tedious and lengthy, and so is not given here. We refer the interested reader to VC. We present here the main results from that reference.

– i.   ML estimation

Define

$$\mathbf{W} \;=\; (\mathbf{I} + \mathbf{Z}'\mathbf{R}^{-1}\mathbf{Z}\mathbf{D})^{-1} = \{\mathbf{W}_{ij}\}_{i,j=1}^{r};$$

$$q_i \;=\; \text{number of levels of the } i\text{th random effect};$$

and

$$\sigma_i^{2(m)} \;=\; \text{the calculated value of } \sigma_i^2 \text{ after the } m\text{th round of iteration}.$$

The superscript parenthesized m is used throughout to indicate iteration number. Then for $i = 1, \ldots, r$ the ML equations (6.60) can be reduced to the iterations

$$\sigma_i^{2(m+1)} \;=\; \frac{\tilde{\mathbf{u}}_i^{\prime(m)}\tilde{\mathbf{u}}_i^{(m)} + \sigma_i^{2(m)}\,\text{tr}\left[\mathbf{W}_{ii}^{(m)}\right]}{q_i} \tag{13.36}$$

or

$$\sigma_i^{2(m+1)} \;=\; \frac{\tilde{\mathbf{u}}_i^{\prime(m)}\tilde{\mathbf{u}}_i^{\prime(m)}}{q_i - \text{tr}\left[\mathbf{W}_{ii}^{(m)}\right]}, \tag{13.37}$$

each along with

$$\sigma_i^{2(m+1)} \;=\; \mathbf{y}'\left[(\mathbf{y} - \mathbf{X}\boldsymbol{\beta}^{0(m)} - \mathbf{Z}\tilde{\mathbf{u}}^{(m)})\right]/N. \tag{13.38}$$

– ii.   *REML estimation*

The iterative procedure is essentially the same for REML as it is for ML but with the following changes. Define

$$\mathbf{S} = \mathbf{R}^{-1} - \mathbf{R}^{-1}\mathbf{X}(\mathbf{X}'\mathbf{R}^{-1}\mathbf{X})^{-}\mathbf{X}'\mathbf{R}^{-1} \qquad (13.39)$$

and use $\mathbf{S}$ instead of $\mathbf{R}^{-1}$ in $\mathbf{W}$. Then, with N-rank($\mathbf{X}$) replacing N as the denominator of (13.38), (13.36) through (13.38) are an iterative procedure for REML estimation of variance components. Making these changes in (13.36), for example, gives

$$\sigma_{i(\text{REML})}^{2(m+1)} = \frac{\tilde{\mathbf{u}}_i^{\prime(m)}\tilde{\mathbf{u}}_i^{(m)} + \sigma_{i(\text{REML})}^{2(m)}\operatorname{tr}\left[\mathbf{W}_{ii(\text{REML})}^{(m)}\right]}{q_i}$$

for $\mathbf{W}_{ii(\text{REML})}$ being the ith diagonal submatrix of

$$\mathbf{W}_{(\text{REML})} = (\mathbf{I} + \mathbf{Z}'\mathbf{SZD})^{-1} = \left\{\mathbf{W}_{ij(\text{REML})}\right\}_{i,j=1}^{r}.$$

VC, pp. 277–285 not only gives details of deriving the preceding results, but also derives the information matrix for both the ML and REML procedures.

13.8   APPENDIX

a.   Verification of (13.5)

In the mean square on the left-hand side of (13.4), to $\tilde{\mathbf{u}} - \mathbf{u}$ add and subtract $\mathrm{E}[\mathbf{u}|\mathbf{y}]$, which, for convenience, will be denoted by $\mathbf{u}_0$; that is, with

$$\mathbf{u}_0 \equiv \mathrm{E}[\mathbf{u}|\mathbf{y}],$$

$$\mathrm{E}[(\tilde{\mathbf{u}}-\mathbf{u})'\mathbf{A}(\tilde{\mathbf{u}}-\mathbf{u})] = \mathrm{E}[(\tilde{\mathbf{u}}-\mathbf{u}_0+\mathbf{u}_0-\mathbf{u})'\mathbf{A}(\tilde{\mathbf{u}}-\mathbf{u}_0+\mathbf{u}_0-\mathbf{u})]. \quad (13.40)$$

To choose a $\tilde{\mathbf{u}}$ that minimizes (13.40), note that in expanding it, the last term, $\mathrm{E}[(\mathbf{u}_0-\mathbf{u})'\mathbf{A}(\mathbf{u}_0-\mathbf{u})]$, does not involve $\tilde{\mathbf{u}}$. And in the cross-product term, using iterated expectation with E_y representing expectation with respect to the distribution of y,

$$\mathrm{E}[(\tilde{\mathbf{u}} - \mathbf{u}_0)'\mathbf{A}(\mathbf{u}_0 - \mathbf{u})] = \mathrm{E}_y\left[\mathrm{E}_{u|y}[(\tilde{\mathbf{u}} - \mathbf{u}_0)'\mathbf{A}(\mathbf{u}_0 - \mathbf{u})|\mathbf{y}]\right]$$

$$= \mathrm{E}_y[(\tilde{\mathbf{u}} - \mathbf{u}_0)'\mathbf{A}(\mathbf{u}_0 - \mathbf{u}_0)] = 0$$

since, for a given $\mathbf{y}$, only $\mathbf{u}$ is not fixed and has $E_{u|y}[\mathbf{u}|\mathbf{y}] = \mathbf{u}_0$. Therefore

$$E[(\tilde{\mathbf{u}} - \mathbf{u})'\mathbf{A}(\tilde{\mathbf{u}} - \mathbf{u})] = E[(\tilde{\mathbf{u}} - \mathbf{u}_0)'\mathbf{A}(\tilde{\mathbf{u}} - \mathbf{u}_0)] + \text{terms without } \tilde{\mathbf{u}}.$$

Since $E[(\tilde{\mathbf{u}} - \mathbf{u}_0)'\mathbf{A}(\tilde{\mathbf{u}} - \mathbf{u}_0)]$ must be non-negative, it is minimized by choosing $\tilde{\mathbf{u}} = \mathbf{u}_0$; i.e., the best predictor is $\tilde{\mathbf{u}} = \mathbf{u}_0 = E[\mathbf{u}|\mathbf{y}]$. Thus the problem of predicting a random variable is simply that of predicting its conditional mean.

b.   Verification of (13.7) and (13.8)

Deriving (13.7) comes from (1.14) with y_{ij} replaced by $\tilde{\mathbf{u}} - \mathbf{u}$ and a_i by $\mathbf{y}$. The two results in (13.8) are established by using (1.16) with $\mathbf{y}$, $\mathbf{w}$ and $\mathbf{u}$ replaced, respectively, by $\tilde{\mathbf{u}}$, $\mathbf{u}'$, and $\mathbf{y}$. This gives

$$\text{cov}(\tilde{\mathbf{u}}, \mathbf{u}') = \text{cov}\left(E[\tilde{\mathbf{u}}|\mathbf{y}], E[\mathbf{u}'|\mathbf{y}]\right) + E_y\left[\text{cov}(\tilde{\mathbf{u}}, \mathbf{u}'|\mathbf{y})\right].$$

The second term here involves the covariance of $\mathbf{u}$ (conditional on $\mathbf{y}$) with its mean $E[\mathbf{u}|\mathbf{y}]$. It is therefore zero. Hence

$$\text{cov}(\tilde{\mathbf{u}}, \mathbf{u}') = \text{cov}(\tilde{\mathbf{u}}, \tilde{\mathbf{u}}') = \text{var}(\tilde{\mathbf{u}}),$$

which is the first result in (13.8). Likewise, for the second result we start with

$$\text{cov}(\mathbf{u}, \mathbf{y}') = \text{cov}\left(E[\mathbf{u}|\mathbf{y}], E[\mathbf{y}'|\mathbf{y}]\right) + E_y\left[\text{cov}(\mathbf{u}, \mathbf{y}'|\mathbf{y})\right].$$

In the second term, the covariance is of $\mathbf{u}$ with $\mathbf{y}'$, which is constant conditional on $\mathbf{y}$. Therefore it is zero and so

$$\text{cov}(\mathbf{u}, \mathbf{y}') = \text{cov}(\tilde{\mathbf{u}}, \mathbf{y}').$$

Thus (13.8) is established.

13.9   EXERCISES

E 13.1 Establish (13.7) and (13.8).

E 13.2 For the random effects one-way classification model with unbalanced data of Section 2.2c, use (13.16) to derive

$$\text{BLP}(a_i) = \frac{n_i\sigma_a^2}{\sigma^2 + n_i\sigma_a^2}(\bar{y}_{i.} - \mu).$$

Under what condition is this the same as $\text{BP}(a_i)$ of Section 2.4b–ii?

E 13.3 Use (13.24) and (13.25) to derive BLUE($\mathbf{X}\boldsymbol{\beta}$).

E 13.4 Derive (13.28) using a partitioned inverse in (13.27).

E 13.5 Derive (13.30) and (13.31).

E 13.6 Derive (13.35) and use it to obtain the solutions in Section 13.7b.

E 13.7 Prove

(a) $\mathbf{V}^{-1} = (\mathbf{ZDZ'} + \mathbf{R})^{-1} = \mathbf{R}^{-1} - \mathbf{R}^{-1}\mathbf{Z}(\mathbf{Z'R}^{-1} + \mathbf{D}^{-1})^{-1}\mathbf{Z'R}^{-1}$.

(b) $\mathbf{DZ'V}^{-1} = (\mathbf{Z'R}^{-1} + \mathbf{D}^{-1})^{-1}\mathbf{Z'R}^{-1}$.

Chapter 14

COMPUTING

14.1 INTRODUCTION

A common theme throughout this book has been the computational difficulties involved in likelihood-based inference. As noted in Chapter 7, computing the likelihood itself is often difficult for GLMMs, requiring the calculation of high-dimensional integrals. In the case of the leaf blight example of Section 7.7, the integral is of more than 200 dimensions. Unfortunately, state-of-the-art statistical software does not include any well-tested and general-purpose routines for performing such calculations. For certain subclasses, for example, linear mixed models, they do exist, but not for the full generality of GLMMs. In this chapter we identify some of the common methods used for likelihood calculation and maximization and briefly describe and give references to some current research topics in computing for GLMMs.

14.2 COMPUTING ML ESTIMATES FOR LMMs

We first consider computing ML estimates for linear mixed models since their structure simplifies the calculations somewhat.

a. The EM algorithm

The EM algorithm (McLachlan and Krishnan, 1996) is an iterative algorithm for calculating ML (or REML) estimates, its name standing for _expectation maximization_: It alternates between calculating conditional _expected_ values and _maximizing_ simplified likelihoods. It only generates parameter estimates and requires extra computations (e.g.,

320

Louis, 1982) for obtaining variance estimates.

The EM algorithm is designed for situations where the recognition or invention of "missing" data simplifies the maximum likelihood calculations. Starting with initial guesses for the model's parameters, the EM algorithm typically fills in the missing data by calculating conditional expected values (given the observed or *incomplete* data $\mathbf{y}$) of the sufficient statistics. The combination of the observed data and the missing data is usually referred to as *complete data*.

For estimating variance components in LMMs with $E[\mathbf{y}|\mathbf{u}] = \mathbf{X}\beta + \sum_i \mathbf{Z}_i \mathbf{u}_i$, the missing data are typically taken to be the realized values of the random effects. Knowledge of the random effects simplifies the calculations from two viewpoints. First, if they were known we could simply estimate $\sigma_i^2 = \text{var}(u_i)$ as:

$$\hat{\sigma}_i^2 = \mathbf{u}_i' \mathbf{u}_i / q_i, \tag{14.1}$$

where q_i is the dimension of $\mathbf{u}_i$. This the ML estimator under the assumption that $\mathbf{u_i} \sim \mathcal{N}(0, \mathbf{I}\sigma_i^2)$. Second, if they were known, we could subtract them from $\mathbf{y}$ leaving the resulting data independent and following a linear model, to which we could apply ordinary least squares:

$$\mathbf{y} - \sum_i \mathbf{Z}_i \mathbf{u}_i \sim \mathcal{N}(bfX\beta, \mathbf{I}\sigma^2). \tag{14.2}$$

All this is very nice, but in reality we do not know the realized values of the $\mathbf{u}_i$. To counter this, the EM algorithm calculates values to use in place of the unknown realized values (the missing data) in order to effect estimation. The conditional expected values of the $\mathbf{u}_i' \mathbf{u}_i$ are used in place of the $\mathbf{u}_i' \mathbf{u}_i$ in (14.1) and the conditional expected values of the $\mathbf{u}_i$ are used in place of the $\mathbf{u}_i$ in (14.2) to form improved estimates of the parameters. This is the maximization step since those equations represent ML estimation from the complete data . The new estimates are then used to recalculate conditional expected values; and so on. This iterative scheme is used until convergence.

Details of applying the preceding ideas to the linear mixed model are given in Searle et al. VC (1992, Sect. 8.3). They result in three procedures, one for ML, a variation of that, and one for REML. The procedure in each case is described by indicating iteratively computed values of (functions of) parameters using parenthesized superscripts. Thus $\sigma_i^{2(m)}$ is the computed value of σ_i^2 after the mth round of iteration; and $\mathbf{V}^{-1(m)}$ is $\mathbf{V}^{-1}$ with σ_i^2 in $\mathbf{V}$ replaced by $\sigma_i^{2(m)}$ for $i = 0, 1, \ldots, r$.

– i. EM for ML

Step 0. Set $m = 0$, and choose starting values $\beta^{(0)}$ and $\sigma_i^{2(0)}$.
Step 1. Calculate

$$\mathbf{X}\beta^{(m+1)} = \mathbf{X}\beta^{(m)} + \sigma_0^{2(m)}\mathbf{X}(\mathbf{X'X})^-\mathbf{X'V}^{-1(m)}(\mathbf{y} - \mathbf{X}\beta^{(m)}). \quad (14.3)$$

and, for $\mathbf{r}^{(m)} = \mathbf{y} - \mathbf{X}\beta^{(m)}$,

$$\sigma_i^{2(m+1)} = \sigma_i^{2(m)} + (\sigma_i^{4(m)}/q_i)[\mathbf{r}^{(m)'}\mathbf{V}^{-1(m)}\mathbf{Z}_i\mathbf{Z}_i'\mathbf{V}^{-1(m)}\mathbf{r}^{(m)}$$

$$- \mathrm{tr}(\mathbf{Z}_i'\mathbf{V}^{-1(m)}\mathbf{Z}_i)]. \quad (14.4)$$

Step 2. If convergence is reached, set $\hat{\sigma}_i^2 = \sigma_i^{(m+1)}$ and $\mathbf{X}\hat{\beta} = \mathbf{X}\beta^{(m+1)}$; otherwise increase m by 1 and return to step 1.

– ii. EM (a variant) for ML

Step 0. Set $m = 0$, and choose starting values $\sigma_i^{2(0)}$.
Step 1. Calculate

$$\sigma_i^{2(m+1)} = \sigma_i^{2(m)} + (\sigma_i^{4(m)}/q_i)[\mathbf{y'P}^{(m)}\mathbf{Z}_i\mathbf{Z}_i'\mathbf{P}^{(m)}\mathbf{y} - \mathrm{tr}(\mathbf{Z}_i'\mathbf{V}^{-1(m)}\mathbf{Z}_i)].$$

$$(14.5)$$

Step 2. If convergence is reached, set $\hat{\sigma}_i^2 = \sigma_i^{(m+1)}$ and then calculate $\mathbf{X}\hat{\beta} = \mathbf{X}(\mathbf{X'V}^{-1(m+1)}\mathbf{X})^-\mathbf{X'V}^{-1(m+1)}\mathbf{y}$; otherwise increase m by 1 and return to step 1.

– iii. EM for REML

Step 0. Set $m = 0$, and choose starting values $\sigma_i^{2(0)}$.
Step 1. Calculate

$$\sigma_i^{2(m+1)} = \sigma_i^{2(m)} + (\sigma_i^{4(m)}/q_i)[\mathbf{y'P}^{(m)}\mathbf{Z}_i\mathbf{Z}_i'\mathbf{P}^{(m)}\mathbf{y} - \mathrm{tr}(\mathbf{Z}_i'\mathbf{P}^{(m)}\mathbf{Z}_i)].$$

$$(14.6)$$

Step 2. If convergence is reached, set $\hat{\sigma}_i^2 = \sigma_i^{(m+1)}$; otherwise increase m by 1 and return to step 1.

b. Using $E[\mathbf{u}|\mathbf{y}]$

The REML equations (6.67) are

$$\left\{ _c \operatorname{tr}(\mathbf{PZ}_i\mathbf{Z}_i') \right\}_{i=0}^{r} = \left\{ _c \mathbf{y}'\mathbf{PZ}_i\mathbf{Z}_i'\mathbf{Py} \right\}_{i=0}^{r}, \tag{14.7}$$

which have to be solved for the σ^2s inherent in $\mathbf{P}$. We now show that those equations can also be written as equating calculated and expected best predicted values:

$$\left\{ _c E[(\tilde{\mathbf{u}}_i^0)'\tilde{\mathbf{u}}_i^0] \right\}_{i=0}^{r} = \left\{ _c (\tilde{\mathbf{u}}_i^0)'\tilde{\mathbf{u}}_i^0 \right\}_{i=0}^{r} \tag{14.8}$$

for

$$\tilde{\mathbf{u}}^0 = \left\{ _c \tilde{\mathbf{u}}_i^0 \right\}_{i=0}^{r} = E[\mathbf{u}|\mathbf{y}]\big|_{\beta=\hat\beta} = \mathbf{DZ}'\mathbf{V}^{-1}(\mathbf{y} - \mathbf{X}\hat\beta) \tag{14.9}$$

of (6.43). First,

$$\tilde{\mathbf{u}}^0 = \mathbf{DZ}'\mathbf{V}^{-1}(\mathbf{y} - \mathbf{X}\hat\beta) = \mathbf{DZ}'\mathbf{Py}$$

$$= \left\{ _d \sigma_i^2\mathbf{I}_{q_i} \right\}\left\{ _c \mathbf{Z}_i' \right\}\mathbf{Py}.$$

Hence

$$\tilde{\mathbf{u}}_i = \sigma_i^2\mathbf{Z}_i'\mathbf{Py} \tag{14.10}$$

and so

$$(\tilde{\mathbf{u}}_i^0)'\tilde{\mathbf{u}}_i^0/\sigma_i^4 = \mathbf{y}'\mathbf{PZ}_i\mathbf{Z}_i'\mathbf{Py}. \tag{14.11}$$

Moreover,

$$E\left[(\tilde{\mathbf{u}}_i^0)'\tilde{\mathbf{u}}_i^0/\sigma_i^4\right] = \operatorname{tr}\left(\mathbf{PZ}_i\mathbf{Z}_i'\mathbf{P}E[\mathbf{yy}']\right)$$

$$= \operatorname{tr}\left[\mathbf{PZ}_i\mathbf{Z}_i'\mathbf{P}(\mathbf{V} + \mathbf{X}\beta\beta'\mathbf{X}')\right]$$

$$= \operatorname{tr}\left[\mathbf{PZ}_i\mathbf{Z}_i'\mathbf{PV}\right] = \operatorname{tr}\left[\mathbf{PVPZ}_i\mathbf{Z}_i'\right]$$

$$= \operatorname{tr}\left[\mathbf{PZ}_i\mathbf{Z}_i'\right]. \tag{14.12}$$

Therefore the REML equations of (14.7) are equivalent to

$$E[(\tilde{\mathbf{u}}_i^0)'\tilde{\mathbf{u}}_i^0] = (\tilde{\mathbf{u}}_i^0)'\tilde{\mathbf{u}}_i^0. \tag{14.13}$$

Moreover, from (14.7), the REML equations can, for $i = 1, 2, \ldots, r$ be written as

$$\sigma_i^4 \text{tr} \left[\mathbf{PZ}_i \mathbf{Z}_i' \right] = \sigma_i^4 \mathbf{y}' \mathbf{PZ}_i \mathbf{Z}_i' \mathbf{Py}$$

$$= (\tilde{\mathbf{u}}_i^0)' \tilde{\mathbf{u}}_i^0 \qquad \text{from (14.11)} \qquad (14.14)$$

or

$$\sigma_i^2 = \frac{(\tilde{\mathbf{u}}_i^0)' \tilde{\mathbf{u}}_i^0}{\sigma_i^2 \text{tr}(\mathbf{PZ}_i \mathbf{Z}_i')}. \qquad (14.15)$$

This suggests a substitution algorithm: starting with initial guesses $\sigma_i^{2(0)}$ find successive iterates of $\sigma_i^2, \sigma_i^{2(m)}$, from

$$\sigma_i^{2(m+1)} = \frac{\tilde{\mathbf{u}}_i^{\prime (m)} \tilde{\mathbf{u}}_i^{(m)}}{\sigma_i^{2(m)} \text{tr}(\mathbf{P}^{(m)} \mathbf{Z}_i \mathbf{Z}_i')}, \qquad (14.16)$$

where

$$\tilde{\mathbf{u}}_i^{(m)} = \sigma_i^{2(m)} \mathbf{Z}_i' \mathbf{P}^{(m)} \mathbf{y} \quad \text{and}$$

$$\mathbf{P}^{(m)} = \mathbf{V}^{-1(m)} - \mathbf{V}^{-1(m)} \mathbf{X} [\mathbf{X}' \mathbf{V}^{-1(m)} \mathbf{X}]^- \mathbf{X}' \mathbf{V}^{-1(m)}.$$

Successive substitutions would be performed in (14.16) until convergence.

c.   Newton-Raphson method

The Newton-Raphson method is an old and celebrated method that can be used for maximization of a nonlinear function. More precisely, it is a root-finding algorithm. Starting from a function $f(\boldsymbol{\theta})$ we wish to find a root of

$$\frac{\partial f(\boldsymbol{\theta})}{\partial \boldsymbol{\theta}} = \mathbf{0}, \qquad (14.17)$$

which we hope is a maximum.

We expand $\partial f(\boldsymbol{\theta}) / \partial \boldsymbol{\theta}$ about $\boldsymbol{\theta}_0$ as

$$\frac{\partial f(\boldsymbol{\theta})}{\partial \boldsymbol{\theta}} = f'(\boldsymbol{\theta}) \approx f'(\boldsymbol{\theta}_0) + \frac{\partial^2 f(\boldsymbol{\theta})}{\partial \boldsymbol{\theta} \, \partial \boldsymbol{\theta}'} (\boldsymbol{\theta} - \boldsymbol{\theta}_0). \qquad (14.18)$$

Equating (14.18) to $\mathbf{0}$, solve for the root as

$$f'(\boldsymbol{\theta}_0) + \frac{\partial^2 f(\boldsymbol{\theta})}{\partial \boldsymbol{\theta} \, \partial \boldsymbol{\theta}'} (\boldsymbol{\theta} - \boldsymbol{\theta}_0) = \mathbf{0}, \qquad (14.19)$$

which gives

$$\boldsymbol{\theta} = \boldsymbol{\theta}_0 - \left[\frac{\partial^2 f(\boldsymbol{\theta})}{\partial \boldsymbol{\theta}\, \partial \boldsymbol{\theta}'} \right]^{-1} f'(\boldsymbol{\theta}_0). \tag{14.20}$$

This can be used iteratively to refine the estimate of the root:

$$\boldsymbol{\theta}^{(m+1)} = \boldsymbol{\theta}^{(m)} - \left[\frac{\partial^2 f(\boldsymbol{\theta})}{\partial \boldsymbol{\theta}\, \partial \boldsymbol{\theta}'} \right]^{-1} \Bigg|_{\theta=\theta^{(m)}} f'(\boldsymbol{\theta}^{(m)}). \tag{14.21}$$

To use (14.21) we need the first and second derivatives of the function.

We illustrate this method on the profile log likelihood, (6.61), of the components-of-variance model:

$$\log l_P(\mathbf{V}) = -\tfrac{1}{2} \mathbf{y}' \mathbf{P} \mathbf{y} - \tfrac{1}{2} \log |\mathbf{V}| - \tfrac{N}{2} \log(2\pi), \tag{14.22}$$

where $\mathbf{V} = \sum_{i=0}^{r} \mathbf{Z}_i \mathbf{Z}_i' \sigma_i^2$ and $\mathbf{P} = \mathbf{V}^{-1} - \mathbf{V}^{-1} \mathbf{X} (\mathbf{X}' \mathbf{V}^{-1} \mathbf{X})^{-} \mathbf{X}' \mathbf{V}^{-1}$. Since this is a profile log likelihood it depends only on the σ^2s and not on $\boldsymbol{\beta}$. We therefore need partial derivatives of $\log l_P$ with respect to σ_i^2 and the mixed partial derivatives with respect to σ_i^2 and σ_j^2.

The ingredients for these are given in Section 6.12. From (6.78) we have

$$\frac{\partial \mathbf{P}}{\partial \sigma_i^2} = -\mathbf{P} \mathbf{Z}_i \mathbf{Z}_i' \mathbf{P}$$

and therefore

$$\frac{\partial^2 \mathbf{P}}{\partial \sigma_i^2\, \partial \sigma_j^2} = \mathbf{P} \mathbf{Z}_j \mathbf{Z}_j' \mathbf{P} \mathbf{Z}_i \mathbf{Z}_i' \mathbf{P} + \mathbf{P} \mathbf{Z}_i \mathbf{Z}_i' \mathbf{P} \mathbf{Z}_j \mathbf{Z}_j' \mathbf{P}.$$

Also, from (M.20)

$$\frac{\partial \log |\mathbf{V}|}{\partial \sigma_i^2} = \operatorname{tr}\left(\mathbf{V}^{-1} \mathbf{Z}_i \mathbf{Z}_i' \right)$$

and using (M.18)

$$\frac{\partial^2 \log |\mathbf{V}|}{\partial \sigma_i^2\, \partial \sigma_j^2} = -\operatorname{tr}\left(\mathbf{V}^{-1} \mathbf{Z}_j \mathbf{Z}_j' \mathbf{V}^{-1} \mathbf{Z}_i \mathbf{Z}_i' \right).$$

From these derivatives it is straightforward to calculate

$$l_{\sigma^2} = \frac{\partial \log l_P}{\partial \sigma^2} = \left\{ \tfrac{1}{2} \mathbf{y}' \mathbf{P} \mathbf{Z}_i \mathbf{Z}_i' \mathbf{P} \mathbf{y} - \tfrac{1}{2} \operatorname{tr}\left(\mathbf{V}^{-1} \mathbf{Z}_i \mathbf{Z}_i' \right) \right\}_c \tag{14.23}$$

and

$$l_{\sigma^2\sigma^2} = \frac{\partial^2 \log l_P}{\partial\sigma^2\,\partial(\sigma^2)'} = \Big\{ \sum_m \tfrac{1}{2}\mathbf{y'PZ}_i\mathbf{Z}_i'\mathbf{PZ}_j\mathbf{Z}_j'\mathbf{Py} + \tfrac{1}{2}\mathbf{y'PZ}_j\mathbf{Z}_j'\mathbf{PZ}_i\mathbf{Z}_i'\mathbf{Py}$$

$$- \tfrac{1}{2}\mathrm{tr}\left(\mathbf{V}^{-1}\mathbf{Z}_j\mathbf{Z}_j'\mathbf{V}^{-1}\mathbf{Z}_i\mathbf{Z}_i'\right) \Big\}. \tag{14.24}$$

The Newton-Raphson algorithm would take the form

$$\sigma^{2(m+1)} = \sigma^{2(m)} - (l_{\sigma^2\sigma^2})^{-1}l_{\sigma^2}, \tag{14.25}$$

where $l_{\sigma^2\sigma^2}$ and l_{σ^2} are evaluated at $\sigma^2 = \sigma^{2(m)}$.

The Newton-Raphson method is not without its drawbacks. First, it does not guarantee convergence, even to a local maximum. It can fail when the linearized approximation in (14.18) is a poor one. Second, it does not necessarily keep iterations in the parameter space. For example, (14.25) could lead to negative $\sigma^2 s$.

Various improvements are possible to remedy these defects. For example, σ_i can be estimated in place of σ_i^2 (and then squared to get σ_i^2) to keep iterative estimates of σ_i^2 positive. Also, taking smaller steps by modifying (14.25) to be of the form

$$\sigma^{2(m+1)} = \sigma^{2(m)} - \alpha(l_{\sigma^2\sigma^2})^{-1}l_{\sigma^2}, \tag{14.26}$$

where $0 < \alpha \leq 1$ is often a good idea. For more details on implementation and for some calculational details for alternative models, see Jennrich and Schluchter (1986), Lindstrom and Bates (1988) and Press et al. (1996).

14.3   COMPUTING ML ESTIMATES FOR GLMMs

a.   Numerical quadrature

Generalized linear mixed models pose special challenges beyond linear mixed models because of the high-dimensional integration required to evaluate (and hence maximize) the likelihood. The direct numerical evaluation of integrals has a long history in mathematics and is a natural first place for considering how to deal with the computational complexity of GLMMs. We start by considering a GLMM with a single, normally distributed random effect. Let y_{ij} be the jth observation corresponding to the ith level of the random effect so that

$$y_{ij}|\mathbf{u} \;\sim\; \text{indep. } f_{y_{ij}|\mathbf{U}}(y_{ij}|\mathbf{u})$$

$$f_{y_{ij}|\mathbf{u}}(y_{ij}|\mathbf{u}) \;=\; \exp\{[y_{ij}\gamma_{ij} - b(\gamma_{ij})]/\tau^2 - c(y_{ij}, \tau)\}$$

$$\mathrm{E}[y_{ij}|\mathbf{u}] \;=\; \mu_{ij} \tag{14.27}$$

$$g(\mu_{ij}) \;=\; \mathbf{x}'_{ij}\boldsymbol{\beta} + u_i$$

$$u_i \;\sim\; \text{i.i.d. } \mathcal{N}(0, \sigma_u^2).$$

The likelihood for this model is

$$L \;=\; \int \prod_{i,j} f_{y_{ij}|U_i}(y_{ij}|u_i) f_{U_i}(u_i)\, du_i,$$

$$=\; \prod_i \int_{-\infty}^{\infty} e^{\sum_j [y_{ij}\gamma_{ij} - b(\gamma_{ij})]/\tau^2 - \sum_j c(y_{ij}, \tau)} \frac{e^{-u_i^2/(2\sigma_u^2)}}{\sqrt{2\pi\sigma_u^2}}\, du_i,$$

$$=\; \prod_i \int_{-\infty}^{\infty} h_i(u_i) \frac{e^{-u_i^2/(2\sigma_u^2)}}{\sqrt{2\pi\sigma_u^2}}\, du_i, \tag{14.28}$$

where $h_i(u_i) = e^{\sum_j [y_{ij}\gamma_{ij} - b(\gamma_{ij})]/\tau^2 - \sum_j c(y_{ij}, \tau)}$ and γ_{ij} is a function of u_i.

It can be seen that the likelihood is the product of one-dimensional integrals of the form

$$\int_{-\infty}^{\infty} h(u) \frac{e^{-u^2/(2\sigma_u^2)}}{\sqrt{2\pi\sigma_u^2}}\, du,$$

which, upon a change of variables of $u = \sqrt{2}\sigma_u v$, can be written as

$$\int_{-\infty}^{\infty} h(\sqrt{2}\sigma_u v) \frac{e^{-v^2}}{\sqrt{\pi}}\, dv \equiv \int_{-\infty}^{\infty} h^*(v) e^{-v^2}\, dv, \tag{14.29}$$

where $h^*(\cdot) \equiv h(\sqrt{2}\,\sigma_u \cdot)/\sqrt{\pi}$.

– i.   *Gauss-Hermite quadrature*

Numerical integration over an unbounded range can be difficult. However, for integrals of smooth functions $h^*(\cdot)$ multiplied by the function

Table 14.1: Constants for Gauss-Hermite Quadrature

	x_k	w_k
$d = 3$	-1.22474487	0.29540898
	0	1.18163590
	1.22474487	0.29540898
$d = 4$	-1.65068012	0.08131284
	-0.52464762	0.80491409
	0.52464762	0.80491409
	1.65068012	0.08131284
$d = 5$	-2.02018287	0.01995324
	-0.95857246	0.39361932
	0	0.94530872
	0.95857246	0.39361932
	2.02018287	0.01995324

e^{-v^2}, the method of Gauss-Hermite quadrature is available. This approximates the integral in (14.29) as a weighted sum:

$$\int_{-\infty}^{\infty} h^*(v)e^{-v^2}\, dv \approx \sum_{k=1}^{d} h^*(x_k)w_k, \qquad (14.30)$$

where the weights, w_k, and the evaluation points, x_k, are designed to provide an accurate approximation in the case where $h^*(\cdot)$ is a polynomial. More specifically, when the sum is from 1 to d, Gauss-Hermite quadrature gives the exact answer for all polynomials up to degree $2d - 1$. Table 14.1 lists the x_k and w_k for $d = 3$, 4, and 5. More extensive tables are available in, for example, Abramowitz and Stegun (1964), or the x_k and w_k can be calculated via mathematical software since

$$x_k = \text{ith zero of } H_n(x)$$

$$w_k = \frac{2^{n-1}n!\sqrt{\pi}}{n^2[H_{n-1}(x_k)]^2}, \qquad (14.31)$$

where $H_n(x)$ is the Hermite polynomial of degree n.

As an illustration consider

$$\int_{-\infty}^{\infty} (1+x^2)e^{-x^2}\, dx = \frac{3}{2}\sqrt{\pi} \doteq 2.6587.$$

This would be approximated using 3-point quadrature as

$$\int_{-\infty}^{\infty} (1 + x^2)e^{-x^2}\, dx \;\doteq\; (1 + [-1.2247]^2)(0.2954) + (1 + 0^2)(1.1816)$$

$$+ (1 + 1.2247^2)(0.2954)$$

$$= (2.5)(0.2954)(2) + 1.1816$$

$$= 2.6587,$$

as expected. On the other hand,

$$\int_{-\infty}^{\infty} (1 + x^6)e^{-x^2}\, dx = \frac{23}{8}\sqrt{\pi} \doteq 5.0958,$$

which is approximated by $(4.375)(0.2954)(2) + 1.1816 = 3.7665$ with 3-point quadrature, a poor approximation. But 4-point quadrature gets the answer exactly right.

By using quadrature of a high-enough degree, accurate approximations can be calculated to integrals of functions that are similar to those of any high-degree polynomial. For the likelihood calculations we have in mind, practical experience shows that quadrature with less than 10 points often gives inaccurate answers, while 20 is usually enough for a good degree of approximation.

Two cautions are in order. First, if the function is not properly "centered," Gauss-Hermite quadrature can give a poor approximation and second, if the function whose integral is to be approximated is not a smooth one, the approximation can also be poor. To illustrate the idea of centering, consider

$$\int_{-\infty}^{\infty} e^{2xa - a^2} e^{-x^2}\, dx,$$

which is easily shown to be $\sqrt{\pi}$ for any value of a. Five-point quadrature gives $\sqrt{\pi}$ as the answer when $a = 0$ but has an error of 0.001 when $a = 1$, and an error of 0.240 when $a = 2$, eventually giving an answer of 0 as $a \to \pm\infty$. This shows that the approximation is more accurate when the values of x_k are near where the function is nonzero and can be inaccurate otherwise. Exercise E 14.4 illustrates that the approximation can be poor for non-smooth functions.

– ii.   *Likelihood calculations*

Gauss-Hermite quadrature can be used to calculate integrals with respect to the normal density as

$$\int_{-\infty}^{\infty} h(x)\frac{e^{-x^2/(2\sigma^2)}}{\sqrt{2\pi\sigma^2}}\,dx \approx \sum_{k=1}^{d} h(\sqrt{2}\,\sigma x_k)w_k/\sqrt{\pi}. \qquad (14.32)$$

To derive an approximation to a likelihood such as (14.28), (14.32) would be used repeatedly. For example, suppose that our model was, for $i = 1, 2, \ldots, m$ and $j = 1, 2, \ldots, n$,

$$y_{ij}|\mathbf{a} \;\sim\; \text{indep. Bernoulli}(p_{ij})$$

$$\text{logit}(p_{ij}) \;=\; \mu + a_i \qquad\qquad\qquad (14.33)$$

$$a_i \;\sim\; \text{i.i.d. } \mathcal{N}(0, \sigma_a^2).$$

The log likelihood would be

$$l \;=\; \sum_{i} \log \int_{-\infty}^{\infty} e^{(\mu+a_i)y_i. \,-n\log(1+e^{\mu+a_i})}\frac{e^{-a_i^2/(2\sigma_a^2)}}{\sqrt{2\pi\sigma_a^2}}\,da_i$$

$$\approx\; \sum_{i} \log\left(\sum_{k} e^{(\mu+x_k)y_i. \,-n\log(1+e^{\mu+x_k})}w_k/\sqrt{\pi}\right). \qquad (14.34)$$

This log likelihood needs to be maximized numerically to get estimates of μ and σ_a. Derivatives of the log likelihood, which are often required by numerical maximization algorithms, can be approximated similarly. Alternatively, quasi-Newton or derivative-free maximization methods (Press et al., 1996) can be used.

A likelihood ratio test, or best predicted values, would require similar numerical calculation. For example, the best predicted values for model (14.33) are

$$E[a_i|\mathbf{y}] \;=\; \int a_i f_{a_i|\mathbf{y}}(a_i|\mathbf{y})\,da_i$$

$$=\; \int a_i f_{\mathbf{y}|a_i}(\mathbf{y}|a_i)f_{a_i}(a_i)/f_{\mathbf{y}}(\mathbf{y})\,da_i$$

$$=\; \frac{\int a_i e^{(\mu+x_k)y_i. \,-n\log(1+e^{\mu+x_k})}f_{a_i}(a_i)\,da_i}{\int e^{(\mu+x_k)y_i. \,-n\log(1+e^{\mu+x_k})}f_{a_i}(a_i)\,da_i}.$$

The denominator is exactly the likelihood, the approximation for which is displayed in (14.34), and the numerator would be approximated similarly. If the MLEs $\hat{\mu}$ and $\hat{\sigma}_a$ were used in the calculation, then the approximation would be for the estimated best predictor.

– iii.   *Limits of numerical quadrature*

Numerical quadrature is limited in its application. The calculations above show that with a single random effect the computations are feasible. It is also possible to use numerical quadrature to approximate the likelihood of a model with two nested random effects (see E 14.5). However, crossed random factors and higher levels of nesting lead to integrals that are not amenable to Gauss-Hermite quadrature.

The possible distributions available for the random effect distribution are also limited. The quadrature techniques described in the preceding sections are appropriate only when integrating products of functions with e^{-x^2}, that is, for normally distributed random effects. To employ other random effects distributions we need alternative quadrature methods. Although these could be derived conceptually they are not readily available, except for integrating products with e^{-x}, which would correspond to exponentially distributed random effects. This methodology is called *Laguerre integration* (Abramowitz and Stegun, 1964, p. 923). Another way to extend the class of random effects distributions is to consider transformations of normally distributed random effects (Piepho and McCulloch, 2004), for example, by using e^{a_i} to introduce lognormally distributed random effects.

b.   EM algorithm

As noted in Section 14.2a, for mixed models a typical missing data configuration is to assume the random effects to be the missing data. Since the random effects introduce correlation in the model, once they are filled in by the EM algorithm and can be treated as fixed known values, the problem often simplifies. The variance components version of the linear mixed model, for example, treated in Section 14.2a, simplifies to the traditional homoscedastic linear model, for which maximum likelihood is ordinary least squares. So, for that model, EM reduces maximum likelihood to a series of least squares problems.

We return to (7.1), the GLMM of Chapter 7, as our most general model:

$$y_i|\mathbf{u} \quad \sim \quad \text{indep. } f_{Y_i|\mathbf{U}}(y_i|\mathbf{u})$$

$$f_{Y_i|\mathbf{u}}(y_i|\mathbf{u}) \quad = \quad \exp\{[y_i\gamma_i - b(\gamma_i)]/\tau^2 - c(y_i, \tau)\}$$

$$\mathrm{E}[y_i|\mathbf{u}] \quad = \quad \mu_i \tag{14.35}$$

$$g(\mu_i) \quad = \quad \mathbf{x}_i'\boldsymbol{\beta} + \mathbf{z}_i'\mathbf{u} \tag{14.36}$$

$$\mathbf{u} \quad \sim \quad f_{\mathbf{U}}(\mathbf{u}|\mathbf{D}),$$

where $\mathbf{D}$ represents the parameters governing the distribution of $\mathbf{u}$ in keeping with the notation in Chapter 6.

To set up the EM algorithm we declare $\mathbf{u}$ to be the missing data so that the *complete data* are $\mathbf{w}' = (\mathbf{y}', \mathbf{u}')$. The EM algorithm proceeds by forming the log likelihood of the complete data, calculating its expectation with respect to the conditional distribution of $\mathbf{u}$ given $\mathbf{y}$ and then maximizing with respect to the parameters. The algorithm is iterative since we now recalculate the log likelihood of the complete data given the new parameter estimates, and so on.

The distribution of the complete data, $\mathbf{w}$, can be factored as $f_{\mathbf{Y},\mathbf{U}} = f_{\mathbf{Y}|\mathbf{U}}f_{\mathbf{U}}$, so that the complete data log likelihood, $\log L_w$, is

$$\log L_w \quad = \quad \log f_{\mathbf{Y}|\mathbf{U}} + \log f_{\mathbf{U}}$$

$$= \quad \sum_{i=1}^{n} \log f_{Y_i|\mathbf{U}} + \log f_{\mathbf{U}}$$

$$= \quad [\sum y_i\gamma_i - b(\gamma_i)]/\tau^2 - \sum c(y_i, \tau) + \log f_{\mathbf{U}}. \tag{14.37}$$

This choice of missing data has two advantages. First, conditional on $\mathbf{u}$, the y_i are independent. Second, $\boldsymbol{\beta}$ and τ enter only the first portion of the log likelihood (the GLM portion) whereas $\mathbf{D}$ enters only through $f_{\mathbf{U}}$, the portion coming from the random effects. The maximization or *M-step* of the algorithm with respect to $\boldsymbol{\beta}$ and τ will be similar to the calculations for GLMs in Section 5.4e. Maximizing with respect to $\mathbf{D}$ is akin to ML using the distribution of $\mathbf{u}$. In fact, if the distribution of $\mathbf{u}$ is a member of the exponential family, then the M-step for $\mathbf{D}$ simplifies to maximum likelihood after replacing the sufficient statistics with their conditional expected values.

The EM algorithm takes the following form:

1. Choose starting values $\beta^{(0)}, \tau^{(0)}$, and $\mathbf{D}^{(0)}$. Set $m = 0$.

2. Calculate (with expectations evaluated under current values)

    (a) $\beta^{(m+1)}$ and $\tau^{(m+1)}$ to maximize $E[\log f_{\mathbf{Y}|\mathbf{U}}(\mathbf{y}|\mathbf{u}, \beta, \tau)|\mathbf{y}]$.

    (b) $\mathbf{D}^{(m+1)}$ to maximize $E[\log f_{\mathbf{U}}(\mathbf{u}|\mathbf{D})|\mathbf{y}]$.

    (c) Set $m = m + 1$.

3. If convergence is achieved, declare the current values to be the MLEs; otherwise return to step 2.

In general, the expectations in neither steps 2(a) nor 2(b) can be computed in closed form for the model. This is because the conditional distribution of $\mathbf{u}|\mathbf{y}$ involves $f_{\mathbf{Y}}$, that is, the likelihood, which we are trying to avoid calculating directly. However, it *is* possible to produce random draws from the conditional distribution of $\mathbf{u}|\mathbf{y}$ without specifying $f_{\mathbf{Y}}$. One can then form Monte Carlo approximations to the required expectations. We describe this approach in the next section.

c.   Markov chain Monte Carlo algorithms

There are a number of ways to produce draws from a difficult-to-calculate density, for example, Gibbs sampling or Markov chain Monte Carlo methods (Robert and Casella, 1999). McCulloch (1994, 1997) uses the Gibbs sampler for probit models, and the Metropolis-Hastings algorithm for general GLMM problems, while Booth and Hobert (1999) use an independence sampler.

– i.   *A Metropolis algorithm*

As an example, we consider a Metropolis algorithm, which generates a Markov chain sequence of values that eventually stabilizes to draws from the candidate distribution. To specify a Metropolis algorithm, a candidate distribution, $h_{\mathbf{U}}(\mathbf{u})$, must be selected, from which potential new values are drawn. The *acceptance function*, which gives the probability of accepting a new value (as opposed to keeping the previous value) is given by

$$A_k(\mathbf{u}^*, \mathbf{u}) = \min\left\{1, \frac{f_{\mathbf{U}|\mathbf{Y}}(\mathbf{u}^*|\mathbf{y}, \beta, \tau, \mathbf{D})h_{\mathbf{U}}(\mathbf{u})}{f_{\mathbf{U}|\mathbf{Y}}(\mathbf{u}|\mathbf{y}, \beta, \tau, \mathbf{D})h_{\mathbf{U}}(\mathbf{u}^*)}\right\}, \qquad (14.38)$$

where $\mathbf{u}^* = (u_1, u_2, \ldots, u_{k-1}, u_k^*, u_{k+1}, \ldots, u_q)'$, which is the candidate new value and has all entries equal to the previous value except the kth.

What can be used for the candidate distribution? Upon choosing $h_{\mathbf{U}} = f_{\mathbf{U}}$, the ratio term in (14.38) simplifies to

$$\frac{f_{\mathbf{U}|\mathbf{Y}}(\mathbf{u}^*|\mathbf{y}, \boldsymbol{\beta}, \tau, \mathbf{D})h_{\mathbf{U}}(\mathbf{u})}{f_{\mathbf{U}|\mathbf{Y}}(\mathbf{u}|\mathbf{y}, \boldsymbol{\beta}, \tau, \mathbf{D})h_{\mathbf{U}}(\mathbf{u}^*)}$$

$$= \frac{\prod_{i=1}^n f_{Y_i|\mathbf{U}}(y_i|\mathbf{u}^*, \boldsymbol{\beta}, \tau)f_{\mathbf{U}}(\mathbf{u}^*|\mathbf{D})f_{\mathbf{U}}(\mathbf{u}|\mathbf{D})}{\prod_{i=1}^n f_{Y_i|\mathbf{U}}(y_i|\mathbf{u}, \boldsymbol{\beta}, \tau)f_{\mathbf{U}}(\mathbf{u}|\mathbf{D})f_{\mathbf{U}}(\mathbf{u}^*|\mathbf{D})}$$

$$= \frac{\prod_{i=1}^n f_{Y_i|\mathbf{U}}(y_i|\mathbf{u}^*, \boldsymbol{\beta}, \tau)}{\prod_{i=1}^n f_{Y_i|\mathbf{U}}(y_i|\mathbf{u}, \boldsymbol{\beta}, \tau)}. \tag{14.39}$$

This calculation involves specifying only the generalized linear model portion of the model, that is, the conditional distribution of $\mathbf{y}$ given $\mathbf{u}$.

Incorporating this Metropolis step into the EM algorithm gives a Monte Carlo EM (MCEM) algorithm as follows:

1. Choose starting values $\boldsymbol{\beta}^{(0)}, \tau^{(0)}$, and $\mathbf{D}^{(0)}$. Set $m = 0$.

2. Generate M values, $\mathbf{u}^{(1)}, \mathbf{u}^{(2)}, \ldots, \mathbf{u}^{(M)}$, from the conditional distribution of $\mathbf{u}$ given $\mathbf{y}$ using the Metropolis algorithm described above.

   (a) Calculate $\boldsymbol{\beta}^{(m+1)}$ and $\tau^{(m+1)}$ to maximize a Monte Carlo estimate of $E[\log f_{\mathbf{Y}|\mathbf{U}}(\mathbf{y}|\mathbf{u}, \boldsymbol{\beta}, \tau)|\mathbf{y}]$, i.e., choose them to maximize $(1/M)\sum_{k=1}^M \log f_{\mathbf{Y}|\mathbf{U}}(\mathbf{y}|\mathbf{u}^{(k)}, \boldsymbol{\beta}, \tau)$.

   (b) Calculate $\mathbf{D}^{(m+1)}$ to maximize $(1/M)\sum_{k=1}^M \log f_{\mathbf{U}}(\mathbf{u}^{(k)}|\mathbf{D})$.

   (c) Set $m = m + 1$.

3. If convergence is achieved, declare the current values to be the MLEs; otherwise return to step 2.

While computationally intensive, this approach remains feasible for a variety of data configurations.

– ii.   *Monte Carlo Newton-Raphson method*

There is also a simulation analog of the working variates or Fisher scoring approach which was used to fit GLMs in Section 5.4e. Whenever the marginal density of $\mathbf{y}$ is formed as a mixture as in (14.35) with separate parameters for $f_{\mathbf{Y}|\mathbf{U}}$ and $f_{\mathbf{U}}$, then the ML equations for $\boldsymbol{\theta} = (\boldsymbol{\beta}', \tau)'$ and $\mathbf{D}$ take the following form (see E 14.6):

$$\mathrm{E}\left[\frac{\partial f_{\mathbf{Y}|\mathbf{U}}(\mathbf{y}|\mathbf{u}, \boldsymbol{\theta})}{\partial \boldsymbol{\theta}}\bigg|\mathbf{y}\right] = \mathbf{0} \tag{14.40}$$

$$\mathrm{E}\left[\frac{\partial f_{\mathbf{U}}(\mathbf{u}|\mathbf{D})}{\partial \mathbf{D}}\bigg|\mathbf{y}\right] = \mathbf{0}. \tag{14.41}$$

Equation (14.41) involves only the distribution of $\mathbf{u}$ and is often fairly easy to solve, for example, when the distribution is normal. On the other hand, (14.40) is amenable to Newton-Raphson or a scoring approach just as in Chapter 5.

Expanding $\partial f_{\mathbf{Y}|\mathbf{U}}(\mathbf{y}|\mathbf{u}, \boldsymbol{\theta})/\partial \boldsymbol{\beta}$ as a function of $\boldsymbol{\beta}$ around a value $\boldsymbol{\theta}_0$ gives

$$\frac{\partial f_{\mathbf{Y}|\mathbf{U}}(\mathbf{y}|\mathbf{u}, \boldsymbol{\theta})}{\partial \boldsymbol{\beta}} \tag{14.42}$$

$$\approx \frac{\partial f_{\mathbf{Y}|\mathbf{U}}(\mathbf{y}|\mathbf{u}, \boldsymbol{\theta})}{\partial \boldsymbol{\beta}}\bigg|_{\boldsymbol{\theta}=\boldsymbol{\theta}_0} + \frac{\partial^2 f_{\mathbf{Y}|\mathbf{U}}(\mathbf{y}|\mathbf{u}, \boldsymbol{\theta})}{\partial \boldsymbol{\beta} \partial \boldsymbol{\beta}'}\bigg|_{\boldsymbol{\theta}=\boldsymbol{\theta}_0} (\boldsymbol{\beta} - \boldsymbol{\beta}_0).$$

Specializing this to our model, and dropping the term with a conditional expected value of zero, the formula for a scoring-type algorithm becomes

$$\frac{\partial f_{\mathbf{Y}|\mathbf{U}}(\mathbf{y}|\mathbf{u}, \boldsymbol{\theta})}{\partial \boldsymbol{\beta}} \approx \frac{1}{\tau^2}\mathbf{X}'\mathbf{W}\boldsymbol{\Delta}(\mathbf{y} - \boldsymbol{\mu}) - \frac{1}{\tau^2}\mathbf{X}'\mathbf{W}\mathbf{X}(\boldsymbol{\beta} - \boldsymbol{\beta}_0), \tag{14.43}$$

where $\mathbf{W} = \left\{_d [v(\mu_i) g_\mu^2(\mu_i)]^{-1}\right\}$ and $\boldsymbol{\Delta} = \left\{_c g_\mu(\mu_i)\right\}$ and it is understood that $\mathbf{W}$, $\boldsymbol{\Delta}$, and $\boldsymbol{\mu} = \mathrm{E}[\mathbf{y}|\mathbf{u}]$ are all functions of $\mathbf{u}$ and that all parameters are evaluated at $\boldsymbol{\theta} = \boldsymbol{\theta}_0$.

Using this approximation in (14.40) leads to an iteration equation of the form

$$\boldsymbol{\beta}^{(m+1)} = \boldsymbol{\beta}^{(m)} + (\mathbf{X}'\mathrm{E}[\mathbf{W}|\mathbf{y}]\mathbf{X})^{-1}\mathbf{X}'\mathrm{E}[\mathbf{W}\boldsymbol{\Delta}(\mathbf{y} - \boldsymbol{\mu})|\mathbf{y}]. \tag{14.44}$$

This analog of scoring would proceed by iteratively solving (14.44), (14.41), and an equation for τ. An advantage of the scoring approach over MCEM is that is makes automatic the maximization step 2(a).

Again, typically the expectations cannot be evaluated in closed form, which leads to a Monte Carlo Newton-Raphson (MCNR) approach. As before, an algorithm like the Metropolis algorithm is used to approximate the expectations in (14.44) since these are expectations with respect to the conditional distribution of $\mathbf{u}$ given $\mathbf{y}$.

d. Stochastic approximation algorithms

A different approach to fitting these models has been suggested by Gu and Kong (1998) through the use of a stochastic approximation (SA) algorithm, although the basic idea of using SA to find MLEs is certainly older (e.g., Moyeed and Baddeley, 1991; Ruppert, 1991). The basic concept is to write $f_{\mathbf{Y},\mathbf{U}}$ as $f_{\mathbf{Y}}f_{\mathbf{U}|\mathbf{Y}}$. It is then straightforward to derive

$$\frac{\partial \log f_{\mathbf{Y},\mathbf{U}}(\mathbf{y},\mathbf{u}|\boldsymbol{\theta})}{\partial \boldsymbol{\theta}} = \frac{\partial \log f_{\mathbf{Y}}(\mathbf{y}|\boldsymbol{\theta})}{\partial \boldsymbol{\theta}} + \frac{\partial \log f_{\mathbf{U}|\mathbf{Y}}(\mathbf{u}|\mathbf{y},\boldsymbol{\theta})}{\partial \boldsymbol{\theta}}. \qquad (14.45)$$

We are interested in finding the root of the likelihood equation, that is, the value of $\boldsymbol{\theta}$ such that $\partial \log f_{\mathbf{Y}}(\mathbf{y}|\boldsymbol{\theta})/\partial \boldsymbol{\theta} = \mathbf{0}$. SA algorithms are methods of finding roots of regression equations, so we need to rewrite (14.45) as a regression equation.

Write $\mathbf{m}(\boldsymbol{\theta})$ for the score function, $\partial \log f_{\mathbf{Y}}(\mathbf{y}|\boldsymbol{\theta})/\partial \boldsymbol{\theta}$, to emphasize we are regarding it as a function of $\boldsymbol{\theta}$ and that it is not a function of $\mathbf{u}$. Next observe that

$$\mathrm{E}\left[\frac{\partial \log f_{\mathbf{U}|\mathbf{Y}}(\mathbf{u}|\mathbf{y},\boldsymbol{\theta})}{\partial \boldsymbol{\theta}}\right] = \mathbf{0} \qquad (14.46)$$

for fixed $\mathbf{y}$ when the expectation is taken with respect to the conditional distribution of $\mathbf{u}$ given $\mathbf{y}$. This is the usual score identity, e.g. (5.9), applied to the conditional distribution. Hence $\partial \log f_{\mathbf{U}|\mathbf{Y}}(\mathbf{u}|\mathbf{y},\boldsymbol{\theta})/\partial \boldsymbol{\theta}$, can be regarded as a mean-zero, "error" term in the following "regression" equation, which is (14.45) rewritten:

$$\frac{\partial \log f_{\mathbf{Y},\mathbf{U}}(\mathbf{y},\mathbf{u}|\boldsymbol{\theta})}{\partial \boldsymbol{\theta}} = \mathbf{m}(\boldsymbol{\theta}) + \text{error}. \qquad (14.47)$$

Thus, inserting random values of $\mathbf{u} \sim f_{\mathbf{U}|\mathbf{Y}}$ into $\partial \log f_{\mathbf{Y},\mathbf{U}}(\mathbf{y},\mathbf{u}|\boldsymbol{\theta})/\partial \boldsymbol{\theta}$ gives "data" for performing the regression.

To implement an SA algorithm, we use the Metropolis algorithm of Section 14.2c to generate a sequence of values $\mathbf{u}^{(k)} \sim f_{\mathbf{U}|\mathbf{Y}}$ and use them to form data $\partial \log f_{\mathbf{Y},\mathbf{U}}(\mathbf{y}, \mathbf{u}^{(k)} | \boldsymbol{\theta}) / \partial \boldsymbol{\theta}$. One can then apply a multivariate version of an SA algorithm in order to find the root of the likelihood equation. Ruppert (1991) provides a nice review.

An SA algorithm applied to maximum likelihood for the GLMM would generally take the form

$$\theta^{(m+1)} = \theta^{(m)} - a_m \frac{\partial \log f_{\mathbf{Y},\mathbf{U}}(\mathbf{y}, \mathbf{u}^{(m)} | \boldsymbol{\theta})}{\partial \boldsymbol{\theta}}, \qquad (14.48)$$

where a_m is chosen to decrease slowly to zero. Ideally, a_m also incorporates information about the derivative of $\partial \log f_{\mathbf{Y}}(\mathbf{y} | \boldsymbol{\theta}) / \partial \boldsymbol{\theta}$ (with respect to $\boldsymbol{\theta}$) at the root, but this is rarely known in practice.

A reasonable choice for a_m allowing it to decrease to zero and using some information about the curvature of the surface to be maximized is

$$a_m = \frac{a}{(m+k)^\alpha} \left(\hat{\mathrm{E}} \left[\frac{\partial^2 \log f_{\mathbf{Y},\mathbf{U}}(\mathbf{y}, \mathbf{U} | \boldsymbol{\theta})}{\partial \boldsymbol{\theta} \, \partial \boldsymbol{\theta}'} \right] \right)^{-1}, \qquad (14.49)$$

where $\hat{\mathrm{E}}$ denotes a Monte Carlo estimate of the expectation. This choice of a_m follows recommendations in the literature [see (Frees and Ruppert, 1990) and Ruppert (1991). There is latitude in the choice of the constants a, k and α; and we have successfully used $a = 3, k = 50$ and $\alpha = 0.75$. Estimates are formed by iterating until convergence.

MCNR and SA are similar, with the main difference being that SA uses a single simulated value at each iteration. The multiplier a_m decreases the step size as the iterations increase in SA. This eventually serves to eliminate the stochastic error involved in the Metropolis step. To achieve a corresponding reduction using MCNR, the simulation size would have to be increased as the iterations increase in order to eliminate the simulation noise.

SA seems to have advantages in that it can use all of the simulated data to calculate estimates and it uses the simulated values one at a time. A theoretical advantage of SA is that convergence proofs are worked out for many cases. Practical details of the implementation of both SA and MCNR have not yet been settled in the literature.

e. Simulated maximum likelihood

While both MCEM and MCNR work on the log of the likelihood, Geyer and Thompson (1992), Gelfand and Carlin (1993), and Durbin and Koopman (1997) have suggested simulation to estimate the value of the likelihood directly. Starting from the likelihood we have

$$l = \int f_{\mathbf{Y}|\mathbf{U}}(\mathbf{y}|\mathbf{u}, \boldsymbol{\beta}, \tau) f_{\mathbf{U}}(\mathbf{u}|\mathbf{D})\, d\mathbf{u}$$

$$= \int \frac{f_{\mathbf{Y}|\mathbf{U}}(\mathbf{y}|\mathbf{u}, \boldsymbol{\beta}, \tau) f_{\mathbf{U}}(\mathbf{u}|\mathbf{D})}{h_{\mathbf{U}}(\mathbf{u})} h_{\mathbf{U}}(\mathbf{u})\, d\mathbf{u}$$

$$= \mathrm{E}\left[\frac{f_{\mathbf{Y}|\mathbf{U}}(\mathbf{y}|\mathbf{U}, \boldsymbol{\beta}, \tau) f_{\mathbf{U}}(\mathbf{U}|\mathbf{D})}{h_{\mathbf{U}}(\mathbf{U})}\right]$$

$$\approx \sum_{k=1}^{M} \frac{f_{\mathbf{Y}|\mathbf{U}}(\mathbf{y}|\mathbf{u}^{(k)}, \boldsymbol{\beta}, \tau) f_{\mathbf{U}}(\mathbf{u}^{(k)}|\mathbf{D})}{h_{\mathbf{U}}(\mathbf{u}^{(k)})}, \qquad (14.50)$$

where $h_{\mathbf{U}}(\mathbf{u})$ is a density with respect to which the expectation is taken, $\mathbf{u}^{(k)}$ are selected from this density, and M is the number of simulated values. This is an unbiased estimate no matter the choice of $h_{\mathbf{U}}(\mathbf{u})$. The simulated likelihood is then numerically maximized, either after a single simulation, or using multiple simulations in an iterative process where the importance sampling distribution is allowed to depend on the current parameter values.

Although unbiased, the approximation is sensitive to the choice of $h_{\mathbf{U}}(\mathbf{u})$ in the sense that it can be highly variable for choices far from the optimal choice (for the optimal choice see E 14.10). So implementation of simulated maximum likelihood must be done with care.

14.4 PENALIZED QUASI-LIKELIHOOD AND LAPLACE

The attractive features of quasi-likelihood, namely model robustness and less restrictive assumptions, have led to a search for generalizations applicable to GLMMs. Central to these is the use of a Laplace approximation (Tierney and Kadane, 1986) for evaluating the high-dimensional integral in the likelihood. The basic form of Laplace's approximation is based on a second-order Taylor series expansion and takes the form

$$\log \int_{\Re^q} e^{h(\mathbf{u})} d\mathbf{u} \approx h(\mathbf{u}_0) + \frac{q}{2}\log 2\pi - \tfrac{1}{2}\log\left| -\frac{\partial^2 h(\mathbf{u})}{\partial \mathbf{u}\partial \mathbf{u}'}\right|_{\mathbf{u}=\mathbf{u}_0}, \quad (14.51)$$

where $\mathbf{u}_0$ is the solution to

$$\left.\frac{\partial h(\mathbf{u})}{\partial \mathbf{u}}\right|_{\mathbf{u}=\mathbf{u}_0} = 0. \qquad (14.52)$$

We utilize this result to approximate the log likelihood of the GLMM via

$$l = \log \int f_{\mathbf{Y}|\mathbf{U}} f_{\mathbf{U}} \, d\mathbf{u}$$

$$= \log \int e^{\log f_{\mathbf{Y}|\mathbf{U}} + \log f_{\mathbf{U}}} \, d\mathbf{u}$$

$$= \log \int e^{h(\mathbf{u})} \, d\mathbf{u}, \tag{14.53}$$

with $h(\mathbf{u}) = \log f_{\mathbf{Y}|\mathbf{U}} + \log f_{\mathbf{U}}$. To construct the Laplace approximation (14.52) must be solved and an expression for $\partial^2 h(\mathbf{u})/\partial \mathbf{u}\, \partial \mathbf{u}'$ is needed.

If we assume that $\mathbf{u} \sim \mathcal{N}(\mathbf{0}, \mathbf{D})$ then

$$\log f_{\mathbf{U}} = -\tfrac{1}{2}\mathbf{u}'\mathbf{D}^{-1}\mathbf{u} - \tfrac{q}{2}\log 2\pi - \tfrac{1}{2}\log|\mathbf{D}|$$

and $h(\mathbf{u})$ becomes

$$\log f_{\mathbf{Y}|\mathbf{U}} + \log f_{\mathbf{U}} = \log f_{\mathbf{Y}|\mathbf{U}} - \tfrac{1}{2}\mathbf{u}'\mathbf{D}^{-1}\mathbf{u} - \tfrac{q}{2}\log 2\pi - \tfrac{1}{2}\log|\mathbf{D}|.$$

Differentiating with respect to $\mathbf{u}$ gives

$$\frac{\partial h(\mathbf{u})}{\partial \mathbf{u}} = \frac{\partial \log f_{\mathbf{Y}|\mathbf{U}}}{\partial \mathbf{u}} - \mathbf{D}^{-1}\mathbf{u}$$

$$= \frac{1}{\tau^2}\mathbf{Z}'\mathbf{W}\boldsymbol{\Delta}(\mathbf{y} - \boldsymbol{\mu}) - \mathbf{D}^{-1}\mathbf{u}, \tag{14.54}$$

where $\mathbf{W}$ and $\boldsymbol{\Delta}$ are defined below (14.43). The second equality comes about from derivations identical to (5.18) with $\mathbf{u}$ replacing $\boldsymbol{\beta}$ and $\mathbf{Z}$ replacing $\mathbf{X}$. To find $\mathbf{u}_0$ it is necessary to solve for $\mathbf{u}$ in

$$\frac{1}{\tau^2}\mathbf{Z}'\mathbf{W}\boldsymbol{\Delta}(\mathbf{y} - \boldsymbol{\mu}) = \mathbf{D}^{-1}\mathbf{u}, \tag{14.55}$$

which is not as simple as it appears since $\mathbf{W}$, $\boldsymbol{\Delta}$, and $\boldsymbol{\mu} = E[\mathbf{y}|\mathbf{u}]$ on the left-hand side of the equation are all functions of $\mathbf{u}$.

We will also need the second derivative:

$$\frac{\partial^2 h(\mathbf{u})}{\partial \mathbf{u}\, \partial \mathbf{u}'} = -\frac{1}{\tau^2}\mathbf{Z}'\mathbf{W}\boldsymbol{\Delta}\frac{\partial \boldsymbol{\mu}}{\partial \mathbf{u}'} + \frac{1}{\tau^2}\mathbf{Z}'\frac{\partial \mathbf{W}\boldsymbol{\Delta}}{\partial \mathbf{u}'}(\mathbf{y} - \boldsymbol{\mu}) - \mathbf{D}^{-1}. \tag{14.56}$$

For some models (e.g., the binomial or Poisson) $\mathbf{W}\boldsymbol{\Delta} = \mathbf{I}$ so the second term is zero. In general, the second term has expectation zero with respect to the conditional distribution of $\mathbf{y}$ given $\mathbf{u}$. So it may be reasonable to consider it as negligible with respect to the other terms. If this is the case, (14.56) becomes

$$-\frac{\partial^2 h(\mathbf{u})}{\partial \mathbf{u} \, \partial \mathbf{u}'} \approx \frac{1}{\tau^2} \mathbf{Z}'\mathbf{W}\boldsymbol{\Delta}\boldsymbol{\Delta}^{-1}\mathbf{Z} + 0 + \mathbf{D}^{-1}$$

$$= \frac{1}{\tau^2} \mathbf{Z}'\mathbf{W}\mathbf{Z} + \mathbf{D}^{-1}$$

$$= \left(\frac{1}{\tau^2} \mathbf{Z}'\mathbf{W}\mathbf{Z}\mathbf{D} + \mathbf{I}\right)\mathbf{D}^{-1}. \qquad (14.57)$$

Using (14.57) in (14.51) gives

$$l \approx \log f_{\mathbf{Y}|\mathbf{U}}(\mathbf{y}|\mathbf{u}_0) - \tfrac{1}{2}\mathbf{u}_0'\mathbf{D}^{-1}\mathbf{u}_0 - \tfrac{q}{2}\log 2\pi - \tfrac{1}{2}\log|\mathbf{D}|$$

$$+ \tfrac{q}{2}\log 2\pi - \tfrac{1}{2}\log|(\mathbf{Z}'\mathbf{W}\mathbf{Z}\mathbf{D}/\tau^2 + \mathbf{I})\mathbf{D}^{-1}| \qquad (14.58)$$

$$= \log f_{\mathbf{Y}|\mathbf{U}}(\mathbf{y}|\mathbf{u}_0) - \tfrac{1}{2}\mathbf{u}_0'\mathbf{D}^{-1}\mathbf{u}_0 - \tfrac{1}{2}\log|\mathbf{Z}'\mathbf{W}\mathbf{Z}\mathbf{D}/\tau^2 + \mathbf{I}|.$$

This still must be maximized with respect to $\boldsymbol{\beta}$ to find the ML estimate. Differentiating with respect to $\boldsymbol{\beta}$ gives an approximate score equation of

$$\frac{\partial l}{\partial \boldsymbol{\beta}} \approx \frac{\partial \log f_{\mathbf{Y}|\mathbf{U}}(\mathbf{y}|\mathbf{u}_0)}{\partial \boldsymbol{\beta}} + \frac{\partial}{\partial \boldsymbol{\beta}}\tfrac{1}{2}\log|\mathbf{Z}'\mathbf{Z}\mathbf{D}/\tau^2 + \mathbf{I}|$$

$$= \frac{1}{\tau^2}\mathbf{X}'\mathbf{W}\boldsymbol{\Delta}(\mathbf{y} - \boldsymbol{\mu}) + \frac{\partial}{\partial \boldsymbol{\beta}}\tfrac{1}{2}\log|\mathbf{Z}'\mathbf{W}\mathbf{Z}\mathbf{D}/\tau^2 + \mathbf{I}|$$

$$\approx \frac{1}{\tau^2}\mathbf{X}'\mathbf{W}\boldsymbol{\Delta}(\mathbf{y} - \boldsymbol{\mu}), \qquad (14.59)$$

where the second equality follows from (5.18) and for the third we have assumed that $\mathbf{W}$ changes negligibly as a function of $\boldsymbol{\beta}$. Thus we jointly solve the equations

$$\frac{1}{\tau^2}\mathbf{X}'\mathbf{W}\boldsymbol{\Delta}(\mathbf{y} - \boldsymbol{\mu}) = \mathbf{0} \qquad (14.60)$$

and

$$\frac{1}{\tau^2}\mathbf{Z}'\mathbf{W}\boldsymbol{\Delta}(\mathbf{y}-\boldsymbol{\mu}) \;=\; \mathbf{D}^{-1}\mathbf{u} \tag{14.61}$$

for $\boldsymbol{\beta}$ and $\mathbf{u}$. Of course, this only gives an estimate of $\boldsymbol{\beta}$; a subsidiary method is needed to estimate $\mathbf{D}$.

Equations (14.60) and (14.61) can also arise from jointly maximizing (with respect to $\boldsymbol{\beta}$ and $\mathbf{u}$)

$$\log f_{\mathbf{Y}|\mathbf{U}} - \tfrac{1}{2}\mathbf{u}'\mathbf{D}^{-1}\mathbf{u}, \tag{14.62}$$

which is similar to a quasi-likelihood (the $f_{\mathbf{Y}|\mathbf{U}}$ term) with a "penalty" function added on (the $\mathbf{u}'\mathbf{D}^{-1}\mathbf{u}$ term). In (14.62) the $\tfrac{1}{2}\mathbf{u}'\mathbf{D}^{-1}\mathbf{u}$ term serves to prevent arbitrary values of $\mathbf{u}$ from being selected and forces them to be closer to zero (a shrinkage effect). Methods to solve these equations are thus frequently called *penalized quasi-likelihood* (PQL) methods. Green (1990), Schall (1991), and Wolfinger (1993) all discuss methods of this type.

In the "derivation" of the PQL equations quite a few approximations of undetermined accuracy are bandied about and the development has an air of ad hocery. How well do these methods work in practice? Unfortunately, not very.

Breslow and Lin (1995), Lin and Breslow (1996), and Neuhaus and Segal (1997) show that PQL methods lead to estimators which are asymptotically biased and hence inconsistent. Of course, inconsistency in itself may not be a worry if the asymptotic bias is small and the the small- or moderate-sized sample performance is good. After all, even full ML is not unbiased in small- or moderate-sized samples. Unfortunately, for situations like paired binary data the PQL estimator can perform quite badly. Its performance improves as the conditional distribution of $\mathbf{y}$ given $\mathbf{u}$ gets closer to normal (and the Laplace approximation becomes more accurate), for example with a Poisson distribution with mean 7 or greater. However, from a practical point of view, we may prefer to transform such data to make them approximately normal and use LMM methods. We thus cannot recommend the use of simple PQL methods in practice.

Recently, there have been improvements in PQL methods (using more accurate Taylor expansions, e.g., Raudenbush et al., 2000) that may lead to better-performing estimators. However, these have not yet been fully tested.

14.5   ITERATIVE BOOTSTRAP BIAS CORRECTION

As the previous sections of this chapter indicate, adding random effects to statistical models leads to computational challenges for parameter estimation. As an alternative to estimators requiring computationally demanding techniques such as numerical quadrature or an EM algorithm, one could base an estimation procedure on an easy-to-compute, but biased estimator, such as obtained using PQL, and then correct the bias to yield asymptotically unbiased estimates. Kuk (1995) proposes a method that iteratively corrects the bias in a biased estimator $\tilde{\xi}$ obtained from an estimating function $\Psi(\xi; y) = 0$ using a parametric bootstrap procedure. Specifically, letting ξ represent the regression coefficients and variance components of a generalized linear mixed model, Kuk's approach calculates updated estimates of bias using

$$b_M^{(k+1)} = g_M(\tilde{\xi} - b^{(k)}) - (\tilde{\xi} - b^{(k)}) \tag{14.63}$$

where

$$g_M(\xi) = (1/M) \sum_{i=1}^{M} \tilde{\xi}(y_i)$$

and one generates $y_1, \ldots, y_M$ as parametric bootstrap samples from the GLMM with parameter ξ. Upon convergence of (14.63), the bias-corrected estimator is

$$\hat{\xi} = \tilde{\xi} - b_M .$$

Kuk (1995) also provides estimates of var$(\hat{\xi})$.

14.6   EXERCISES

E 14.1   If $\mathbf{u}_i$ of order q_i is distributed $\mathcal{N}(\mathbf{0}, \mathbf{I}\sigma_i^2)$ show that the ML estimator of σ_i^2 is $\hat{\sigma}_i^2 = \mathbf{u}_i'\mathbf{u}_i/q_i$.

E 14.2   For w_k of (14.30) show that $\sum_k w_k = \sqrt{\pi}$ for any order Gauss-Hermite quadrature.

E 14.3   Calculate $\int_{-\infty}^{\infty} (1 + x^2) \dfrac{e^{-x^2/2}}{\sqrt{2\pi}}\, dx$ both analytically and using 3-point Gauss-Hermite quadrature. What relationship is there between w_1, w_2, and w_3?

E 14.4   Calculate P$\{Z > 1.7\}$ when $Z \sim \mathcal{N}(0, 1)$ using 3-, 4- and 5-point quadrature and compare to the value from a table. Is the approximation likely to improve by using a slightly higher-order quadrature? Why or why not?

E 14.5 Consider a nested logit-normal model:

$$y_{ijk}|\mathbf{a}, \mathbf{b} \sim \text{indep. Bernoulli}(p_{ijk})$$

$$\text{logit}(p_{ijk}) = \mu + a_i + b_{ij}$$

$$a_i \sim \text{i.i.d. } \mathcal{N}(0, \sigma_a^2) \text{ independently of}$$

$$b_{ij} \sim \text{i.i.d. } \mathcal{N}(0, \sigma_b^2).$$

Write the likelihood in as simple a form as possible with regard to the integrations involved.

E 14.6 Derive (14.40) and (14.41).

E 14.7 For $u_i \sim$ i.i.d. $\mathcal{N}(0, \sigma^2)$ and hence $\mathbf{D} = \mathbf{I}\sigma^2$, write out (14.41).

E 14.8 Derive (14.44).

E 14.9 Show that (14.50) is an unbiased estimator of the likelihood independent of the choice of $h_U(\mathbf{u})$.

E 14.10 Show that $h_U(\mathbf{u}) = f_{U|Y}(\mathbf{u}|\mathbf{y})$ is the optimal choice of $h_U(\mathbf{u})$ in the sense that it gives a zero variance estimator for (14.50).

E 14.11 For the case of a scalar u, derive (14.51) by approximating $h(u)$ in a second-order Taylor series about the point u_0 for which $h'(u_0) = 0$.

E 14.12 Show that the Laplace approximation is exact for the case of a linear mixed model.

E 14.13 Why might we expect the Laplace approximation method of Section 14.4 to work for models such as (11.7) when it fails for some of the models discussed previously?

Appendix M: Some Matrix Results

We assume that readers of this book have a working knowledge of matrix algebra. Nevertheless, we provide a few reminders in this appendix.

M.1 VECTORS AND MATRICES OF ONES

Vectors having every element equal to unity are denoted by $\mathbf{1}$: Thus $\mathbf{1}_3' = [\, 1 \quad 1 \quad 1 \,]$. With $\mathbf{x}' = [\, x_1 \quad x_2 \quad x_3 \,]$, $\mathbf{1}'\mathbf{x} = \sum_{i=1}^{3} x_i$. The inner product of $\mathbf{1}_n$ with itself is n : $\mathbf{1}_n'\mathbf{1}_n = n$ and outer products of these vectors with each other are matrices having every element unity. They are denoted by $\mathbf{J}$. For example,

$$\mathbf{1}_2\mathbf{1}_3' = \begin{bmatrix} 1 \\ 1 \end{bmatrix} \begin{bmatrix} 1 & 1 & 1 \end{bmatrix} = \begin{bmatrix} 1 & 1 & 1 \\ 1 & 1 & 1 \end{bmatrix} = \mathbf{J}_{2\times 3}.$$

Square $\mathbf{J}$-matrices are the most common form: $\mathbf{1}_n\mathbf{1}_n' = \mathbf{J}_n$. Products of $\mathbf{J}$s with each other and with $\mathbf{1}$s are, respectively, $\mathbf{J}$s and $\mathbf{1}$s multiplied by scalars. For square $\mathbf{J}$

$$\mathbf{J}_n^2 = n\mathbf{J}_n \quad \text{and} \quad \mathbf{J}_n\mathbf{1}_n = n\mathbf{1}_n; \quad \text{also} \quad \text{tr}(\mathbf{J}_n) = n.$$

Illustration The mean and variance of data $x_1, x_2, \ldots, x_n$ are easily expressed in terms of the preceding matrices. Thus

$$\bar{x} = \sum_{i=1}^{n} \frac{x_i}{n} = \frac{\mathbf{1}'\mathbf{x}}{n} = \frac{\mathbf{x}'\mathbf{1}}{n}, \quad \text{and}$$

$$s^2 = \frac{1}{n-1}\sum_{i=1}^{n}(x_i - \bar{x})^2 = \frac{1}{n-1}\mathbf{x}'\left(\mathbf{I}_n - \frac{1}{n}\mathbf{J}_n\right)\mathbf{x}.$$

344

Linear combinations of $\mathbf{I}$ (an identity matrix) and $\mathbf{J}$ often arise in a variety of circumstances, for which the following results are often found useful.

1. $(a\mathbf{I}_n + b\mathbf{J}_n)(\alpha\mathbf{I}_n + \beta\mathbf{J}_n) = a\alpha\mathbf{I}_n + (a\beta + b\alpha + b\beta n)\mathbf{J}_n$.

2. $(a\mathbf{I}_n + b\mathbf{J}_n)^{-1} = \frac{1}{a}\left(\mathbf{I}_n - \frac{b}{a+nb}\mathbf{J}_n\right)$, for $a \neq 0$ and $a \neq -nb$.

3. $|a\mathbf{I}_n + b\mathbf{J}_n| = a^{n-1}(a + nb)$.

4. Eigenroots of $a\mathbf{I}_n + b\mathbf{J}_n$ are a, with multiplicity $n-1$, and $a+nb$.

M.2   KRONECKER (OR DIRECT) PRODUCTS

The Kronecker product of two matrices $\mathbf{A} = \{a_{ij}\}$ and $\mathbf{B} = \{b_{ij}\}$ is

$$\mathbf{A} \otimes \mathbf{B} = \{a_{ij}\mathbf{B}\}.$$

Examples arising in linear models are

$$
\mathbf{I}_2 \otimes \mathbf{1}_3 =
\begin{bmatrix} 1(\mathbf{1}_3) & 0(\mathbf{1}_3) \\ 0(\mathbf{1}_3) & 1(\mathbf{1}_3) \end{bmatrix}
=
\begin{bmatrix}
1 & \cdot \\
1 & \cdot \\
1 & \cdot \\
\cdot & 1 \\
\cdot & 1 \\
\cdot & 1
\end{bmatrix}
\quad \text{and} \quad
\mathbf{1}_2 \otimes \mathbf{I}_3 =
\begin{bmatrix}
1 & \cdot & \cdot \\
\cdot & 1 & \cdot \\
\cdot & \cdot & 1 \\
1 & \cdot & \cdot \\
\cdot & 1 & \cdot \\
\cdot & \cdot & 1
\end{bmatrix}.
$$

Kronecker products have many properties. Assuming conformability

$$
\begin{aligned}
(\mathbf{A} \otimes \mathbf{B})' &= \mathbf{A}' \otimes \mathbf{B}' \\[4pt]
(\mathbf{A} \otimes \mathbf{B})^{-1} &= \mathbf{A}^{-1} \otimes \mathbf{B}^{-1} \\[4pt]
(\mathbf{A} \otimes \mathbf{B})(\mathbf{X} \otimes \mathbf{Y}) &= \mathbf{A}\mathbf{X} \otimes \mathbf{B}\mathbf{Y} \\[4pt]
\operatorname{rank}(\mathbf{A} \otimes \mathbf{B}) &= \operatorname{rank}(\mathbf{A})\operatorname{rank}(\mathbf{B}) \\[4pt]
\operatorname{tr}(\mathbf{A} \otimes \mathbf{B}) &= \operatorname{tr}(\mathbf{A})\operatorname{tr}(\mathbf{B}) \\[4pt]
|\mathbf{A}_{a\times a} \otimes \mathbf{B}_{b\times b}| &= |\mathbf{A}|^b|\mathbf{B}|^a.
\end{aligned}
$$

M.3   A MATRIX NOTATION IN TERMS OF ELEMENTS

Familiar notation for $\mathbf{A}$ of order $p \times q$ is

$$\mathbf{A} = \{a_{ij}\} \text{ for } i = 1, \ldots, p \text{ and } j = 1, \ldots, q,$$

where a_{ij} is the element in row i and column j of $\mathbf{A}$. We abbreviate this to

$$\mathbf{A} = \left\{ {}_m a_{ij} \right\}_{i=1, j=1}^{p \quad q} = \left\{ {}_m a_{ij} \right\}_{ij} = \left\{ {}_m a_{ij} \right\}$$

using only as much detail concerning i and j as is needed for context. Similarly we use

$$\mathbf{u} = \left\{ {}_c u_i \right\}_{i=1}^{q} \text{ and } \mathbf{u}' = \left\{ {}_r u_i \right\}_{i=1}^{q}$$

for a column and a row, respectively, of elements u_i. Also we use $\left\{ {}_d x_i \right\}_{i=1}^{t}$ for a diagonal matrix with t diagonal elements x_i.

The advantage of this notation is, for example, that instead of writing $\mathbf{A} = \{a_{ij}\}$ for $i = 1, \ldots, p$ and $j = 1, \ldots, q$, and $\mathbf{u}$ as a column vector of elements u_i for $i = 1, \ldots, q$ with $\mathbf{Au} = \sum_{j=1}^{q} a_{ij} u_j$ for $i = 1, \ldots, p$, one has no need of the symbols $\mathbf{A}$ and $\mathbf{u}$ but simply writes

$$\left\{ {}_m a_{ij} \right\}_{i=1, j=1}^{p \quad q} \left\{ {}_c u_j \right\}_{j=1}^{q} = \left\{ {}_c \sum_{j=1}^{q} a_{ij} u_j \right\}_{i=1}^{p}.$$

M.4   GENERALIZED INVERSES

a.   Definition

Readers will be familiar with a nonsingular matrix $\mathbf{T}$ being a square matrix that has an inverse $\mathbf{T}^{-1}$ such that $\mathbf{TT}^{-1} = \mathbf{T}^{-1}\mathbf{T} = \mathbf{I}$. More generally, for any non-null matrix $\mathbf{A}$, be it rectangular, or square and singular, there are always matrices $\mathbf{A}^-$ satisfying

$$\mathbf{AA}^-\mathbf{A} = \mathbf{A}. \tag{M.1}$$

When $\mathbf{A}$ is non-singular, (M.1) leads to $\mathbf{A}^- = \mathbf{A}^{-1}$, but otherwise there is an infinite number of matrices $\mathbf{A}^-$ that, for each $\mathbf{A}$, satisfy (M.1). Each such $\mathbf{A}^-$ is called a *generalized inverse* of $\mathbf{A}$.

Example: For

$$
\mathbf{A} = \begin{bmatrix} 1 & 2 & 3 & 2 \\ 3 & 7 & 11 & 4 \\ 4 & 9 & 14 & 6 \end{bmatrix}, \quad \mathbf{A}^- = \begin{bmatrix} 7-t & -2-t & t \\ -3+2t & 1+2t & -2t \\ -t & -t & t \\ 0 & 0 & 0 \end{bmatrix}.
$$

Calculation of $\mathbf{A}\mathbf{A}^-\mathbf{A}$ yields $\mathbf{A}$ no matter what value of t is used, thus illustrating the existence of infinitely many matrices $\mathbf{A}^-$ satisfying (M.1).

A great deal has been written about generalized inverse matrices, with much of what is useful for linear models being available in books such as Rao (1962) and Searle (1997) and many others. We direct attention here solely to generalized inverses of $\mathbf{X}'\mathbf{X}$ and their properties, which are extremely useful in solving the normal equations $\mathbf{X}'\mathbf{X}\beta^0 = \mathbf{X}'\mathbf{y}$ of (4.18) or their more general form $\mathbf{X}'\mathbf{V}^{-1}\mathbf{X}\beta^0 = \mathbf{X}'\mathbf{V}^{-1}\mathbf{y}$ of (6.19).

b. Generalized inverses of $\mathbf{X}'\mathbf{X}$

Clearly $\mathbf{X}'\mathbf{X}$ is square and symmetric; its generalized inverses are denoted by $(\mathbf{X}'\mathbf{X})^-$ and $\mathbf{G}$ interchangeably. Thus $\mathbf{G}$ is defined as

$$\mathbf{X}'\mathbf{X}\mathbf{G}\mathbf{X}'\mathbf{X} = \mathbf{X}'\mathbf{X}. \tag{M.2}$$

Note that although $\mathbf{X}'\mathbf{X}$ is symmetric, $\mathbf{G}$ need not be symmetric. For example,

$$
\mathbf{X}'\mathbf{X} = \begin{bmatrix} 7 & 3 & 2 & 2 \\ 3 & 3 & \cdot & \cdot \\ 2 & \cdot & 2 & \cdot \\ 2 & \cdot & \cdot & 2 \end{bmatrix} \quad \text{has} \quad \mathbf{G} = \begin{bmatrix} 9 & 0 & 0 & 3 \\ 5 & -13\frac{2}{3} & -14 & 17 \\ 1 & -10 & -9\frac{1}{2} & -13 \\ 0 & -9 & -9 & -11\frac{1}{2} \end{bmatrix} \tag{M.3}
$$

as a non-symmetric generalized inverse. Despite this, transposing (M.2) shows that when $\mathbf{G}$ is a generalized inverse of $\mathbf{X}'\mathbf{X}$, then so also is $\mathbf{G}'$. As a consequence, as may easily be verified,

$$(\mathbf{X}'\mathbf{X})^- = \mathbf{G}\mathbf{X}'\mathbf{X}\mathbf{G}' \tag{M.4}$$

is a symmetric generalized inverse of $\mathbf{X}'\mathbf{X}$.

The following theorem is vital for linear model theory.

Theorem M.1. When $\mathbf{G}$ is a generalized inverse of $\mathbf{X'X}$:

$\mathbf{G'}$ is also a generalized inverse of $\mathbf{X'X}$,   (M.5)

$\mathbf{XGX'X} = \mathbf{X}$,   (M.6)

$\mathbf{XGX'}$ is invariant to $\mathbf{G}$; i.e., is the same for

every $\mathbf{G}$,   (M.7)

$\mathbf{XGX'}$ is symmetric, whether or not $\mathbf{G}$ is,   (M.8)

$\mathbf{XGX'1} = \mathbf{1}$ when $\mathbf{1}$ is a column of $\mathbf{X}$.   (M.9)

Proof. Condition (M.5) comes from transposing (M.2). Result (M.6) is true because, for real matrices, there is a theorem (e.g., Searle, 1982, p. 63) indicating that if $\mathbf{PX'X} = \mathbf{QX'X}$ then $\mathbf{PX'} = \mathbf{QX'}$; applying this to the transpose of (M.2) and then transposing it yields (M.6); and applying it to $\mathbf{XGX'X} = \mathbf{X} = \mathbf{XFX'X}$ for $\mathbf{F}$ being any other generalized inverse of $\mathbf{X'X}$ yields (M.7). Using $(\mathbf{X'X})^-$ of (M.4) in place of $\mathbf{G}$ in $\mathbf{X'GX}$ demonstrates the symmetry of (M.8) which, by (M.7), therefore holds for any $\mathbf{G}$. Finally, (M.9) follows from considering an individual column of $\mathbf{X}$ in (M.6).   *Q.E.D.*

Notice that (M.5) and (M.6) spawn three other results similar to (M.6): $\mathbf{XG'X'X} = \mathbf{X}$, $\mathbf{X'XGX'} = \mathbf{X'}$, and $\mathbf{X'XG'X'} = \mathbf{X'}$.

A particularly useful matrix is $\mathbf{M} = \mathbf{I} - \mathbf{XGX'}$. Theorem M.1 provides the means for verifying that $\mathbf{M}$ has the following properties: $\mathbf{M}$ is symmetric, idempotent, invariant to $\mathbf{G}$, of rank $N - r_\mathbf{X}$ when $\mathbf{X}$ has N rows, and its products with $\mathbf{X}$ and $\mathbf{X'}$ are null. Thus

$$\mathbf{M} = \mathbf{M'} = \mathbf{M}^2, r_\mathbf{M} = N - r_\mathbf{X}, \mathbf{MX} = \mathbf{0} \text{ and } \mathbf{X'M} = \mathbf{0}. \quad (\text{M.10})$$

c.   Two results involving $\mathbf{X}(\mathbf{X'V}^{-1}\mathbf{X})^-\mathbf{X'V}^{-1}$

For $\mathbf{V}$ being symmetric and positive definite [as it usually is when it is var$(\mathbf{y})$]

$$\mathbf{X}(\mathbf{X'V}^{-1}\mathbf{X})^-\mathbf{X'V}^{-1} \text{ is invariant to } (\mathbf{X'V}^{-1}\mathbf{X})^- \quad (\text{M.11})$$

and

$$\mathbf{X}(\mathbf{X'V}^{-1}\mathbf{X})^-\mathbf{X'V}^{-1}\mathbf{X} = \mathbf{X}.$$

Proof of these results stems from the nature of $\mathbf{V}$ ($= \mathbf{V}'$ and p.s.d.) enabling us to write $\mathbf{V}^{-1} = \mathbf{L}'\mathbf{L}$ for some $\mathbf{L}$. Then

$$\mathbf{X}(\mathbf{X}'\mathbf{V}^{-1}\mathbf{X})^{-}\mathbf{X}'\mathbf{V}^{-1} = \mathbf{V}\mathbf{V}^{-1}\mathbf{X}(\mathbf{X}'\mathbf{V}^{-1}\mathbf{X})^{-}\mathbf{X}'\mathbf{V}^{-1}$$

$$= \mathbf{V}\mathbf{L}'\mathbf{L}\mathbf{X}(\mathbf{X}'\mathbf{L}'\mathbf{L}\mathbf{X})^{-}\mathbf{X}'\mathbf{L}'\mathbf{L}$$

$$= \mathbf{V}\mathbf{L}'\mathbf{T}(\mathbf{T}'\mathbf{T})^{-}\mathbf{T}'\mathbf{L} \text{ for } \mathbf{T} = \mathbf{L}\mathbf{X},$$

which by (M.6) is invariant to the generalized inverse.

Also

$$\mathbf{X}(\mathbf{X}'\mathbf{V}^{-1}\mathbf{X})^{-}\mathbf{X}'\mathbf{V}^{-1}\mathbf{X} = \mathbf{V}\mathbf{L}'\mathbf{T}(\mathbf{T}'\mathbf{T})^{-}\mathbf{T}'\mathbf{T}$$

$$= \mathbf{V}\mathbf{L}'\mathbf{T} \qquad \text{by (M.6)}$$

$$= \mathbf{X}. \qquad\qquad (\text{M.12})$$

d. Solving linear equations

Rao (1962) shows that equations $\mathbf{A}\mathbf{x} = \mathbf{y}$ have solutions $\mathbf{A}^{-}\mathbf{y} + (\mathbf{I} - \mathbf{A}^{-}\mathbf{A})\mathbf{z}$ for any $\mathbf{z}$ of the appropriate order. The simplest application of this to

$$\mathbf{X}'\mathbf{X}\beta^0 = \mathbf{X}'\mathbf{y} \text{ is } \beta^0 = (\mathbf{X}'\mathbf{X})^{-}\mathbf{X}'\mathbf{y}$$

for any $(\mathbf{X}'\mathbf{X})^{-}$. Equation (M.7) ensures that $\mathbf{X}\beta^0 = \mathbf{X}(\mathbf{X}'\mathbf{X})^{-}\mathbf{X}'\mathbf{y}$ is invariant to $(\mathbf{X}'\mathbf{X})^{-}$. Extension to $\mathbf{X}'\mathbf{V}^{-1}\mathbf{X}\beta^0 = \mathbf{X}'\mathbf{V}^{-1}\mathbf{y}$ is obvious.

e. Rank results

The standard result for the rank of a product matrix is $r_{\mathbf{A}\mathbf{B}} \le r_{\mathbf{B}}$. Thus using $r(\mathbf{X})$ and $r_{\mathbf{X}}$ interchangeably to represent the rank of $\mathbf{X}$, we have $r(\mathbf{A}\mathbf{A}^{-}) \le r_{\mathbf{A}}$; and from $\mathbf{A} = \mathbf{A}\mathbf{A}^{-}\mathbf{A}$ we have $r_{\mathbf{A}} \le r(\mathbf{A}\mathbf{A}^{-})$. Therefore $r(\mathbf{A}\mathbf{A}^{-}) = r_{\mathbf{A}}$. And so, because (M.5) shows that $(\mathbf{X}'\mathbf{X})^{-}\mathbf{X}'$ is a generalized inverse of $\mathbf{X}$, $r[\mathbf{X}(\mathbf{X}'\mathbf{X})^{-}\mathbf{X}'] = r_{\mathbf{X}}$.

f. Vectors orthogonal to columns of X

Suppose $\mathbf{k}'$ is such that $\mathbf{k}'\mathbf{X} = \mathbf{0}$. Then $\mathbf{X}'\mathbf{k} = 0$ and, from the theory of solving linear equations (e.g., Searle, 1982, Sec. 9.4b), $\mathbf{k} = [\mathbf{I} - (\mathbf{X}')^{-}\mathbf{X}']\mathbf{c}$ for any vector $\mathbf{c}$, of the appropriate order. Therefore, since $(\mathbf{X}')^{-}$ is a generalized inverse of $\mathbf{X}'$ we can write $\mathbf{k}' = \mathbf{c}'(\mathbf{I} - \mathbf{X}\mathbf{X}^{-})$.

Moreover, because $(\mathbf{X'X})^-\mathbf{X'}$ is a generalized inverse of $\mathbf{X}$, another form for $\mathbf{k'}$ is $\mathbf{k'} = \mathbf{c'}[\mathbf{I} - \mathbf{X}(\mathbf{X'X})^-\mathbf{X'}]$. Thus two forms of $\mathbf{k}$ are

$$\mathbf{k'} = \mathbf{c'}[\mathbf{I} - \mathbf{XX}^-] \quad \text{or} \quad \mathbf{k'} = \mathbf{c'}[\mathbf{I} - \mathbf{X}(\mathbf{X'X})^-\mathbf{X'}] = \mathbf{c'M},$$

for $\mathbf{M} = \mathbf{I} - \mathbf{X}(\mathbf{X'X})^-\mathbf{X'}$ of Section M.4b.

With $\mathbf{X}$ of order $N \times p$ of rank r, there are only $N - r$ linearly independent vectors $\mathbf{k'}$ satisfying $\mathbf{k'X} = \mathbf{0}$ (e.g., Searle, 1982, Sec. 9.7a). Using a set of such $N - r$ linearly independent vectors $\mathbf{k'}$ as rows of $\mathbf{K'}$, we then have the following theorem, for $\mathbf{K'X} = \mathbf{0}$ with $\mathbf{K'}$ having maximum row rank $N - r$, and with $\mathbf{K'} = \mathbf{C'M}$ for some $\mathbf{C}$.

g. A theorem for $\mathbf{K'}$ with $\mathbf{K'X}$ being null

Theorem M.2. If $\mathbf{K'X} = \mathbf{0}$, where $\mathbf{K'}$ has maximum row rank and $\mathbf{V}$ is positive definite then

$$\mathbf{K}(\mathbf{K'VK})^{-1}\mathbf{K'} = \mathbf{P}$$

for

$$\mathbf{P} \equiv \mathbf{V}^{-1} - \mathbf{V}^{-1}\mathbf{X}(\mathbf{X'V}^{-1}\mathbf{X})^-\mathbf{X'V}^{-1}.$$

The proof of this is lengthy and technical and we do not show it here; it can be found in VC p. 452.

M.5 DIFFERENTIAL CALCULUS

a. Definition

Differentiation with respect to elements of a vector $\mathbf{x} = \left\{{}_c\, x_i\right\}_{i=1}^{k}$ is defined by the notation

$$\frac{\partial}{\partial \mathbf{x}} = \begin{bmatrix} \dfrac{\partial}{\partial x_1} \\ \dfrac{\partial}{\partial x_2} \\ \vdots \\ \dfrac{\partial}{\partial x_k} \end{bmatrix}.$$

b. Scalars

Thus

$$\frac{\partial}{\partial \mathbf{x}}(\mathbf{a'x}) = \frac{\partial}{\partial \mathbf{x}}(\mathbf{x'a}) = \mathbf{a}. \tag{M.13}$$

c. Vectors

For $\mathbf{y}' = [\begin{array}{cccc} y_1 & y_2 & \cdots & y_p \end{array}]$

$$\frac{\partial \mathbf{y}'}{\partial \mathbf{x}} = \left\{ {}_m \frac{\partial y_j}{\partial x_i} \right\}_{i=1,j=1}^{k \quad p}, \quad \text{a matrix of order } k \times p.$$

Then

$$\frac{\partial \mathbf{x}'}{\partial \mathbf{x}} = \mathbf{I} \tag{M.14}$$

and for $\mathbf{A}$ not involving $\mathbf{x}$

$$\frac{\partial}{\partial \mathbf{x}}(\mathbf{x}'\mathbf{A}) = \frac{\partial \mathbf{x}'}{\partial \mathbf{x}}\mathbf{A} = \mathbf{A}. \tag{M.15}$$

d. Inner products

Consider $\mathbf{u}$ and $\mathbf{v}$, of the same order, each having elements that are functions of the elements of $\mathbf{x}$. Then $\mathbf{u}'\mathbf{v}$ is a scalar, and so by (M.13) $\partial(\mathbf{u}'\mathbf{v})/\partial \mathbf{x}$ is a column. Therefore, because differentiating the $\mathbf{u}'$ part of $\mathbf{u}'\mathbf{v}$ gives $(\partial \mathbf{u}'/\partial \mathbf{x})\mathbf{v}$ and because $\mathbf{u}'\mathbf{v} = \mathbf{v}'\mathbf{u}$, we have

$$\frac{\partial \mathbf{u}'\mathbf{v}}{\partial \mathbf{x}} = \frac{\partial \mathbf{u}'}{\partial \mathbf{x}}\mathbf{v} + \frac{\partial \mathbf{v}'}{\partial \mathbf{x}}\mathbf{u}. \tag{M.16}$$

e. Quadratic forms

To differentiate $\mathbf{x}'\mathbf{A}\mathbf{x}$ with respect to $\mathbf{x}$, use (M.16) with $\mathbf{u}'$ and $\mathbf{v}$ being $\mathbf{x}'$ and $\mathbf{A}\mathbf{x}$ respectively. This gives

$$\begin{aligned}\frac{\partial}{\partial \mathbf{x}}\mathbf{x}'\mathbf{A}\mathbf{x} &= \frac{\partial \mathbf{x}'}{\partial \mathbf{x}}\mathbf{A}\mathbf{x} + \frac{\partial \mathbf{A}\mathbf{x}}{\partial \mathbf{x}}\mathbf{x} \\[2mm] &= \mathbf{A}\mathbf{x} + \mathbf{A}'\mathbf{x} \\[2mm] &= 2\mathbf{A}\mathbf{x} \text{ when } \mathbf{A} \text{ is symmetric,} \end{aligned} \tag{M.17}$$

which it usually is.

f. Inverse matrices

If $\mathbf{V}$ is non-singular of order n and has elements which are functions of a scalar w, differentiating $\mathbf{V}^{-1}$ with respect to w comes from differentiating the identity $\mathbf{V}^{-1}\mathbf{V} = \mathbf{I}$. Thus

$$\frac{\partial \mathbf{V}^{-1}}{\partial w}\mathbf{V} + \mathbf{V}^{-1}\frac{\partial \mathbf{V}}{\partial w} = 0$$

and so

$$\frac{\partial \mathbf{V}^{-1}}{\partial w} = -\mathbf{V}^{-1}\frac{\partial \mathbf{V}}{\partial w}\mathbf{V}^{-1}, \tag{M.18}$$

where

$$\frac{\partial \mathbf{V}}{\partial w} = \left\{_m \frac{\partial v_{ij}}{\partial w}\right\}_{i,j=1}^{n}.$$

Note that (M.18) is a special case of (6.75) for generalized inverses.

Finally, using $\mathbf{P} = \mathbf{K}(\mathbf{K}'\mathbf{V}\mathbf{K})^{-1}\mathbf{K}'$ note that

$$\frac{\partial \mathbf{P}}{\partial w} = -\mathbf{K}(\mathbf{K}'\mathbf{V}\mathbf{K})^{-1}\mathbf{K}'\frac{\partial \mathbf{V}}{\partial w}\mathbf{K}(\mathbf{K}'\mathbf{V}\mathbf{K})^{-1}\mathbf{K}'$$

$$= -\mathbf{P}\frac{\partial \mathbf{V}}{\partial w}\mathbf{P}. \tag{M.19}$$

g. Determinants

$$\frac{\partial \log |\mathbf{V}|}{\partial w} = \frac{1}{|\mathbf{V}|}\frac{\partial |\mathbf{V}|}{\partial w} = \sum_i \sum_j \frac{|\mathbf{V}_{ij}|}{|\mathbf{V}|}\frac{\partial v_{ij}}{\partial w}$$

$$= \sum_i \sum_j v^{ij}\frac{\partial v_{ij}}{\partial w} = \mathrm{tr}\left(\mathbf{V}^{-1}\frac{\partial \mathbf{V}}{\partial w}\right), \tag{M.20}$$

with v_{ij} being an element of $\mathbf{V}$, $|\mathbf{V}_{ij}|$ its cofactor, and $v^{ij} = |\mathbf{V}_{ij}|/|\mathbf{V}|$ representing an element of $\mathbf{V}^{-1}$. The last step arises from $\mathbf{V}$ being symmetric. Some intermediate details are given in VC pp. 456–457.

Appendix S: Some Statistical Results

As with Appendix M, we assume that a reader's background knowledge includes familiarity with basic mathematical statistics. Nevertheless, here are a few reminders.

S.1   MOMENTS

a.   Conditional moments

For random variables $\mathbf{y}$ and $\mathbf{u}$, let $f(\mathbf{y}, \mathbf{u})$ and $f(\mathbf{y}|\mathbf{u})$ denote, respectively, the joint density of $\mathbf{y}$ and $\mathbf{u}$, and the density of $\mathbf{y}$ conditional on $\mathbf{u}$. Also, let E_Y and var_Y denote expectation and variance with respect to the distribution of $\mathbf{y}$. There are three well-established results (Searle et al., 1992):

$$\mathrm{E}[\mathbf{y}] \quad = \quad \mathrm{E}_{Y,U}[\mathbf{y}] = \mathrm{E}_U\left[\mathrm{E}[\mathbf{y}|\mathbf{u}]\right] \tag{S.1}$$

$$\mathrm{cov}(\mathbf{y}, \mathbf{w}) \quad = \quad \mathrm{E}_U[\mathrm{cov}(\mathbf{y}, \mathbf{w}|\mathbf{u})] + \mathrm{cov}_U\left(\mathrm{E}[\mathbf{y}|\mathbf{u}], \mathrm{E}[\mathbf{w}|\mathbf{u}])\right), \tag{S.2}$$

and using the latter with $\mathbf{w} = \mathbf{y}$,

$$\mathrm{var}(\mathbf{y}) = \mathrm{E}_U[\mathrm{var}(\mathbf{y}|\mathbf{u})] + \mathrm{var}_U\left(\mathrm{E}[\mathbf{y}|\mathbf{u}])\right). \tag{S.3}$$

(S.1) is established as follows:

$$\mathrm{E}[\mathbf{y}] \quad = \quad \int \int \mathbf{y} f(\mathbf{y}|\mathbf{u}) f(\mathbf{u}) \, d\mathbf{y} \, d\mathbf{u}$$

$$= \quad \int \mathrm{E}_Y[\mathbf{y}|\mathbf{u}] d\mathbf{u} = \mathrm{E}_U\left[\mathrm{E}_Y[\mathbf{y}|\mathbf{u}]\right]. \tag{S.4}$$

When $\mathbf{y}$ of $\mathrm{E}[\mathbf{y}]$ is replaced by $(\mathbf{y} - \mathrm{E}[\mathbf{y}])(\mathbf{w} - \mathrm{E}[\mathbf{w}])$, the left-hand side of (S.1) becomes $\mathrm{cov}(\mathbf{y},\mathbf{w})$. That same replacement on the right-hand side of (S.1) followed by some tedious algebra (see VC, p. 462), yields (S.2). And then replacing $\mathbf{w}$ by $\mathbf{y}$ in (S.2) gives it as

$$\mathrm{var}(\mathbf{y}) \;=\; \mathrm{E}_U[\mathrm{cov}(\mathbf{y},\mathbf{y}|\mathbf{u})] + \mathrm{cov}_U\left(\mathrm{E}[\mathbf{y}|\mathbf{u}], \mathrm{E}[\mathbf{y}|\mathbf{u}]\right)$$

$$=\; \mathrm{E}_U[\mathrm{var}(\mathbf{y}|\mathbf{u})] + \mathrm{var}_U\left(\mathrm{E}[\mathbf{y}|\mathbf{u}]\right), \tag{S.5}$$

which is (S.3).

b. Mean of a quadratic form

Suppose $\mathrm{E}[\mathbf{y}] = \boldsymbol{\mu}$ and $\mathrm{var}(\mathbf{y}) = \mathbf{V}$, that is, $\mathbf{y} \sim (\boldsymbol{\mu}, \mathbf{V})$, not necessarily normally distributed. Then, for a quadratic form in $\mathbf{y}$, we have

$$\mathrm{E}[\mathbf{y}'\mathbf{A}\mathbf{y}] \;=\; \mathrm{E}[\mathrm{tr}(\mathbf{y}'\mathbf{A}\mathbf{y})] = \mathrm{E}[\mathrm{tr}(\mathbf{A}\mathbf{y}\mathbf{y}')]$$

$$=\; \mathrm{tr}(\mathbf{A}\,\mathrm{E}[\mathbf{y}\mathbf{y}']) = \mathrm{tr}(\mathbf{A}[\mathbf{V} + \boldsymbol{\mu}\boldsymbol{\mu}'])$$

$$=\; \mathrm{tr}(\mathbf{A}\mathbf{V}) + \boldsymbol{\mu}'\mathbf{A}\boldsymbol{\mu}. \tag{S.6}$$

c. Moment generating function

The moment generating function (m.g.f.) of a random variable $\mathbf{y}$ is a function [carefully defined; see Casella and Berger (1990, p. 61)] of a mathematical variable t. For y having a density $f(y)$ the m.g.f. is

$$M_Y(t) = \mathrm{E}[e^{ty}] = \int_{-\infty}^{\infty} e^{ty} f(y) \, dy. \tag{S.7}$$

This yields the rth moment (about zero) of y as

$$\mu_Y^{(r)} = \left. \frac{\partial^r M_Y(t)}{\partial t^r} \right|_{t=0}. \tag{S.8}$$

Similarly, for a function $h(y)$ of y

$$M_{h(Y)}(t) = \mathrm{E}[e^{th(y)}] \qquad \text{and} \qquad \mu_{h(Y)}^{(r)} = \left. \frac{\partial^r M_{h(Y)}(t)}{\partial t^r} \right|_{t=0}. \tag{S.9}$$

For a vector random variable $\mathbf{y}$, the m.g.f. is

$$M_Y(\mathbf{t}) = \mathrm{E}[e^{\mathbf{t}'\mathbf{y}}]. \tag{S.10}$$

S.2 NORMAL DISTRIBUTIONS

a. Univariate

The scalar random variable y is said to be normally distributed with mean μ and variance σ^2 when it has probability density function

$$f(y) = \frac{e^{-\frac{1}{2}(y-\mu)^2/\sigma^2}}{\sqrt{2\pi\sigma^2}}.$$

We represent this as $y \sim \mathcal{N}(\mu, \sigma^2)$.

b. Multivariate

The vector of n random variables $\mathbf{y}' = [\ y_1\ \ y_2\ \ \cdots\ \ y_n\ ]$ is said to have a multivariate normal distribution with mean vector $\boldsymbol{\mu}$ and non-singular variance-covariance matrix $\mathbf{V}$ when it has probability density function

$$f(\mathbf{y}) = \frac{e^{-\frac{1}{2}(\mathbf{y}-\boldsymbol{\mu})'\mathbf{V}^{-1}(\mathbf{y}-\boldsymbol{\mu})}}{(2\pi)^{n/2}|\mathbf{V}|^{\frac{1}{2}}}. \tag{S.11}$$

This is represented as $\mathbf{y} \sim \mathcal{N}_n(\boldsymbol{\mu}, \mathbf{V})$, often with the subscript n omitted when it is evident from the context. Many texts (e.g., Searle, 1997) have numerous details about these distributions, so we summarize just some of the properties of the multivariate normal, mostly those which are useful to the purposes of this book.

For $\mathbf{y} \sim \mathcal{N}_n(\boldsymbol{\mu}, \mathbf{V})$:

1. $E[\mathbf{y}] = \boldsymbol{\mu}$ and $\text{var}(\mathbf{y}) = \mathbf{V}$.

2. $\mathbf{Ky} \sim \mathcal{N}(\mathbf{K}\boldsymbol{\mu}, \mathbf{KVK}')$.

On writing

$$\mathbf{y} = \begin{bmatrix} \mathbf{y}_1 \\ \mathbf{y}_2 \end{bmatrix} \sim \mathcal{N}\left(\begin{bmatrix} \boldsymbol{\mu}_1 \\ \boldsymbol{\mu}_2 \end{bmatrix}, \begin{bmatrix} \mathbf{V}_{11} & \mathbf{V}_{12} \\ \mathbf{V}_{21} & \mathbf{V}_{22} \end{bmatrix} \right),$$

3. the marginal distribution of $\mathbf{y}_1$ is

$$\mathbf{y}_1 \sim \mathcal{N}(\boldsymbol{\mu}_1, \mathbf{V}_{11}).$$

4. the conditional distribution of $\mathbf{y}_1|\mathbf{y}_2$ is

$$\mathbf{y}_1|\mathbf{y}_2 \sim \mathcal{N}\left(\boldsymbol{\mu}_1 + \mathbf{V}_{12}\mathbf{V}_{22}^{-1}(\mathbf{y}_2 - \boldsymbol{\mu}_2), \mathbf{V}_{11} - \mathbf{V}_{12}\mathbf{V}_{22}^{-1}\mathbf{V}_{21} \right).$$

5. The moment generating function is

$$M_Y(\mathbf{t}) = \mathrm{E}[e^{\mathbf{t}'\mathbf{y}}] = e^{\mathbf{t}'\boldsymbol{\mu} + \frac{1}{2}\mathbf{t}'\mathbf{V}\mathbf{t}}.$$

6. $\mathrm{var}(\mathbf{y}'\mathbf{A}\mathbf{y}) = 2\mathrm{tr}[(\mathbf{A}\mathbf{V})^2] + 4\boldsymbol{\mu}'\mathbf{A}\mathbf{V}\mathbf{A}\boldsymbol{\mu}.$

Extensive details of deriving these, especially properties 4, 5 and 6, are available in Searle (1971, Chap. 2).

c.   Quadratic forms in normal variables

Section S.1b shows the derivation of $\mathrm{E}[\mathbf{y}'\mathbf{A}\mathbf{y}]$ no matter what the distribution of $\mathbf{y}$ is. When that distribution is normal $\mathbf{y}'\mathbf{A}\mathbf{y}$ has three very useful properties, the first of which requires the prelude of describing the non-central chi-square (χ^2) distribution.

– i.   *The non-central χ^2*

For $\mathbf{y}_{n\times 1} \sim \mathcal{N}(\mathbf{0}, \mathbf{I})$, we have the well-known result that $\sum_i y_i^2 = \mathbf{y}'\mathbf{y}$ is distributed as chi-square on n degrees of freedom, that is, $\mathbf{y}'\mathbf{y} \sim \chi_n^2$. A common variant of this is that when $\mathbf{y}_{n\times 1} \sim \mathcal{N}(\mu\mathbf{1}, \mathbf{I})$ then $\sum_{i=1}^n (y_i - \bar{y})^2 \sim \chi_{n-1}^2$. An extension of these two cases is when $\mathbf{y} \sim \mathcal{N}(\boldsymbol{\mu}, \mathbf{I})$. Then the resulting distribution of $\mathbf{y}'\mathbf{y}$ is known as the non-central chi-square. It is akin to the customary χ^2 (now called the central χ^2), with degrees of freedom n, but with a second parameter $\lambda = \frac{1}{2}\boldsymbol{\mu}'\boldsymbol{\mu}$, known as the non-centrality parameter. And when $\boldsymbol{\mu} = \mathbf{0}$ the non-central chi-square [denoted by $\chi^{2'}(n, \lambda)$] reduces to being the central χ^2.

– ii.   *Properties of $\mathbf{y}'\mathbf{A}\mathbf{y}$ when $\mathbf{y} \sim \mathcal{N}(\boldsymbol{\mu}, \mathbf{V})$*

When $\mathbf{y} \sim \mathcal{N}(\boldsymbol{\mu}, \mathbf{V})$

$$\mathbf{y}'\mathbf{A}\mathbf{y} \quad \sim \quad \chi^{2'}(r_\mathbf{A}, \tfrac{1}{2}\boldsymbol{\mu}'\mathbf{A}\boldsymbol{\mu}) \text{ if and only if } \mathbf{A}\mathbf{V} \text{ is idempotent;}$$

$$\mathbf{y}'\mathbf{A}\mathbf{y} \quad \text{and} \quad \mathbf{y}'\mathbf{B}\mathbf{y} \text{ are independent if and only if } \mathbf{A}\mathbf{V}\mathbf{B} = \mathbf{0}.$$

Details and proofs of these widely known results can be found in Searle (1997, Chap. 2). The sufficient condition in each is easily proven, whereas the necessity conditions are not. Driscoll and Gundberg (1986) and Driscoll and Krasnicka (1995) have an interesting history of these necessity conditions.

Two further results for $\mathbf{y} \sim \mathcal{N}(\boldsymbol{\mu}, \mathbf{V})$ are

$$\text{var}(\mathbf{y}'\mathbf{A}\mathbf{y}) = 2\,\text{tr}[(\mathbf{A}\mathbf{V})^2] + 4\boldsymbol{\mu}'\mathbf{A}\mathbf{V}\mathbf{A}\boldsymbol{\mu} \text{ and}$$

$$\text{cov}(\mathbf{y}'\mathbf{A}\mathbf{y}, \mathbf{y}'\mathbf{B}\mathbf{y}) = 2\,\text{tr}(\mathbf{A}\mathbf{V}\mathbf{B}\mathbf{V}) + 4\boldsymbol{\mu}'\mathbf{A}\mathbf{V}\mathbf{B}\boldsymbol{\mu}.$$

The first of these two is a special case of the more general result that the kth cumulant of $\mathbf{y}'\mathbf{A}\mathbf{y}$ is $2^{k-1}(k-1)![\text{tr}(\mathbf{A}\mathbf{V})^k + k\boldsymbol{\mu}'\mathbf{A}(\mathbf{V}\mathbf{A})^{k-1}\boldsymbol{\mu}]$. And the second of the two comes from applying the first to $\text{var}[\mathbf{y}'(\mathbf{A}+\mathbf{B})\mathbf{y}]$.

S.3   EXPONENTIAL FAMILIES

Probability densities which can be written in the form

$$f(\mathbf{y}; \boldsymbol{\theta}) = h(\mathbf{y})d(\boldsymbol{\theta})\exp\{\boldsymbol{\nu}(\boldsymbol{\theta})'\mathbf{T}(\mathbf{y})\}$$

$$= h(\mathbf{y})d(\boldsymbol{\theta})\exp\left\{\sum_{i=1}^{k}\nu_i(\boldsymbol{\theta})'T_i(\mathbf{y})\right\}, \qquad \text{(S.12)}$$

are said to constitute an *exponential family*. Many of the commonly used distributions are of this form: for example, normal, gamma, beta, binomial, and Poisson. An important consequence of the form (S.12) is that the sufficient statistics are $[T_1(\mathbf{y}), T_2(\mathbf{y}), \dots, T_k(\mathbf{y})]'$.

S.4   MAXIMUM LIKELIHOOD

a.   The likelihood function

Suppose a vector of random variables, $\mathbf{y}$, has density function $f(\mathbf{y})$. Let $\boldsymbol{\theta}$ be the vector of parameters involved in $f(\mathbf{y})$. Then $f(\mathbf{y})$ is a function of both $\mathbf{y}$ and $\boldsymbol{\theta}$. As a result, it can be viewed in two different ways. The first is (as above) as a density function, in which case $\boldsymbol{\theta}$ is usually assumed to be known. With this in mind we use the symbol $f(\mathbf{y}|\boldsymbol{\theta})$ in place of $f(\mathbf{y})$ to emphasize that $\boldsymbol{\theta}$ is being taken as known.

A second viewpoint is where $\mathbf{y}$ represents a known vector of data and where $\boldsymbol{\theta}$ is unknown. Then $f(\mathbf{y})$ will be a function of just $\boldsymbol{\theta}$. It is called the *likelihood function* for the data $\mathbf{y}$; and because in this context $\boldsymbol{\theta}$ is unknown and $\mathbf{y}$ is known we use the notation $L(\boldsymbol{\theta}|\mathbf{y})$ or just $L(\boldsymbol{\theta})$ or even just L. Thus, although $f(\mathbf{y}|\boldsymbol{\theta})$ and $L(\boldsymbol{\theta}|\mathbf{y})$ represent the same thing mathematically, that is,

$$f(\mathbf{y}|\boldsymbol{\theta}) = L(\boldsymbol{\theta}|\mathbf{y}),$$

it is convenient to use each in its appropriate context.

b.   Maximum likelihood estimation

The likelihood function $L(\theta|\mathbf{y})$ is the foundation of the widely used method of estimation known as *maximum likelihood estimation*. It yields estimators that have many good properties. ML is used as an abbreviation for maximum likelihood and MLE for maximum likelihood estimate — with whatever suffix is appropriate to the context: estimate, estimator (and their plurals) or estimation.

The essence of the ML method is to view $L(\theta|\mathbf{y})$ as a function of the mathematical variable θ and to derive $\hat{\theta}$ as the value of θ that maximizes $L(\theta|\mathbf{y})$. The only proviso is that this maximization must be carried out within the range of permissible values for θ. For example, if one element of θ is a variance then permissible values for that variance are non-negative values. This aspect of ML estimation is very important in estimating variances of random effects.

Under widely existing regularity conditions on $f(\mathbf{y}|\theta)$, a general method of establishing equations that yield MLEs is to differentiate $L(\theta)$ with respect to θ and equate the derivative to $\mathbf{0}$. But finding the values of θ that maximize L is equivalent to maximizing $\log L$, which we denote by l, and it is often easier to use l rather than L. Thus for $l = \log L(\theta|\mathbf{y})$ the equations

$$\frac{\partial l}{\partial \theta}\bigg|_{\theta=\dot{\theta}} = \mathbf{0} \qquad (S.13)$$

are known as the ML equations, with $\dot{\theta}$, their solution, being called the ML solution. When this solution is the global maximum and is in the parameter space it is also the maximum likelihood estimator, $\hat{\theta}$. When it is not within the permissible range, then adjustments must be made to the solution to find the MLE; these adjustments depend on the context and form of $f(\mathbf{y}|\theta)$.

c.   Asymptotic variance-covariance matrix

A useful property of the ML estimator, $\hat{\theta}$, is that its large-sample, or asymptotic, variance-covariance matrix is easy to calculate. From $\mathbf{I}(\theta)$, known as the information (or Fisher information) matrix, and defined as

$$\mathbf{I}(\theta) = \mathrm{E}\left[\frac{\partial l}{\partial \theta}\frac{\partial l}{\partial \theta'}\right] = \mathrm{E}\left[\left\{\,_m\frac{\partial l}{\partial \theta_i}\frac{\partial l}{\partial \theta_j}\right\}_{i,j}\right], \qquad (S.14)$$

the asymptotic variance-covariance matrix of $\hat{\boldsymbol{\theta}}$ is

$$\text{var}(\hat{\boldsymbol{\theta}}) \approx [\mathbf{I}(\boldsymbol{\theta})]^{-1}. \tag{S.15}$$

Note that this is available without even needing a formula or the sampling distribution of $\hat{\boldsymbol{\theta}}$. An alternative form of the information matrix that is valid in many situations is

$$\mathbf{I}(\boldsymbol{\theta}) = -\text{E}\left[\frac{\partial^2 l}{\partial \boldsymbol{\theta}\, \partial \boldsymbol{\theta}'}\right] = -\text{E}\left[\left\{\frac{\partial^2 l}{m\, \partial \theta_i\, \partial \theta_j}\right\}_{i,j}\right]. \tag{S.16}$$

Proof of this is given in a number of books (e.g., Searle et al., 1992, p. 473).

d.   Asymptotic distribution of MLEs

No matter what the distribution of one's data vector $\mathbf{y}$, it is ordinarily the case for an MLE that, as the sample size increases, the MLE of $\boldsymbol{\theta}$ is consistent and asymptotically normally distributed with mean $\boldsymbol{\theta}$ and variance $[\mathbf{I}(\boldsymbol{\theta})]^{-1}$; we summarize this by writing

$$\hat{\boldsymbol{\theta}} \sim \mathcal{AN}\left(\boldsymbol{\theta}, [\mathbf{I}(\boldsymbol{\theta})]^{-1}\right), \tag{S.17}$$

where $\mathbf{I}(\boldsymbol{\theta})$ is given in (S.14) or (S.16).

S.5   LIKELIHOOD RATIO TESTS

The likelihood ratio test is a standard test for composite hypotheses. It has the advantage of an easily derived large sample distribution. Suppose the parameter vector $\boldsymbol{\theta}$ is partitioned into two components $\boldsymbol{\theta}' = [\boldsymbol{\theta}'_1, \boldsymbol{\theta}'_2]$ and suppose interest focuses on $\boldsymbol{\theta}_1$ while $\boldsymbol{\theta}_2$ is left unspecified. $\boldsymbol{\theta}_2$ is often called a *nuisance parameter*. Either or both of $\boldsymbol{\theta}_1$ and $\boldsymbol{\theta}_2$ could be vector-valued and, if the entire parameter vector is of interest, $\boldsymbol{\theta}_2$ would be null.

Suppose our hypothesis is of the form $H_0 : \boldsymbol{\theta}_1 = \boldsymbol{\theta}_{1,0}$, where $\boldsymbol{\theta}_{1,0}$ is a specified value of $\boldsymbol{\theta}_1$, and let $\hat{\boldsymbol{\theta}}_{2,0}$ be the MLE of $\boldsymbol{\theta}_2$ under the restriction that $\boldsymbol{\theta}_1 = \boldsymbol{\theta}_{1,0}$.

With $L(\boldsymbol{\theta}) = L([\boldsymbol{\theta}'_1, \boldsymbol{\theta}'_2]')$ being the likelihood, the likelihood ratio statistic is

$$\Lambda = \frac{L([\boldsymbol{\theta}'_{1,0}, \hat{\boldsymbol{\theta}}'_{2,0}]')}{L(\hat{\boldsymbol{\theta}})}. \tag{S.18}$$

The test is to reject H_0 when $\Lambda \leq k$ and with k determined such that

$$\sup_{\theta_2} P\{\Lambda \leq k | \boldsymbol{\theta}_1 = \boldsymbol{\theta}_{1,0}\} \leq \alpha.$$

An equivalent rejection region is when

$$-2\log \Lambda \geq -2\log k \equiv k^*.$$

Under regularity conditions, a notable one being that $\boldsymbol{\theta}_{1,0}$ is not on the boundary of the parameter space, the large-sample distribution of $-2\log \Lambda$ is χ^2 with degrees of freedom equal to ν, the dimension of $\boldsymbol{\theta}_1$. The value of k^* is then given by $\chi^2_{\nu,1-\alpha}$. This can be written in terms of the log likelihood, l, as follows:

$$-2\log \Lambda = -2\left[l(\boldsymbol{\theta}_{1,0}, \hat{\boldsymbol{\theta}}_{2,0}) - l(\hat{\boldsymbol{\theta}}_1, \hat{\boldsymbol{\theta}}_2)\right] \tag{S.19}$$

with the large-sample critical region of the test given by

$$-2\log \Lambda > \chi^2_{\nu,1-\alpha}. \tag{S.20}$$

If $\boldsymbol{\theta}_{1,0}$ *is* on the boundary of the parameter space then special care must be taken. This can arise in the analysis of random effects since we may be interested in testing the null hypothesis that the variance of the random effect is zero. See Self and Liang (1987) for details.

S.6   MLE UNDER NORMALITY

In this section we consider maximum likelihood estimation under the linear model: $\mathbf{y} \sim \mathcal{N}(\mathbf{X}\boldsymbol{\beta}, \mathbf{V})$.

a.   Estimation of β

If we assume $\mathbf{V}$ to be known then, from (S.11), with $\boldsymbol{\mu} = \mathbf{X}\boldsymbol{\beta}$

$$l = \log L = -\frac{N}{2}\log 2\pi - \frac{1}{2}\log|\mathbf{V}| - \frac{1}{2}(\mathbf{y} - \mathbf{X}\boldsymbol{\beta})'\mathbf{V}^{-1}(\mathbf{y} - \mathbf{X}\boldsymbol{\beta}). \tag{S.21}$$

In Section 6.3 we use the general results of Section 6.12 to differentiate l with respect to $\boldsymbol{\beta}$. Now we confirm that result by differentiating l of (S.21) using the rules in Section M.5:

$$\frac{\partial l}{\partial \boldsymbol{\beta}} = -\mathbf{X}'\mathbf{V}^{-1}\mathbf{X}\boldsymbol{\beta} + \mathbf{X}'\mathbf{V}^{-1}\mathbf{y}. \tag{S.22}$$

Equating this to $\mathbf{0}$ and using β^0 for the solution gives β^0 of (6.19):

$$\beta^0 = (\mathbf{X}'\mathbf{V}^{-1}\mathbf{X})^-\mathbf{X}'\mathbf{V}^{-1}\mathbf{y}. \tag{S.23}$$

And then, as in (6.20),

$$\mathrm{ML}(\mathbf{X}\beta) = \mathbf{X}\beta^0 = \mathbf{X}(\mathbf{X}'\mathbf{V}^{-1}\mathbf{X})^-\mathbf{X}'\mathbf{V}^{-1}\mathbf{y}, \tag{S.24}$$

which, by (M.11), is invariant to the choice of $(\mathbf{X}'\mathbf{V}^{-1}\mathbf{X})^-$.

It is to be noted in passing when $\mathbf{y} \sim (\mathbf{X}\beta, \mathbf{V})$, whether normally distributed or not, that (S.24) is the generalized least squares estimator (GLSE) of $\mathbf{X}\beta$. Moreover, if $\mathbf{V} = \sigma^2\mathbf{I}$, then (S.24) simplifies to

$$\mathbf{X}\beta^0 = \mathbf{X}(\mathbf{X}'\mathbf{X})^-\mathbf{X}'\mathbf{y}, \tag{S.25}$$

which is known as the ordinary least squares estimator (OLSE) of $\mathbf{X}\beta$.

b. Estimation of variance components

Equation (6.59) for obtaining $\mathrm{ML}(\sigma^2)$ is

$$\left\{{}_c\,\mathrm{tr}(\mathbf{V}^{-1}\mathbf{Z}_i\mathbf{Z}_i')\right\}_{i=0}^r = \left\{{}_c\,\mathbf{y}'\mathbf{P}\mathbf{Z}_i\mathbf{Z}_i'\mathbf{P}\mathbf{y}\right\}_{i=0}^r. \tag{S.26}$$

c. Asymptotic variance-covariance matrix

The asymptotic variance of the ML estimators $\mathbf{X}\hat{\beta}$ and $\hat{\sigma}^2$ is shown in Section 6.8c. Using (S.16) the terms for $\mathbf{I}\begin{pmatrix}\mathbf{X}\beta \\ \sigma^2\end{pmatrix}$ are

$$-\mathrm{E}\left[\frac{\partial^2 l}{\partial\beta\,\partial\beta'}\right] = \mathrm{E}\left[\mathbf{X}'\mathbf{V}^{-1}\mathbf{X}\right] = \mathbf{X}'\mathbf{V}^{-1}\mathbf{X} \tag{S.27}$$

and based on $\partial l/\partial\sigma_i^2$ preceding (6.57)

$$-\mathrm{E}\left[\frac{\partial^2 l}{\partial\beta\,\partial\sigma_i^2}\right] = \mathrm{E}\left[\mathbf{X}'\mathbf{V}^{-1}\mathbf{Z}_i\mathbf{Z}_i'\mathbf{V}^{-1}\mathbf{X}\beta - \mathbf{X}'\mathbf{V}^{-1}\mathbf{Z}_i\mathbf{Z}_i'\mathbf{V}^{-1}\mathbf{y}\right]$$

$$= \mathbf{0}, \quad \text{since } \mathrm{E}[\mathbf{y}] = \mathbf{X}\beta. \tag{S.28}$$

And

$$-\mathrm{E}\left[\frac{\partial^2 l}{\partial\sigma_j^2\,\partial\sigma_i^2}\right] = \mathrm{E}\left[\tfrac{1}{2}\mathrm{tr}(\mathbf{V}^{-1}\mathbf{Z}_j\mathbf{Z}_j'\mathbf{V}^{-1}\mathbf{Z}_i\mathbf{Z}_i')\right]$$

$$-\tfrac{1}{2}(\mathbf{y} - \mathbf{X}\boldsymbol{\beta})'\mathbf{V}^{-1}\mathbf{Z}_j\mathbf{Z}_j'\mathbf{V}^{-1}\mathbf{Z}_i\mathbf{Z}_i'\mathbf{V}^{-1}(\mathbf{y} - \mathbf{X}\boldsymbol{\beta})$$

$$-\tfrac{1}{2}(\mathbf{y} - \mathbf{X}\boldsymbol{\beta})'\mathbf{V}^{-1}\mathbf{Z}_i\mathbf{Z}_i'\mathbf{V}^{-1}\mathbf{Z}_j\mathbf{Z}_j'\mathbf{V}^{-1}(\mathbf{y} - \mathbf{X}\boldsymbol{\beta})\Big].$$

The last term is a scalar and so equals its transpose, which is the penultimate term. Moreover, a scalar equals its own trace. Thus the expected value of the sum of those two (equal) terms is

$$-\text{tr}\left(\mathbf{V}^{-1}\mathbf{Z}_j\mathbf{Z}_j'\mathbf{V}^{-1}\mathbf{Z}_i\mathbf{Z}_i'\mathbf{V}^{-1}\text{E}\left[(\mathbf{y} - \mathbf{X}\boldsymbol{\beta})(\mathbf{y} - \mathbf{X}\boldsymbol{\beta})'\right]\right) \qquad (\text{S.29})$$

$$= -\text{tr}\left(\mathbf{V}^{-1}\mathbf{Z}_j\mathbf{Z}_j'\mathbf{V}^{-1}\mathbf{Z}_i\mathbf{Z}_i'\mathbf{V}^{-1}\mathbf{V}\right) = -\text{tr}\left(\mathbf{V}^{-1}\mathbf{Z}_j\mathbf{Z}_j'\mathbf{V}^{-1}\mathbf{Z}_i\mathbf{Z}_i'\right).$$

Thus

$$-\text{E}\left[\frac{\partial^2 l}{\partial\sigma_j^2\,\partial\sigma_j^2}\right] = -\tfrac{1}{2}\text{tr}\left(\mathbf{V}^{-1}\mathbf{Z}_j\mathbf{Z}_j'\mathbf{V}^{-1}\mathbf{Z}_i\mathbf{Z}_i'\right). \qquad (\text{S.30})$$

From these results we get (6.62), (6.63), and (6.64).

d.   Restricted maximum likelihood (REML)

The underlying concept of restricted maximum likelihood (REML) is described at the beginning of Section 6.9. In that section we give a wholly technical derivation of the REML methodology whereas here we describe the derivation from basic principles. That involves estimating variance components from linear combinations of the data that do not involve $\boldsymbol{\beta}$. This is achieved by using maximum likelihood on $\mathbf{K}'\mathbf{y}$ where $\mathbf{K}'$ is chosen to have as many linearly independent rows as possible satisfying $\mathbf{K}'\mathbf{X} = 0$. This results in $\mathbf{K}'$ having row rank $r_\mathbf{K} = N - r_\mathbf{X}$. Then with

$$\mathbf{y} \sim \mathcal{N}(\mathbf{X}\boldsymbol{\beta}, \mathbf{V}) \text{ we have } \mathbf{K}'\mathbf{y} \sim \mathcal{N}(\mathbf{0}, \mathbf{K}'\mathbf{V}\mathbf{K}). \qquad (\text{S.31})$$

– i.   *Estimation*

Using (S.31) the log likelihood of $\mathbf{K}'\mathbf{y}$ is

$$l = \tfrac{1}{2}r_\mathbf{K}\log 2\pi - \tfrac{1}{2}\log|\mathbf{K}'\mathbf{V}\mathbf{K}| - \tfrac{1}{2}\mathbf{y}'\mathbf{K}(\mathbf{K}'\mathbf{V}\mathbf{K})^{-1}\mathbf{K}'\mathbf{y}. \qquad (\text{S.32})$$

To apply maximum likelihood we need

$$\frac{\partial l}{\partial\sigma_i^2} = -\tfrac{1}{2}\text{tr}\left[(\mathbf{K}'\mathbf{V}\mathbf{K})^{-1}\mathbf{K}'\mathbf{Z}_i\mathbf{Z}_i'\mathbf{K}\right]$$

$$+ \tfrac{1}{2}\mathbf{y}'\mathbf{K}(\mathbf{K}'\mathbf{V}\mathbf{K})^{-1}\mathbf{K}'\mathbf{Z}_i\mathbf{Z}_i'\mathbf{K}(\mathbf{K}'\mathbf{V}\mathbf{K})^{-1}\mathbf{K}'\mathbf{y}. \qquad (\text{S.33})$$

Equating this to zero, and using $\mathbf{P} = \mathbf{K}(\mathbf{K}'\mathbf{V}\mathbf{K})^{-1}\mathbf{K}'$ of Section M.4g in doing so, gives

$$\operatorname{tr}(\mathbf{P}\mathbf{Z}_i\mathbf{Z}_i') = \mathbf{y}'\mathbf{P}\mathbf{Z}_i\mathbf{Z}_i'\mathbf{P}\mathbf{y}.$$

This equation written for each $i = 0, 1, \ldots, r$ is equation (6.66) which is the REML procedure.

− ii. *Asymptotic variance*

Using (S.33) together with (M.19) of Section M.5f gives

$$\frac{\partial^2 l}{\partial\sigma_j^2\,\partial\sigma_i^2} = -\tfrac{1}{2}\operatorname{tr}(\mathbf{P}\mathbf{Z}_j\mathbf{Z}_j'\mathbf{P}\mathbf{Z}_i\mathbf{Z}_i') \tag{S.34}$$

$$- \tfrac{1}{2}\mathrm{E}\left[\mathbf{y}'\mathbf{P}\mathbf{Z}_j\mathbf{Z}_j'\mathbf{P}\mathbf{Z}_i\mathbf{Z}_i'\mathbf{P}\mathbf{y} + \mathbf{y}'\mathbf{P}\mathbf{Z}_i\mathbf{Z}_i'\mathbf{P}\mathbf{Z}_j\mathbf{Z}_j'\mathbf{P}\mathbf{y}\right].$$

Each quadratic in $\mathbf{y}$ is a scalar and so equals its transpose; hence the last two terms are equal. Using the fact that, for any $\mathbf{A}$,

$$\mathrm{E}\left[\operatorname{tr}(\mathbf{y}'\mathbf{P}\mathbf{A}\mathbf{P}\mathbf{y})\right] = \operatorname{tr}(\mathbf{A}\mathbf{P}\mathrm{E}\left[\mathbf{y}\mathbf{y}'\right]\mathbf{P})$$

$$= \operatorname{tr}(\mathbf{A}\mathbf{P}[\mathbf{V} + \mathbf{X}\beta\beta'\mathbf{X}']\mathbf{P}) \tag{S.35}$$

$$= \operatorname{tr}(\mathbf{A}\mathbf{P}) \text{ because } \mathbf{P}\mathbf{V}\mathbf{P} = \mathbf{P} \text{ and } \mathbf{P}\mathbf{X} = \mathbf{0}.$$

Thus for $\mathbf{A}$ being $\mathbf{Z}_i\mathbf{Z}_i'\mathbf{P}\mathbf{Z}_j\mathbf{Z}_j'$, (S.35) leads to

$$\mathbf{I}(\sigma^2) = \tfrac{1}{2}\left\{_m \operatorname{tr}(\mathbf{Z}_i\mathbf{Z}_i'\mathbf{P}\mathbf{Z}_j\mathbf{Z}_j'\mathbf{P})\right\}_{i,j=0}^r$$

$$= \tfrac{1}{2}\left\{_m \operatorname{tr}(\mathbf{Z}_j'\mathbf{P}\mathbf{Z}_i[\mathbf{Z}_j'\mathbf{P}\mathbf{Z}_i]')\right\}_{i,j=0}^r \tag{S.36}$$

for which

$$\operatorname{var}_\infty(\hat{\sigma}_{\mathrm{REML}}^2) = [I(\sigma^2)]^{-1}. \tag{S.37}$$

References

Abramowitz, M. and I. Stegun (eds.) (1964). *Handbook of Mathematical Functions*. National Bureau of Standards, Washington, D.C.

Abu-Libdeh, H., B. W. Turnbull, and L. C. Clark (1990). Analysis of multi-type recurrent events in longitudinal studies: Application to a skin cancer prevention trial. *Biometrics*, **46**:1017–1034.

Andersen, E. (1970). Asymptotic properties of conditional maximum-likelihood estimators. *Journal of the Royal Statistical Society. Series B*, **32**:283–301.

Arnold, S. (1981). *The Theory of Linear Models and Multivariate Analysis*. Wiley, New York.

Atwill, E., H. O. Mohammed, J. W. Lopez, C. E. McCulloch, and E. J. Dubovi (1996). Cross-sectional evaluation of environmental, host, and management factors associated with the risk of seropositivity to *Ehrlichia risticii* in horses of New York State. *American Journal of Veterinary Research*, **57**:278–285.

Benhin, E., J. N. K. Rao, and S. A. J. (2005). Mean estimating equation approach to analysing cluster-correlated data with nonignorable cluster sizes. *Biometrika*, **92**:435–450.

Binder, D. A. (1983). On the variances of asymptotically normal estimators from complex surveys. *International Statistical Review*, **51**:279–292.

Bishop, Y. M. M., S. E. Fienberg, and P. W. Holland (1975). *Discrete Multivariate Analysis: Theory and Practice*. MIT Press, Cambridge, MA.

Bliss, C. (1934). The method of probits. *Science*, **79**:38–39.

Bliss, C. (1935). The calculation of the dose-mortality curve. *Annals of Applied Biology*, **22**:134–167.

Blyth, C. R. and H. A. Still (1983). Binomial confidence intervals. *Journal of the American Statistical Association*, **78**:108–116.

Booth, J. G. and J. P. Hobert (1999). Maximizing generalized linear mixed model likelihoods with an automated Monte Carlo EM algorithm. *Journal of the Royal Statistical Society, Series B*, **61**:265–285.

Box, G. E. P. and D. R. Cox (1962). An analysis of transformations. *Journal of the Royal Statistical Society, Series B*, **26**:211–252.

Breslow, N. E. and D. G. Clayton (1993). Approximate inference in generalized linear mixed models. *Journal of the American Statistical Association*, **88**:9–25.

Breslow, N. E. and X. Lin (1995). Bias correction in generalized linear mixed models with a single component of dispersion. *Biometrika*, **82**:81–91.

Butler, S. M. and T. A. Louis (1997). Consistency of maximum likelihood estimators in general random effects models for binary data. *Annals of Statistics*, **25**:351–377.

Casella, G. and R. L. Berger (1990). *Statistical Inference*. Wadsworth and Brooks/Cole, Pacific Grove, CA.

Chakravorti, S. R. and J. E. Grizzle (1975). Analysis of data from multiclinic experiments. *Biometrics*, **31**:325–338.

Churchill, G. (1995). Personal communication. Cornell University, Ithaca, NY.

Cochran, W. G. (1951). Improvement by means of selection. In *Proceedings of the Second Berkeley Symposium on Mathematical Statistics and Probability, 1950*. University of California Press, Berkeley and Los Angeles, CA.

Cortesi, P. and M. Milgroom (1998). Genetics of vegetative incompatibility in *Cryphonectria parasitica*. *Applied Environmental Microbiology*, **64**:2988–2994.

Cortesi, P., M. Milgroom, and M. Bisiach (1996). Distribution and diversity of vegetative incompatibility types in subpopulations of *Cryphonectria parasitica* in Italy. *Mycological Research*, **100**:1087–1093.

Crainiceanu, C. and D. Ruppert (2004). Likelihood ratio tests in linear mixed models with one variance component. *Journal of the Royal Statistical Society, Series B*, **66**:165–185.

Czado, C. and T. J. Santner (1992). The effect of link misspecification in binary regression inference. *Journal of Statistical Planning and Inference*, **33**:213–231.

David, H. A. (1995). First (?) occurrence of common terms in mathematical statistics. *The American Statistician*, **29**:21–31.

Devore, J. and R. Peck (1993). *Statistics: The Exploration and Analysis of Data, 2nd ed.* Brooks/Cole, Pacific Grove, CA.

Diggle, P., K.-Y. Liang, and S. L. Zeger (1994). *Longitudinal Data Analysis*. Oxford University Press, Oxford.

Driscoll, M. F. and W. R. Gundberg (1986). The history of the development of Craig's theorem. *The American Statistician*, **40**:65–71.

Driscoll, M. F. and B. Krasnicka (1995). An accessible proof of Craig's theorem in the general case. *The American Statistician*, **49**:59–62.

Durbin, J. and S. J. Koopman (1997). Monte Carlo maximum likelihood estimation for non-Gaussian state space models. *Biometrika*, **84**:669–684.

Ekholm, A., P. Smith, and J. McDonald (1995). Marginal analysis of a multivariate binary response. *Biometrika*, **82**:847–854.

Finney, D. J. (1952). *Probit Analysis*. Cambridge University Press, Cambridge.

Firth, D. (1987). On the efficiency of quasi–likelihood estimation. *Biometrika*, **74**:233–245.

Fisher, R. A. (1935). Appendix to the calculation of the dose-mortality curve (by Bliss). *Annals of Applied Biology*, **22**:164–165.

Fitzmaurice, G. and N. Laird (1997). Regression models for a mixed discrete and continuous responses with potentially missing valuess. *Biometrics*, **53**:110–122.

Fitzmaurice, G. M. (1995). A caveat concerning independence estimating equations with multivariate binary data. *Biometrics*, **51**:309–317.

Follman, D. and D. Lambert (1989). Generalizing logistic regression by nonparametric mixing. *Journal of the American Statistical Association*, **84**:295–300.

Frees, E. W. and D. Ruppert (1990). Estimation following a sequentially designed experiment. *Journal of the American Statistical Association*, **85**:1123–1129.

Gail, M., S. Wieand, and S. Piantadosi (1984). Biased estimates of treatment effect in randomized experiments with non-linear regressions and omitted covariates. *Biometrika*, **71**:807–815.

Gelfand, A. and B. Carlin (1993). Maximum-likelihood estimation for constrained- or missing-data models. *Canadian Journal of Statistics*, **21**:303–311.

Geyer, C. J. and E. A. Thompson (1992). Constrained Monte Carlo maximum likelihood for dependent data. *Journal of the Royal Statistical Society, Series B*, **54**:657–699.

Gilks, W., S. Richardson, and S. D.J. (1996). *Markov chain Monte Carlo in practice*. Chapman and Hall, New York.

Giltinan, D. and M. Davidian (1995). *Nonlinear Models for Repeated Measurement Data*. Chapman & Hall, London.

Goldberger, A. S. (1962). Best linear unbiased prediction in the generalized regression model. *Journal of the American Statistical Association*, **57**:369–375.

Graybill, F. A. (1976). *Theory and Application of the Linear Model*. Duxbury Press, Pacific Grove, CA.

Green, P. J. (1990). On use of the EM algorithm for penalized likelihood estimation. *Journal of the Royal Statistical Society, Series B*, **52**:443–452.

Green, P. J. and B. W. Silverman (1994). *Nonparametric Regression and Generalized Linear Models*. Chapman & Hall, London.

Griffiths, D. A. (1973). Maximum likelihood estimation for the beta-binomial distribution and an application to the household distribution of the total number of cases of a disease. *Biometrics*, **29**:637–648.

Gu, M. G. and F. H. Kong (1998). A stochastic approximation algorithm with Markov chain Monte-Carlo method for incomplete data estimation problems. *Proceedings of the National Academy of Sciences*, **95**:7270–7274.

Gueorguieva, R. and A. Agresti (2001). Correlated probit model for joint modeling of clustered binary and continuous responses. *Journal of the American Statistical Association*, **96**:1102–1112.

Gueorguieva, R. and G. Sanacora (2006). Missing data in longitudinal studies. *Statistics in Medicine*, **25**:1307–1322.

Hartley, H. O. and J. N. K. Rao (1967). Maximum-likelihood estimation for the mixed analysis of variance model. *Biometrika*, **54**:93–108.

Hausman, J. A. (1978). Specification tests in econometrics. *Econometrica*, **46**:1251–1272.

Hayes, K. and J. Haslett (1999). Simplifying general least squares. *The American Statistician*, **53**:376–381.

Heagerty, P. (1999). Marginally specified logistic-normal models for longitudinal binary data. *Biometrics*, **55**:688–698.

Heagerty, P. J. and B. F. Kurland (2001). Misspecified maximum likelihood estimates and generalised linear mixed models. *Biometrika*, **88**:973–985.

Heagerty, P. J. and S. L. Zeger (2000). Marginalized multilevel models and likelihood inference (with comments and a rejoinder by the authors). *Statistical Science*, **15**:1–26.

Hedeker, D. and R. D. Gibbons (1994). A random-effects ordinal regression model for multilevel analysis. *Biometrics*, **50**:933–944.

Henderson, C. R. (1953). Estimation of variance and covariance components. *Biometrics*, **9**:226–252.

Henderson, C. R. (1963). Selection index and expected genetic advance. In *Statistical Genetics and Plant Breeding* (edited by W. D. Hansen and H. F. Robinson). Publication 982, National Academy of Sciences and National Research Council. Washington, D.C.

Henderson, C. R., O. Kempthorne, S. R. Searle, and C. N. von Krosigk (1959). Estimation of environmental and genetic trends from records subject to culling. *Biometrics*, **15**:192–218.

Heyde, C. C. (1997). *Quasi–likelihood and Its Application: A General Approach to Optimal Parameter Application.* Springer-Verlag, New York.

Hoaglin, D. (1985). *Summarizing shape numerically: The g- and h-distributions,* In *Exploring Data Tables, Trends and Shapes (D.C. Hoaglin, F. Mosteller and J.W. Tukey, eds.).* Wiley, New York.

Hocking, R. R. (1985). *The Analysis of Linear Models.* Brooks/Cole, Pacific Grove, CA.

Jennrich, R. I. and M. D. Schluchter (1986). Unbalanced repeated-measures models with structured covariance matrices. *Biometrics*, **42**:805–820.

Joe, H. (1997). *Multivariate Models and Multivariate Dependence Concepts.* Chapman and Hall, London.

Johnson, N. L. and S. Kotz (1970). *Continuous Univariate Distributions - Vol. 2.* Wiley, New York.

Kackar, R. N. and D. A. Harville (1984). Approximations for standard errors of estimators of fixed and random effects in mixed linear models. *Journal of the American Statistical Association*, **79**:853–862.

Kenward, M. G. and J. H. Roger (1997). Small sample inference for fixed effects from restricted maximum likelihood. *Biometrics*, **53**:983–997.

Khuri, A. I., T. Mathew, and B. K. Sinha (1998). *Statistical Tests for Mixed Linear Models.* Wiley, New York.

Kiefer, J. and J. Wolfowitz (1956). Consistency of the maximum likelihood estimator in the presence of infinitely many incidental parameters. *Annals of Mathematical Statistics*, **27**:886–906.

Kuk, A. Y. C. (1995). Asymptotically unbiased estimation in generalized linear models with random effects. *Journal of the Royal Statistical Society, Series B*, **57**:395–407.

Kullback, S. (1959). *Information Theory and Statistics*. Wiley, New York.

Laird, N. (1978). Nonparametric maximum likelihood estimation of a mixing distribution. *Journal of the American Statistical Association*, **73**:805–811.

Laird, N. (1988). Missing data in longitudinal studies. *Statistics in Medicine*, **7**:305–315.

Laird, N. M. and J. H. Ware (1982). Random-effects models for longitudinal data. *Biometrics*, **38**:963–974.

Lee, Y. and J. A. Nelder (1996). Hierarchical generalized linear models. With discussion. *Journal of the Royal Statistical Society, Series B*, **58**:619–678.

Lehmann, E. L. (1986). *Testing Statistical Hypotheses*. Wiley, New York.

Lesaffre, E. and G. Molenberghs (2001). Multivariate probit analysis: A neglected procedure in medical statistics. *Statistics in Medicine*, **10**:1391–1403.

Lesperance, M. and J. Kalbfleisch (1992). An algorithm for computing the nonparametric mle of a mixing distribution. *Journal of the American Statistical Association*, **87**:120–126.

Li, K.-C. and N. Duan (1989). Regression analysis under link violation. *Annals of Statistics*, **17**:1009–1052.

Liang, K.-Y., S. Zeger, and B. Qaqish (1992). Multivariate regression analyses for categorical data. *Journal of the Royal Statistical Society, Series B*, **54**:673–687.

Liang, K.-Y. and S. L. Zeger (1986). Longitudinal data analysis using generalized linear models. *Biometrika*, **73**:13–22.

Lin, X. and N. E. Breslow (1996). Bias correction in generalized linear mixed models with multiple components of dispersion. *Journal of the American Statistical Association*, **91**:1007–1016.

Lindsay, C., B.and Clogg and J. Grego (1991). Semiparametric estimation in the rasch model and related exponential response models, including a simple latent class model for item analysis. *Journal of the American Statistical Association*, **86**:96–107.

Lindsey, J. K. and P. Lambert (1998). On the appropriateness of marginal models for repeated measurements in clinical trials. *Statistics in Medicine*, **17**:447–469.

Lindstrom, M. J. and D. M. Bates (1988). Newton–Raphson and EM algorithms for linear mixed-effects models for repeated-measures data. *Journal of the American Statistical Association*, **83**:1014–1022.

Lindstrom, M. J. and D. M. Bates (1990). Nonlinear mixed effects models for repeated measures data. *Biometrics*, **46**:673–687.

Liu, Q. and D. A. Pierce (1994). A note on Gauss-Hermite quadrature. *Biometrika*, **81**:624–629.

Louis, T. A. (1982). Finding the observed information matrix when using the EM algorithm. *Journal of the Royal Statistical Society, Series B*, **44**:226–233.

Lu, M. and B. Tilley (2001). Use of odds ratio or relative risk to measure a treatment effect in clinical trials with multiple correlated binary outcomes: Data from the ninds t-pa stroke trial. *Statistics in Medicine*, **20**:1891–1901.

Magder, L. S. and S. L. Zeger (1996). A smooth nonparametric estimate of a mixing distribution using mixtures of Gaussians. *Journal of the American Statistical Association*, **91**:1141–1151.

McCullagh, P. (1983). Quasi–likelihood functions. *Annals of Statistics*, **11**:59–67.

McCullagh, P. and J. A. Nelder (1989). *Generalized Linear Models, 2nd Ed.* Chapman & Hall, London.

McCulloch, C. (2008). Joint modeling of mixed outcome types using latent variables. *Statistical Methods in Medical Research*, **17**:1–10.

McCulloch, C. E. (1997). Maximum likelihood algorithms for generalized linear mixed models. *Journal of the American Statistical Association*, **92**:162–170.

McLachlan, G. J. and T. Krishnan (1996). *The EM Algorithm and Extensions*. Wiley, New York.

Mehta, C. and N. Patel (1992). *StatXact-Turbo User's Manual*. Cytel Software, Cambridge, MA.

Mood, A. M., F. A. Graybill, and D. C. Boes (1974). *Introduction to the Theory of Statistics, 3rd Ed*. McGraw-Hill, New York.

Moyeed, R. A. and A. J. Baddeley (1991). Stochastic approximation of the MLE for a spatial point pattern. *Scandinavian Journal of Statistics*, **18**:39–50.

Nelder, J. A. and R. W. M. Wedderburn (1972). Generalized linear models. *Journal of the Royal Statistical Society, Series A*, **135**:370–384.

Neuhaus, J. M. (1998). Estimation efficiency with omitted covariates in generalized linear models. *Journal of the American Statistical Association*, **93**:1124–1129.

Neuhaus, J. M. (1999). Bias and efficiency loss due to misclassified responses in binary regression. *Biometrika*, **86**:843–855.

Neuhaus, J. M. (2002). Analysis of clustered and longitudinal binary data subject to response misclassification. *Biometrics*, **58**:675–683.

Neuhaus, J. M., W. W. Hauck, and J. D. Kalbfleisch (1992). The effects of mixture distribution misspecification when fitting mixed-effects logistic models. *Biometrika*, **79**:755–762.

Neuhaus, J. M. and N. P. Jewell (1993). A geometric approach to assess bias due to omitted covariates in generalized linear models. *Biometrika*, **80**:807–815.

Neuhaus, J. M., J. D. Kalbfleisch, and W. W. Hauck (1994). Conditions for consistent estimation in mixed-effects models for binary matched pairs data. *Canadian Journal of Statistics*, **22**:139–148.

Neuhaus, J. M. and C. E. McCulloch (2006). Separating between and within-cluster covariate effects using conditional and partitioning methods. *Journal of the Royal Statistical Society, Series B*, **68**:859–872.

Neuhaus, J. M. and M. R. Segal (1997). An assessment of approximate maximum likelihood estimators in generalized linear mixed models. In *Modelling longitudinal and spatially correlated data* (edited by T. G. Gregoire, D. R. Brillinger, P. J. Diggle, E. Russek-Cohen, W. G. Warren, and R. D. Wolfinger). Springer, New York.

Neyman, J. and E. S. Pearson (1928). On the use and interpretation of certain test criteria for purposes of statistical inference: Part I. *Biometrika*, **20A**:175–240.

Neyman, J. and E. L. Scott (1948). Consistent estimates based on partially consistent observations. *Econometrica*, **16**:1–32.

Parker, H. (1995). Personal communication. Cornell University, Ithaca, NY.

Patterson, H. D. and R. Thompson (1971). Recovery of inter-block information when block sizes are unequal. *Biometrika*, **58**:545–554.

Piepho, H.-P. and C. E. McCulloch (2004). Transformations in mixed models: Application to risk analysis for a multienvironment trial. *Journal of Agricultural, Biological and Environmental Statistics*, **9**, 2:123–137.

Pinheiro, J. C. and D. M. Bates (1995). Approximations to the log-likelihood function in nonlinear mixed-effects models. *Journal of Computational and Graphical Statistics*, **4**:12–35.

Portnoy, S. (1982). Maximizing the probability of correctly ordering random variables using linear predictors. *Journal of Multivariate Analysis*, **12**:256–269.

Prasad, N. G. N. and J. N. K. Rao (1990). The estimation of the mean squared error of small-area estimators. *Journal of the American Statistical Association*, **85**:163–171.

Pregibon, D. (1980). Goodness of link tests for generalized linear models. *Applied Statistics*, **29**:15–24.

Prentice, R. (1986). Binary regression using an extended beta-binomial distribution, with discussion of correlation induced by covariate measurement errors. *Journal of the American Statistical Association*, **81**:321–327.

Press, W. H., S. A. Teukolsky, W. T. Vetterling, and B. P. Flannery (1996). *Numerical Recipes in Fortran 90, 2nd ed.*. Cambridge University Press, Cambridge.

Rao, C. R. (1962). A note on a generalized inverse of a matrix with applications to problems in mathematical statistics. *Journal of the Royal Statistical Society, Series B*, **24**:152–158.

Rao, C. R. (1965). *Linear Statistical Inference and Its Applications*. Wiley, New York.

Raudenbush, S., M. Yang, and Y. M. (2000). Maximum likelihood for generalized linear models with nested random effects via high-order, multivariate Laplace approximation. *Journal of Computational and Graphical Statistics*, **9**:141–157.

Robert, C. P. and G. Casella (1999). *Monte Carlo Statistical Methods*. Springer-Verlag, New York.

Ruppert, D. (1991). Stochastic approximation. In *Handbook of Sequential Analysis*. Marcel Dekker, New York, 503–529.

Ruppert, D., N. Cressie, and R. J. Carroll (1989). A transformation/weighting model for estimating Michaelis-Menten parameters. *Biometrics*, **45**:637–656.

Santner, T. J. and D. E. Duffy (1990). *The Statistical Analysis of Discrete Data*. Springer-Verlag, New York.

Santner, T. J. and M. K. Snell (1980). Small-sample confidence intervals for $p_1 - p_2$ and p_1/p_2 in 2×2 contingency tables. *Journal of the American Statistical Association*, **75**:386–394.

Sartori, N. and T. Severini (2004). Conditional likelihood inference in generalized linear mixed models. *Statistica Sinica*, **14**:349–360.

Satterthwaite, F. E. (1946). An approximate distribution of estimates of variance components. *Biometrics Bulletin*, **2**:110–114.

Schall, R. (1991). Estimation in generalized linear models with random effects. *Biometrika*, **78**:719–727.

Searle, S. R. (1971). *Linear Models*. Wiley, New York.

Searle, S. R. (1974). Prediction, mixed models and variance components. In *Reliability and Biometry* (edited by F. Proschan and R. Ser-
fling). Society for Industrial and Applied Mathematics, Philadelphia.

Searle, S. R. (1982). *Matrix Algebra Useful for Statistics*. Wiley, New
York.

Searle, S. R. (1987). *Linear Models for Unbalanced Data*. Wiley, New
York.

Searle, S. R. (1997). *Linear Models*. Classic Edition, Wiley, New York.
(Reprinted from 1971).

Searle, S. R. (1999). On Linear Models with Restrictions on Parameters. Department of Biometrics Technical Report BU-1450-M, Cornell University, Ithaca, NY.

Searle, S. R., G. Casella, and C. E. McCulloch (1992). *Variance Components*. Wiley, New York.

Seber, G. A. F. (1977). *Linear Regression Analysis*. Wiley, New York.

Self, S. and K.-Y. Liang (1987). Asymptotic properties of maximum
likelihood estimators and likelihood ratio tests under nonstandard
conditions. *Journal of the American Statistical Association*, **82**:605–
610.

Sheiner, L. B., S. L. Beal, and A. Dunne (1997). Analysis of nonran-
domly censored ordered categorical longitudinal data from analgesic
trials (c/r: pp. 1245–1255). *Journal of the American Statistical Association*, **92**:1235–1244.

Shiboski, C., J. M. Neuhaus, G. D, and G. J (1999). Effect of receptive
oral sex and smoking on the incidence of hairy leukoplakia and gender: A longitudinal analysis. *JAIDS Journal of the Acquired Immune
Deficiency Syndromes*, **21**:236–242.

Snedecor, G. W. and W. G. Cochran (1989). *Statistical Methods, 8th
ed.* Iowa State University Press, Ames, IA.

Tchetgen, E. J. and B. A. Coull (2006). A diagnostic test for the
mixing distribution in a generalised linear mixed model. *Biometrika*,
93:1003–1010.

Tierney, L. and J. B. Kadane (1986). Accurate approximations for posterior moments and marginal densities. *Journal of the American Statistical Association*, **81**:82–86.

Verbyla, A. P. (1993). Modelling variance heterogeneity: Residual maximum likelihood and diagnostics. *Journal of the Royal Statistical Society, Series B*, **55**:493–508.

Wald, A. (1941). Asymptotically most powerful tests of statistical hypotheses. *Annals of Mathematical Statistics*, **12**:1–19.

Wang, Y. (2007). On fast computation of the non-parametric maximum likelihood estimate of a mixing distribution. *Journal of the Royal Statistical Society, Series B*, **69**:185–198.

Wang, Z. and T. Louis (2003). Matching conditional and marginal shapes in binary random intercept models using a bridge distribution function. *Biometrika*, **90**:765–775.

Wedderburn, R. W. M. (1974). Quasi–likelihood functions, generalized linear models, and the Gauss–Newton method. *Biometrika*, **61**:439–447.

Weisberg, S. (1980). *Applied Linear Regression*. Wiley, New York.

White, H. (1994). *Estimation, Inference, and Specification Analysis*. Cambridge University Press, Cambridge.

Williams, D. A. (1975). The analysis of binary response from toxicological experiments involving reproduction and teratogenicity. *Biometrics*, **31**:949–952.

Wolfinger, R. W. (1993). Laplace's approximation for nonlinear mixed models. *Biometrika*, **80**:791–795.

Zeger, S. L. and M. R. Karim (1991). Generalized linear models with random effects: a gibbs sampling approach. *Journal of the American Statistical Association*, **86**:79–86.

Zeger, S. L. and K.-Y. Liang (1986). Longitudinal data analysis for discrete and continuous outcomes. *Biometrics*, **42**:121–130.

Zhang, D. and M. Davidian (2001). Linear mixed models with flexible distribution of random effects for longitudinal data. *Biometrics*, **57**:795–802.

Zhao, L. and R. Prentice (1990). Correlated binary regression using a quadratic exponential model. *Biometrika*, **77**:642–648.

Index

WILEY SERIES IN PROBABILITY AND STATISTICS
ESTABLISHED BY WALTER A. SHEWHART AND SAMUEL S. WILKS

Editors: *David J. Balding, Noel A. C. Cressie, Garrett M. Fitzmaurice, Iain M. Johnstone, Geert Molenberghs, David W. Scott, Adrian F. M. Smith, Ruey S. Tsay, Sanford Weisberg*
Editors Emeriti: *Vic Barnett, J. Stuart Hunter, Jozef L. Teugels*

The **Wiley Series in Probability and Statistics** is well established and authoritative. It covers many topics of current research interest in both pure and applied statistics and probability theory. Written by leading statisticians and institutions, the titles span both state-of-the-art developments in the field and classical methods.

Reflecting the wide range of current research in statistics, the series encompasses applied, methodological and theoretical statistics, ranging from applications and new techniques made possible by advances in computerized practice to rigorous treatment of theoretical approaches.

This series provides essential and invaluable reading for all statisticians, whether in academia, industry, government, or research.

† ABRAHAM and LEDOLTER · Statistical Methods for Forecasting
AGRESTI · Analysis of Ordinal Categorical Data
AGRESTI · An Introduction to Categorical Data Analysis, *Second Edition*
AGRESTI · Categorical Data Analysis, *Second Edition*
ALTMAN, GILL, and McDONALD · Numerical Issues in Statistical Computing for the Social Scientist
AMARATUNGA and CABRERA · Exploration and Analysis of DNA Microarray and Protein Array Data
ANDĚL · Mathematics of Chance
ANDERSON · An Introduction to Multivariate Statistical Analysis, *Third Edition*
* ANDERSON · The Statistical Analysis of Time Series
ANDERSON, AUQUIER, HAUCK, OAKES, VANDAELE, and WEISBERG · Statistical Methods for Comparative Studies
ANDERSON and LOYNES · The Teaching of Practical Statistics
ARMITAGE and DAVID (editors) · Advances in Biometry
ARNOLD, BALAKRISHNAN, and NAGARAJA · Records
* ARTHANARI and DODGE · Mathematical Programming in Statistics
* BAILEY · The Elements of Stochastic Processes with Applications to the Natural Sciences
BALAKRISHNAN and KOUTRAS · Runs and Scans with Applications
BALAKRISHNAN and NG · Precedence-Type Tests and Applications
BARNETT · Comparative Statistical Inference, *Third Edition*
BARNETT · Environmental Statistics
BARNETT and LEWIS · Outliers in Statistical Data, *Third Edition*
BARTOSZYNSKI and NIEWIADOMSKA-BUGAJ · Probability and Statistical Inference
BASILEVSKY · Statistical Factor Analysis and Related Methods: Theory and Applications
BASU and RIGDON · Statistical Methods for the Reliability of Repairable Systems
BATES and WATTS · Nonlinear Regression Analysis and Its Applications
BECHHOFER, SANTNER, and GOLDSMAN · Design and Analysis of Experiments for Statistical Selection, Screening, and Multiple Comparisons

*Now available in a lower priced paperback edition in the Wiley Classics Library.
†Now available in a lower priced paperback edition in the Wiley–Interscience Paperback Series.

BELSLEY · Conditioning Diagnostics: Collinearity and Weak Data in Regression
† BELSLEY, KUH, and WELSCH · Regression Diagnostics: Identifying Influential
  Data and Sources of Collinearity
BENDAT and PIERSOL · Random Data: Analysis and Measurement Procedures,
  Third Edition
BERRY, CHALONER, and GEWEKE · Bayesian Analysis in Statistics and
  Econometrics: Essays in Honor of Arnold Zellner
BERNARDO and SMITH · Bayesian Theory
BHAT and MILLER · Elements of Applied Stochastic Processes, *Third Edition*
BHATTACHARYA and WAYMIRE · Stochastic Processes with Applications
BILLINGSLEY · Convergence of Probability Measures, *Second Edition*
BILLINGSLEY · Probability and Measure, *Third Edition*
BIRKES and DODGE · Alternative Methods of Regression
BISWAS, DATTA, FINE, and SEGAL · Statistical Advances in the Biomedical Sciences:
  Clinical Trials, Epidemiology, Survival Analysis, and Bioinformatics
BLISCHKE AND MURTHY (editors) · Case Studies in Reliability and Maintenance
BLISCHKE AND MURTHY · Reliability: Modeling, Prediction, and Optimization
BLOOMFIELD · Fourier Analysis of Time Series: An Introduction, *Second Edition*
BOLLEN · Structural Equations with Latent Variables
BOLLEN and CURRAN · Latent Curve Models: A Structural Equation Perspective
BOROVKOV · Ergodicity and Stability of Stochastic Processes
BOULEAU · Numerical Methods for Stochastic Processes
BOX · Bayesian Inference in Statistical Analysis
BOX · R. A. Fisher, the Life of a Scientist
BOX and DRAPER · Response Surfaces, Mixtures, and Ridge Analyses, *Second Edition*
* BOX and DRAPER · Evolutionary Operation: A Statistical Method for Process
  Improvement
BOX and FRIENDS · Improving Almost Anything, *Revised Edition*
BOX, HUNTER, and HUNTER · Statistics for Experimenters: Design, Innovation,
  and Discovery, *Second Editon*
BOX and LUCEÑO · Statistical Control by Monitoring and Feedback Adjustment
BRANDIMARTE · Numerical Methods in Finance: A MATLAB-Based Introduction
† BROWN and HOLLANDER · Statistics: A Biomedical Introduction
BRUNNER, DOMHOF, and LANGER · Nonparametric Analysis of Longitudinal Data in
  Factorial Experiments
BUCKLEW · Large Deviation Techniques in Decision, Simulation, and Estimation
CAIROLI and DALANG · Sequential Stochastic Optimization
CASTILLO, HADI, BALAKRISHNAN, and SARABIA · Extreme Value and Related
  Models with Applications in Engineering and Science
CHAN · Time Series: Applications to Finance
CHARALAMBIDES · Combinatorial Methods in Discrete Distributions
CHATTERJEE and HADI · Regression Analysis by Example, *Fourth Edition*
CHATTERJEE and HADI · Sensitivity Analysis in Linear Regression
CHERNICK · Bootstrap Methods: A Guide for Practitioners and Researchers,
  Second Edition
CHERNICK and FRIIS · Introductory Biostatistics for the Health Sciences
CHILÈS and DELFINER · Geostatistics: Modeling Spatial Uncertainty
CHOW and LIU · Design and Analysis of Clinical Trials: Concepts and Methodologies,
  Second Edition
CLARKE and DISNEY · Probability and Random Processes: A First Course with
  Applications, *Second Edition*
* COCHRAN and COX · Experimental Designs, *Second Edition*
CONGDON · Applied Bayesian Modelling
CONGDON · Bayesian Models for Categorical Data

*Now available in a lower priced paperback edition in the Wiley Classics Library.
†Now available in a lower priced paperback edition in the Wiley–Interscience Paperback Series.

*Now available in a lower priced paperback edition in the Wiley Classics Library.

†Now available in a lower priced paperback edition in the Wiley–Interscience Paperback Series.

---

*Now available in a lower priced paperback edition in the Wiley Classics Library.

†Now available in a lower priced paperback edition in the Wiley–Interscience Paperback Series.

MYERS, MONTGOMERY, and VINING · Generalized Linear Models. With Applications in Engineering and the Sciences
† NELSON · Accelerated Testing, Statistical Models, Test Plans, and Data Analyses
† NELSON · Applied Life Data Analysis
NEWMAN · Biostatistical Methods in Epidemiology
OCHI · Applied Probability and Stochastic Processes in Engineering and Physical Sciences
OKABE, BOOTS, SUGIHARA, and CHIU · Spatial Tesselations: Concepts and Applications of Voronoi Diagrams, *Second Edition*
OLIVER and SMITH · Influence Diagrams, Belief Nets and Decision Analysis
PALTA · Quantitative Methods in Population Health: Extensions of Ordinary Regressions
PANJER · Operational Risk: Modeling and Analytics
PANKRATZ · Forecasting with Dynamic Regression Models
PANKRATZ · Forecasting with Univariate Box-Jenkins Models: Concepts and Cases
* PARZEN · Modern Probability Theory and Its Applications
PEÑA, TIAO, and TSAY · A Course in Time Series Analysis
PIANTADOSI · Clinical Trials: A Methodologic Perspective
PORT · Theoretical Probability for Applications
POURAHMADI · Foundations of Time Series Analysis and Prediction Theory
POWELL · Approximate Dynamic Programming: Solving the Curses of Dimensionality
PRESS · Bayesian Statistics: Principles, Models, and Applications
PRESS · Subjective and Objective Bayesian Statistics, *Second Edition*
PRESS and TANUR · The Subjectivity of Scientists and the Bayesian Approach
PUKELSHEIM · Optimal Experimental Design
PURI, VILAPLANA, and WERTZ · New Perspectives in Theoretical and Applied Statistics
† PUTERMAN · Markov Decision Processes: Discrete Stochastic Dynamic Programming
QIU · Image Processing and Jump Regression Analysis
* RAO · Linear Statistical Inference and Its Applications, *Second Edition*
RAUSAND and HØYLAND · System Reliability Theory: Models, Statistical Methods, and Applications, *Second Edition*
RENCHER · Linear Models in Statistics
RENCHER · Methods of Multivariate Analysis, *Second Edition*
RENCHER · Multivariate Statistical Inference with Applications
* RIPLEY · Spatial Statistics
* RIPLEY · Stochastic Simulation
ROBINSON · Practical Strategies for Experimenting
ROHATGI and SALEH · An Introduction to Probability and Statistics, *Second Edition*
ROLSKI, SCHMIDLI, SCHMIDT, and TEUGELS · Stochastic Processes for Insurance and Finance
ROSENBERGER and LACHIN · Randomization in Clinical Trials: Theory and Practice
ROSS · Introduction to Probability and Statistics for Engineers and Scientists
ROSSI, ALLENBY, and McCULLOCH · Bayesian Statistics and Marketing
† ROUSSEEUW and LEROY · Robust Regression and Outlier Detection
* RUBIN · Multiple Imputation for Nonresponse in Surveys
RUBINSTEIN and KROESE · Simulation and the Monte Carlo Method, *Second Edition*
RUBINSTEIN and MELAMED · Modern Simulation and Modeling
RYAN · Modern Engineering Statistics
RYAN · Modern Experimental Design
RYAN · Modern Regression Methods
RYAN · Statistical Methods for Quality Improvement, *Second Edition*
SALEH · Theory of Preliminary Test and Stein-Type Estimation with Applications
* SCHEFFE · The Analysis of Variance
SCHIMEK · Smoothing and Regression: Approaches, Computation, and Application

---

*Now available in a lower priced paperback edition in the Wiley Classics Library.
†Now available in a lower priced paperback edition in the Wiley–Interscience Paperback Series.

WEERAHANDI · Generalized Inference in Repeated Measures: Exact Methods in MANOVA and Mixed Models

WEISBERG · Applied Linear Regression, *Third Edition*

WELSH · Aspects of Statistical Inference

WESTFALL and YOUNG · Resampling-Based Multiple Testing: Examples and Methods for *p*-Value Adjustment

WHITTAKER · Graphical Models in Applied Multivariate Statistics

WINKER · Optimization Heuristics in Economics: Applications of Threshold Accepting

WONNACOTT and WONNACOTT · Econometrics, *Second Edition*

WOODING · Planning Pharmaceutical Clinical Trials: Basic Statistical Principles

WOODWORTH · Biostatistics: A Bayesian Introduction

WOOLSON and CLARKE · Statistical Methods for the Analysis of Biomedical Data, *Second Edition*

WU and HAMADA · Experiments: Planning, Analysis, and Parameter Design Optimization

WU and ZHANG · Nonparametric Regression Methods for Longitudinal Data Analysis

YANG · The Construction Theory of Denumerable Markov Processes

YOUNG, VALERO-MORA, and FRIENDLY · Visual Statistics: Seeing Data with Dynamic Interactive Graphics

ZELTERMAN · Discrete Distributions—Applications in the Health Sciences

 * ZELLNER · An Introduction to Bayesian Inference in Econometrics

ZHOU, OBUCHOWSKI, and McCLISH · Statistical Methods in Diagnostic Medicine